Greta Ricarda Patzke and Pierre-Emmanuel Car (Eds.)

Polyoxometalates

MDPI

This book is a reprint of the Special Issue that appeared in the online, open access journal, *Inorganics* (ISSN 2304-6740) in 2015 (available at: http://www.mdpi.com/journal/inorganics/special_issues/polyoxometalates).

Guest Editors
Greta Ricarda Patzke
University of Zurich
Switzerland

Pierre-Emmanuel Car
University of Zurich
Switzerland

Editorial Office
MDPI AG
Klybeckstrasse 64
Basel, Switzerland

Publisher
Shu-Kun Lin

Managing Editor
Min Su

1. Edition 2016

MDPI • Basel • Beijing • Wuhan

ISBN 978-3-03842-161-0 (Hbk)
ISBN 978-3-03842-162-7 (PDF)

Table of Contents

IV

List of Contributors

Pavel A. Abramov: Nikolaev Institute of Inorganic Chemistry SB RAS, Novosibirsk 630090, Russia; Novosibirsk State University, Novosibirsk 630090, Russia.

Gregory Absillis: Department of Chemistry, KU Leuven, Celestijnenlaan 200F, 3001 Heverlee, Belgium.

Revathi Bacsa: Laboratoire de Chimie de Coordination UPR CNRS 8241, Composante ENSIACET, Université Toulouse, 4 allée Emile Monso, 31030 Toulouse, France.

Bassem S. Bassil: Department of Life Sciences and Chemistry, Jacobs University, P.O. Box 750 561, 28725 Bremen, Germany; Department of Chemistry, Faculty of Sciences, University of Balamand, P.O. Box 100, Tripoli, Lebanon.

Tarik Benali: Institut Lavoisier de Versailles, UMR8180, Université de Versailles St Quentin en Yvelines, 45 Avenue des Etats Unis, 78035 Versailles cedex, France.

Stephen P. Best: School of Chemistry, University of Melbourne, Melbourne 3010, Australia.

Kerry Lee Buchwalder: Department of Inorganic and Analytical Chemistry, University of Geneva, 30 quai E. Ansermet, Geneva CH-1211, Switzerland.

Emmanuel Cadot: Institut Lavoisier de Versailles, UMR 8180, University of Versailles, 45 avenue des Etats-Unis, Versailles 78035, France.

Pierre-Emmanuel Car: Department of Chemistry, University of Zurich, Winterthurerstrasse 190, CH-8057 Zurich, Switzerland.

Ricardo J. Carvalho: REQUIMTE/LAQV, Departamento de Química e Bioquímica, Faculdade de Ciências, Universidade do Porto, 4169-007 Porto, Portugal.

William H. Casey: Department of Geology; Department of Chemistry, University of California, Davis, CA 95616, USA.

Pedro de Oliveira: Laboratoire de Chimie Physique, UMR 8000 CNRS, Université Paris-Sud, 91405 Orsay Cedex, France; Laboratoire de Chimie Physique, Equipe d'Electrochimie et de Photo-électrochimie, Université Paris-Sud, UMR 8000 CNRS, Orsay, F-91405, France.

Philippe Deniard: Institut des Matériaux Jean-Rouxel, Université de Nantes, CNRS, 2 Rue de la Houssinière, BP 32229, 44322 Nantes cedex, France.

Rémi Dessapt: Institut des Matériaux Jean-Rouxel, Université de Nantes, CNRS, 2 Rue de la Houssinière, BP 32229, 44322 Nantes cedex, France.

Anne Dolbecq: Institut Lavoisier de Versailles, UMR8180, Université de Versailles St Quentin en Yvelines, 45 Avenue des Etats Unis, 78035 Versailles cedex, France.

Stefano Fabris: CNR-IOM DEMOCRITOS, Istituto Officina dei Materiali, c/o SISSA, Via Bonomea 265, Trieste 34136, Italy.

Leire San Felices: Servicios Generales de Investigación SGIker, Universidad del País Vasco UPV/EHU, P. O. Box 644, Bilbao 48080, Spain.

Diana M. Fernandes: REQUIMTE/LAQV, Departamento de Química e Bioquímica, Faculdade de Ciências, Universidade do Porto, 4169-007 Porto, Portugal.

Sébastien Floquet: Institut Lavoisier de Versailles, UMR 8180, University of Versailles, 45 avenue des Etats-Unis, Versailles 78035, France.

Tori Z. Forbes: Department of Chemistry, University of Iowa, Iowa City, IA 52242, USA.

Cristina Freire: REQUIMTE/LAQV, Departamento de Química e Bioquímica, Faculdade de Ciências, Universidade do Porto, 4169-007 Porto, Portugal.

Eric Le Fur: ENSCR, UMR 6226, 11, allée de Beaulieu - CS 50837-35708 Rennes Cedex 07, France.

José Ramón Galán-Mascarós: Institute of Chemical Research of Catalonia (ICIQ), Av. Països Catalans 16, E-43007 Tarragona, Spain; Catalan Institution for Research and Advanced Studies (ICREA), Passeig Lluis Companys, 23, E-08010 Barcelona, Spain.

Sara Goberna-Ferrón: Institute of Chemical Research of Catalonia (ICIQ), Av. Països Catalans 16, E-43007 Tarragona, Spain.

Lars Goerigk: School of Chemistry, University of Melbourne, Melbourne 3010, Australia.

Goovaerts: Department of Chemistry, KU Leuven, Celestijnenlaan 200F, 3001 Heverlee, Belgium.

Laure Guénée: Department of Inorganic and Analytical Chemistry, University of Geneva, 30 quai E. Ansermet, Geneva CH-1211, Switzerland.

Juan M. Gutiérrez-Zorrilla: Departamento de Química Inorgánica, Facultad de Ciencia y Tecnología, Universidad del País Vasco UPV/EHU, P. O. Box 644, Bilbao 48080, Spain; BCMaterials, Parque Científico y Tecnológico de Bizkaia, Edificio 500, Derio 48160, Spain.

Mohamed Haouas: Institut Lavoisier de Versailles, Université de Versailles St. Quentin, UMR 8180 CNRS, Versailles, F-78035, France.

Yoshihito Hayashi: Department of Chemistry, Graduate School of Natural Science and Technology, Kanazawa University, Kakuma, Kanazawa 920-1192, Japan.

Caitlyn R. Hazlett: Department of Chemistry, University of Oregon, Eugene, OR 97043, USA.

Merinda R. Healey: School of Chemistry, University of Melbourne, Melbourne 3010, Australia.

Patricio Hermosilla-Ibáñez: Facultad de Química y Biología, Universidad de Santiago de Chile, USACH, Av. Libertador Bernardo O'Higgins 3363, 9170022, Santiago, Chile; Centro para el Desarrollo de la Nanociencia y Nanotecnología, CEDENNA, 9170022, Santiago, Chile.

Akram Hijazi: Laboratoire de Chimie de Coordination Inorganique et Organométallique LCIO, Université Libanaise, Faculté des Sciences I, Hadath, Lebanon; Ecole Doctorale des Sciences et Technologie EDST, PRASE, Université Libanaise, Hadath, Lebanon.

Masooma Ibrahim: Department of Life Sciences and Chemistry, Jacobs University, P.O. Box 750 561, 28725 Bremen, Germany; Present address: Institute of Nanotechnology, Karlsruhe Institute of Technology (KIT), Hermann-von-Helmholtz Platz 1, 76344 Eggenstein-Leopoldshafen, Germany.

Amaia Iturrospe: Departamento de Química Inorgánica, Facultad de Ciencia y Tecnología, Universidad del País Vasco UPV/EHU, P. O. Box 644, Bilbao 48080, Spain.

Yuji Kikukawa: Department of Chemistry, Graduate School of Natural Science and Technology, Kanazawa University, Kakuma, Kanazawa 920-1192, Japan.

Ulrich Kortz: Department of Life Sciences and Chemistry, Jacobs University, P.O. Box 750 561, 28725 Bremen, Germany.

Chelsey Lamar: Department of Chemistry, Howard University, Washington, D.C. 20059, USA.

Luis Lezama: Departamento de Química Inorgánica, Facultad de Ciencia y Tecnología, Universidad del País Vasco UPV/EHU, P. O. Box 644, Bilbao 48080, Spain; BCMaterials, Parque Científico y Tecnológico de Bizkaia, Edificio 500, Derio 48160, Spain.

Jérome Marrot: Institut Lavoisier de Versailles, UMR8180, Université de Versailles St Quentin en Yvelines, 45 Avenue des Etats Unis, 78035 Versailles cedex, France.

Israël M. Mbomekallé: Laboratoire de Chimie Physique, Equipe d'Electrochimie et de Photo-électrochimie, Université Paris-Sud, UMR 8000 CNRS, Orsay F-91405, France.

Pierre Mialane: Institut Lavoisier de Versailles, UMR8180, Université de Versailles St Quentin en Yvelines, 45 Avenue des Etats Unis, 78035 Versailles cedex, France.

Karina Muñoz-Becerra: Facultad de Química y Biología, Universidad de Santiago de Chile, USACH, Av. Libertador Bernardo O'Higgins 3363, 9170022, Santiago, Chile; Centro para el Desarrollo de la Nanociencia y Nanotecnología, CEDENNA, 9170022, Santiago, Chile.

Jun-ichi Nambu: Department of Applied Science, Faculty of Science, Kochi University, Kochi 780-8520, Japan.

Daoud Naoufal: Laboratoire de Chimie de Coordination Inorganique et Organométallique LCIO, Université Libanaise, Faculté des Sciences I, Hadath, Lebanon; Ecole Doctorale des Sciences et Technologie EDST, PRASE, Université Libanaise, Hadath, Lebanon.

Yuriko Nishimoto: Department of Applied Science, Faculty of Science, Kochi University, Kochi 780-8520, Japan.

Marta Nunes: REQUIMTE/LAQV, Departamento de Química e Bioquímica, Faculdade de Ciências, Universidade do Porto, 4169-007 Porto, Portugal.

Kazuhiro Ogihara: Department of Chemistry, Graduate School of Natural Science and Technology, Kanazawa University, Kakuma, Kanazawa 920-1192, Japan.

Miho Ohnishi: Department of Applied Science, Faculty of Science, Kochi University, Kochi 780-8520, Japan.

Marilyn M. Olmstead: Department of Chemistry, University of California, Davis, CA 95616, USA.

Olivier Oms: Institut Lavoisier de Versailles, UMR8180, Université de Versailles St Quentin en Yvelines, 45 Avenue des Etats Unis, 78035 Versailles cedex, France.

Aroa Pache: Departamento de Química Inorgánica, Facultad de Ciencia y Tecnología, Universidad del País Vasco UPV/EHU, P. O. Box 644, Bilbao 48080, Spain.

Tatjana N. Parac-Vogt: Department of Chemistry, KU Leuven, Celestijnenlaan 200F, 3001 Heverlee, Belgium.

Verónica Paredes-García: Centro para el Desarrollo de la Nanociencia y Nanotecnología, CEDENNA, 9170022 Santiago, Chile; Departamento de Ciencias Químicas, Universidad Andres Bello, Republica 275, 8370146, Santiago, Chile.

Loïc Parent: Institut Lavoisier de Versailles, Université de Versailles St. Quentin, UMR 8180 CNRS, Versailles, F-78035, France.

Greta R. Patzke: Department of Chemistry, University of Zurich, Winterthurerstrasse 190, CH-8057 Zurich, Switzerland.

Simone Piccinin: CNR-IOM DEMOCRITOS, Istituto Officina dei Materiali, c/o SISSA, Via Bonomea 265, Trieste 34136, Italy.

Claude Piguet: Department of Inorganic and Analytical Chemistry, University of Geneva, 30 quai E. Ansermet, Geneva CH-1211, Switzerland.

Marin Puget: Institut des Matériaux Jean-Rouxel, Université de Nantes, CNRS, 2 Rue de la Houssinière, BP 32229, 44322 Nantes cedex, France.

Santiago Reinoso: Departamento de Química Inorgánica, Facultad de Ciencia y Tecnología, Universidad del País Vasco UPV/EHU, P. O. Box 644, Bilbao 48080, Spain.

Chris Ritchie: School of Chemistry, University of Melbourne, Melbourne 3010, Australia.

Saito: Department of Applied Science, Faculty of Science, Kochi University, Kochi 780-8520, Japan.

William Salomon: Institut Lavoisier de Versailles, UMR 8180, University of Versailles, 45 avenue des Etats-Unis, Versailles 78035, France.

Hélène Serier-Brault: Institut des Matériaux Jean-Rouxel, Université de Nantes, CNRS, 2 Rue de la Houssinière, BP 32229, 44322 Nantes cedex, France.

Philippe Serp: Laboratoire de Chimie de Coordination UPR CNRS 8241, Composante ENSIACET, Université Toulouse, 4 allée Emile Monso, 31030 Toulouse, France.

Maxim N. Sokolov: Nikolaev Institute of Inorganic Chemistry SB RAS, Novosibirsk 630090, Russia; Novosibirsk State University, Novosibirsk 630090, Russia.

Joaquín Soriano-López: Institute of Chemical Research of Catalonia (ICIQ), Av. Països Catalans 16, E-43007 Tarragona, Spain.

Evgenia Spodine: Centro para el Desarrollo de la Nanociencia y Nanotecnología, CEDENNA, 9170022, Santiago, Chile; Facultad de Ciencias Químicas y Farmacéuticas, Universidad de Chile, Sergio Livingstone 1007, Independencia, 8380492, Santiago, Chile.

Karen Stroobants: Department of Chemistry, KU Leuven, Celestijnenlaan 200F, 3001 Heverlee, Belgium.

Anne-Lucie Teillout: Laboratoire de Chimie Physique, Equipe d'Electrochimie et de Photo-électrochimie, Université Paris-Sud, UMR 8000 CNRS, Orsay, F-91405, France.

Emmanuel Terazzi: Department of Inorganic and Analytical Chemistry, University of Geneva, 30 quai E. Ansermet, Geneva CH-1211, Switzerland.

Tadaharu Ueda: Department of Applied Science, Faculty of Science, Kochi University, Kochi 780-8520, Japan.

Diego Venegas-Yazigi: Facultad de Química y Biología, Universidad de Santiago de Chile, USACH, Av. Libertador Bernardo O'Higgins 3363, 9170022, Santiago, Chile; Centro para el Desarrollo de la Nanociencia y Nanotecnología, CEDENNA, 9170022, Santiago, Chile.

Cristian Vicent: Serveis Centrals d'Instrumentació Científica, Universitat Jaume I, Av. Sos Baynat s/n, 12071 Castelló, Spain.

Nancy Watfa: Institut Lavoisier de Versailles, UMR 8180, University of Versailles, 45 avenue des Etats-Unis, Versailles 78035, France; Ecole Doctorale des Sciences et Technologie EDST, PRASE; Laboratoire de Chimie de Coordination Inorganique et Organométallique LCIO, Université Libanaise, Faculté des Sciences I, Hadath, Lebanon.

About the Guest Editors

Greta R. Patzke received her Ph.D. *summa cum laude* from the University of Hannover in 1999, and she worked on the synthesis, characterization and properties of mixed oxides with special emphasis on crystal growth methods from the gas phase. She then moved to the ETH, Zurich and joined the group of Prof. Reinhard Nesper to work on her Habilitation. During these years, she developed a wide range of research interests including nanomaterials synthesis, polyoxometalates (POMs) and the mechanistic investigation of hydrothermal techniques, crossing the border between nano- and molecular chemistry. In October 2006, she received the *Venia Legendi* for Inorganic Chemistry from the ETH, Zurich. Her work was recognized by an offer of the *Alfred Werner Assistant Professorship* soon afterwards, but in summer 2007, Greta Patzke started to work as Assistant Professor of Inorganic Chemistry at the University of Zurich, Switzerland, endowed with a *Förderungsprofessur* by the *Swiss National Science Foundation*. Over the following years, she and her research team focused on the targeted development of oxide-based materials for environmental applications. In spring 2013, Greta Patzke was promoted to Associate Professor (tenured). Recent highlights of her work range from new POM-based catalysts for visible-light-driven water oxidation to innovative monitoring and carrier strategies for the interaction of bio-active POMs with cells. Current activities are focused on the exploration of abundant transition metal clusters and oxides for water splitting in search of design concepts for artificial photosynthesis concepts. She is a board member of the University Research Priority Program "Light to Chemical Energy Conversion" (LightChEC, http://www.lightchec.uzh.ch/).

Pierre-Emmanuel Car obtained his Ph.D. degree from the University of Rennes 1 (France) in 2008, undertaking his Ph.D. research under the supervision of Eric Le Fur and Jean-Yves Pivan, working on polyoxovanadates. After post-doctoral experience with Roberta Sessoli and Andrea Caneschi at the University of Florence (Italy), and with Greta R. Patzke at the University of Zurich (Switzerland), he started his independent career as an "Oberassistent" (Habilitation) in 2014. His research interests are currently focused on polyoxometalate chemistry and molecular materials relative to the fields of molecular magnetism and photocatalytic water splitting.

Preface

The Fascination of Polyoxometalate Chemistry

Pierre-Emmanuel Car and Greta R. Patzke

Reprinted from *Inorganics.* Cite as: Car, P.-E.; Patzke, G.R. The Fascination of Polyoxometalate Chemistry. *Inorganics* **2015**, *3*, 511-515.

We are delighted to introduce this special issue of *Inorganics*. This themed issue is dedicated to polyoxometalates (POMs) as an outstanding class of oxo-cluster materials. Polyoxometalates have fascinated generations of researchers since the mid-18th-century; and they continue to attract promising young scientists all over the world. Since the first pioneering studies; the manifold structures and properties of POMs; have been the focus of interdisciplinary research synthetic/structural chemistry; biology; physics and theoretical chemistry. Moreover; polyoxometalates excel through outstanding compositional and structural diversity; which enables fine-tuning of their electronic properties; redox properties; and chemical stability along with robustness for the design of future applied devices. The growing family of polyoxometalates can be divided into two classes: transition-metal-substituted polyoxometalates (TMSPs) and lanthanide-substituted polyoxometalates (LnSPs). They currently attract particular interest due to their strong potential in the most challenging forefront research areas; e.g.; water splitting; catalysis; magnetism; electronic materials and bio-medical applications. In the present themed issue; several research domains of polyoxometalate chemistry are covered by internationally renowned research groups in this topical field; ranging from synthesis/characterization and biological properties; through water oxidation catalysis and photochromic properties; to liquid crystal properties.

The present special issue starts with a new report by William H. Casey and co-workers on the synthesis and crystal structure determination of a new polynuclear Ga(III)-oxyhydroxo cluster [1]. This compound extends the hitherto very short list of group 13 clusters (with Al^{3+} and Ga^{3+} cations). The newly discovered oxo-cluster contains 30 Ga(III) centers, and its structure was determined from single crystal X-ray diffraction techniques at the Advanced Light Source.

The contribution of Juan M. Gutierrez-Zorrilla and co-workers [2], sheds light on a fascinating and versatile domain of polyoxometalate chemistry, namely the use of Keggin-type polyoxometalates $[XM_{12}O_{40}]^{n-}$ (X = Si, Ge) as building-blocks and precursors for the design and synthesis of new inorganic-metalorganic materials. Juan M. Gutierrez-Zorrilla *et al.* report on three new hybrids that were fully characterized. Single crystal X-ray diffraction techniques revealed the formation of 2D networks for all three compounds, highlighting the structure-directing role of the guanidinium ions.

Next, the up-coming field of lanthanide substituted polyoxometalates is represented by the contributions of Ulrich Kortz and co-workers [3] and of Israël M. Mbomekallé and co-workers [4]. Ulrich Kortz and his team present the synthesis and characterization of a new polynuclear lanthanide-substituted polyoxometalate with the formula $[Y_8(CH_3COO)(H_2O)_{18}(As_2W_{19}O_{68})_4(W_2O_6)_2(WO_4)]^{43-}$. Crystallographic studies revealed that this lanthanide polyoxometalate is formed by four $\{Y_2As_2W_{19}O_{68}\}$ units linked to each other via two $\{W_2O_{10}\}$ groups and one $\{WO_6\}$ fragment. Israël M. Mbomekallé et al. synthesized and characterized a mononuclear europium hetero-polyoxometalate. The central europium(III) ion displays a square anti-prismatic coordination environment, and it is surrounded by two monovacant heterometallic W/Mo Keggin moieties $[\alpha\text{-}(SiW_9Mo_2O_{39})]^{8-}$. The new compound was fully characterized with a wide range of analytical methods, and it displays promising electro-catalytic activity for O_2 and H_2O_2 reduction.

Among the wide range of properties of POMs, bio-medical applications continue to attract the interest of international research groups. Two chemical approaches are reported in the present themed issue. The first contribution of Tatjana N. Parac-Vogt and co-workers [5] employs Zr(IV)-substituted polyoxometalates for the investigation of the regioselective hydrolysis of human serum albumin. Tryptophan fluorescence spectroscopy studies revealed for the first time a direct correlation between the metal incorporated in the POM and the rate of protein hydrolysis, as well as the strength of their interaction. The second article by Christina Freire and co-workers [6] features a novel hybrid nanocomposite formed from the combination of a vanadium substituted phosphomolybdate $\{PMo_{11}V\}$ and N-doped few layer graphene (N-FLG), which was newly prepared and fully characterized. The efficiency of the novel hybrid material for the electrochemical sensing of biomolecules, as well as for electro-catalytic and sensing properties, was investigated in detail. The polyoxometalate compound was successfully immobilized on N-FLG, and it exhibited excellent electrocatalytic and sensing properties towards acetaminophen and theophylline oxidations.

The present special issue also contains selected contributions focused on polyoxometalates with group 4 (Zr) and 5 (V, Nb, Ta) metal ions: two review articles by Pavel A. Abramov and co-workers on polyoxonobiates and polyoxotantalates [7], and by Diego Venegas-Yazigi and co-workers on polyoxovanadates [8]. Pavel A. Abramov et al. summarize their contributions to the coordination chemistry of noble metals (Rh, Ir, Ru, Pt(IV)) and polyoxometalates of Nb(V) and Ta(V). In a complementary manner, Diego Venegas-Yazigi et al. give an account of the structural and electronic properties of the less explored polyoxovanadoborate system. This review covers the different existing vanadium (V_6, V_{10}, V_{12}) and borate fragments ($B_{10}O_{22}^{14-}$, ...) as well as the use of the polyoxovanadoborate as building-blocks for the design and the synthesis of extended structures (1D to 3D). This section is rounded off with an article from Yoshihito Hayashi and co-workers [9] who report on the transformation of three different polyoxovanadates and their efficiency as catalysts for the oxidation of thioanisole.

Over the past decade, transition metal substituted polyoxometalates (TMSPs) have been intensely studied as promising catalyst types for water oxidation and water reduction processes. In the present themed issue, two contributions from Simone Piccinin and co-workers [10] and from José R. Galan-Mascaros and co-workers [11] shed new and interesting light on TMSP compounds as water oxidation catalysts (WOCs). Simone Piccinin et al. compared different ruthenium substituted polyoxometalates as WOCs containing one or four ruthenium centers as

WOCs by means of density functional theory (DFT). Theoretical studies showed that the oxidation state of the active Ru sites was found to be more important than their nuclearity. José R. Galan-Mascaros *et al.* explore the challenging field of POM-WOCs inspired by the {CaMn$_4$O$_5$} cluster of photosystem II. They investigated the activity and the stability of a tetranuclear manganese-substituted polyoxometalate $[Mn_4(H_2O)_2(PW_9O_{34})_2]^{10-}$ as an electrocatalyst for the water oxidation reaction, and compared its activity to the well-known {Co$_4$} analogue. Electrocatalytic studies revealed that the {Mn$_4$} title POM shows lower efficiency and stability under catalytic conditions in comparison with its {Co$_4$} analogue, thereby illustrating the challenges associated with bio-inspired POM-WOC design.

Finally, this special issue of *Inorganics* on the most recent advancements of the polyoxometalate chemistry is concluded with several contributions on recent progress in the rewarding field of polyoxomolybdates. The featured articles cover a wide range of investigations, from the synthesis and characterization of hexamolybdate $[Mo_6O_{19}]^{2-}$ functionalized by a heteroaromatic thiophene molecule containing an organoimido group to form the novel polyoxomolybdate $[Mo_6O_{18}(\textbf{L4})]^{2-}$ (**L4** = 4(4-bromo-5-methylthiophen-2-yl)-2,6-dimethylaniline) by Chris Ritchie and co-workers [12], to the investigation of the vanadium(V) substitution reaction in Wells–Dawson type polyoxometalates, by Jun-ichi Nambu and co-workers [13]. In this work, several analytical methods, such as cyclic voltammetry, ^{31}P NMR and Raman spectroscopy shed new light on the vanadium(V)-substitution processes in the $[X_2M_{18}O_{62}]^{6-}$ to $[X_2VM_{17}O_{62}]^{7-}$ (X = P, As; and M = Mo, W) transformation reactions. These two articles are complemented by two contributions on new polyoxomolybdate materials. Emmanuel Cadot and co-workers [14] report on the synthesis and characterization of eight new Keplerates obtained through the combination of $[Mo_{132}O_{372}(CH_3COO)_{30}(H_2O)_{72}]^{42-}$ polyoxoanions and 1-methyl-3-alkylimidazolium cations. The obtained complexes were fully characterized by a wide range of analytical methods, and the liquid crystal properties of the newly reported materials were investigated. Next, Anne Dolbecq and co-workers [15] synthesized and characterized two new hybrid organic–inorganic Mo(VI) and mixed Mo(V/VI) polyoxomolybdates. The fully oxidized Mo(VI) POM $[(Mo^{VI}_3O_8)_2(O_3PC(O)(C_3H_6NH_2CH_2C_5H_4NH)PO_3)_2]^{4-}$ was obtained as sodium salt and as sodium/potassium salt, while the mixed Mo(V/VI) POM $[(Mo^V_2O_4)(Mo^{VI}_2O_6)_2$ $\{O_3PC(O)(C_3H_6N(CH_2C_5H_4N)_2)PO_3\}_2]^{4-}$ was obtained as ammonium salt in the presence of hydrazine. The pH stability domain of the three new hybrids was evaluated by ^{31}P NMR spectroscopic studies, and they were found to exhibit solid state photochromic properties with rapid color-change under UV excitation.

Finally, we are very much indebted to all of the authors for their inspirational, exciting and interdisciplinary contributions, which cover a wide range of contemporary POM chemistry. We enjoyed editing this issue, and we hope that our audience will share the fascination of polyoxometalate chemistry.

References

1. Casey, W.H.; Olmstead, M.M.; Hazlett, C.R.; Lamar, C.; Forbes, T.Z. A new nanometer-sized Ga(III)-oxyhydroxide cation. *Inorganics* **2015**, *3*, 21–26.

2. Pache, A.; Reinoso, S.; San Felices, L.; Iturrospe, A.; Lezama, L.; Gutierrez-Zorrilla, J.M. Single-crystal to single-crystal reversible transformations induced by thermal dehydration in Keggin-type polyoxometalates decorated with copper(II)-picolinate complexes: The structure directing role of guanidinium. *Inorganics* **2015**, *3*, 194–218.

3. Ibrahim, M.; Bassil, B.S.; Kortz, U. Synthesis and characterization of 8-yttrium(III)-containing 81-tungsto-8-arsenate(III), $[Y_8(CH_3COO)(H_2O)_{18}(As_2W_{19}O_{68})_4(W_2O_6)_2(WO_4)]^{43-}$. *Inorganics* **2015**, *3*, 267–278.

4. Parent, L.; de Oliveira, P.; Teillout, A.-L.; Dolbecq, A.; Haouas, M.; Cadot, E.; Mbomekallé, I.M. Synthesis and characterisation of the europium (III) dimolybdo-enneatungsto-silicate dimer, $[Eu(\alpha\text{-}SiW_9Mo_2O_{39})_2]^{13-}$. *Inorganics* **2015**, *3*, 341–354.

5. Goovaerts, V.; Stroobants, K.; Absillis, G.; Parac-Vogt, T.N. Understanding the regioselective hydrolysis of human serum albumin by Zr(IV)-substituted polyoxotungstates using tryptophan fluorescence spectroscopy. *Inorganics* **2015**, *3*, 230–245.

6. Fernandes, D.M.; Nunes, M.; Carvalho, R.J.; Bacsa, R.; Mbomekalle, I.-M.; Serp, P.; de Oliveira, P.; Freire, C. Biomolecules electrochemical sensing properties of a $PMo_{11}V@N$-doped few layer graphene nanocomposite. *Inorganics* **2015**, *3*, 178–193.

7. Abramov, P.A.; Sokolov, M.N.; Vicent, C. Polyoxonobiates and polyoxotantalates as ligands-revisited. *Inorganics* **2015**, *3*, 160–177.

8. Hermosilla-Ibáñez, P.; Muñoz-Becerra, K.; Paredes-García, V.; Le Fur, E.; Spodine, E.; Venegas-Yazigi, D. Structural and electronic properties of polyoxovanadoborates containing the $[V_{12}B_{18}O_{60}]$ core in different mixed valence states. *Inorganics* **2015**, *3*, 309–331.

9. Kikukawa, Y.; Ogihara, K.; Hayashi, Y. Structure transformation among deca-, dodeca- and tridecavanadates and their properties for thioanisole oxidation. *Inorganics* **2015**, *3*, 295–308.

10. Piccinin, S.; Fabris, S. Water oxidation by Ru-polyoxometalate catalysts: Overpotential dependency on the number and charge of the metal centers. *Inorganics* **2015**, *3*, 374–387.

11. Goberna-Ferrón, S.; Soriano-López, J.; Galán-Mascarós, J.R. Activity and stability of the tetramanganese polyanion $[Mn_4(H_2O)_2(PW_9O_{34})_2]^{10-}$ during electrocatalytic water oxidation. *Inorganics* **2015**, *3*, 332–340.

12. Healey, M.R.; Best, S.P.; Goerigk, L.; Ritchie, C. A heteroaromatically functionalized hexamolybdate. *Inorganics* **2015**, *3*, 82–100.

13. Ueda, T.; Nishimoto, Y.; Saito, R.; Ohnishi, M.; Nambu, J.-I. Vanadium(V)-substitution reactions of Wells–Dawson-type polyoxometalates: From $[X_2M_{18}O_{62}]^{6-}$ (X = P, As; M = Mo, W) to $[X_2VM_{17}O_{62}]^{7-}$. *Inorganics* **2015**, *3*, 355–369.

14. Watfa, N.; Floquet, S.; Terazzi, E.; Salomon, W.; Guénée, L.; Buchwalder, K.L.; Hijazi, A.; Naoufal, D.; Piguet, C.; Cadot, E. Synthesis, characterization and study of liquid crystals based on the ionic association of the Keplerate anion $[Mo_{132}O_{372}(CH_3COO)_{30}(H_2O)_{72}]^{42-}$ and imidazolium cations. *Inorganics* **2015**, *3*, 246–266.

15. Oms, O.; Benali, T.; Marrot, J.; Mialane, P.; Puget, M.; Serier-Brault, H.; Deniard, P.; Dessapt, R.; Dolbecq, A. Fully oxidized and mixed-valent polyoxomolybdates structured by bisphosphonates with pendant pyridine groups: Synthesis, structure and photochromic properties. *Inorganics* **2015**, *3*, 279–294.

A New Nanometer-Sized Ga(III)-Oxyhydroxide Cation

William H. Casey, Marilyn M. Olmstead, Caitlyn R. Hazlett, Chelsey Lamar and
Tori Z. Forbes

Abstract: A new 30-center Ga(III)-oxy-hydroxide cation cluster was synthesized by hydrolysis of an aqueous $GaCl_3$ solution near pH = 2.5 and crystallized using 2,6-napthalene disulfonate (NDS). The cluster has 30 metal centers and a nominal stoichiometry: $[Ga_{30}(\mu_4\text{-}O)_{12}(\mu_3\text{-}O)_4(\mu_3\text{-}OH)_4(\mu_2\text{-}OH)_{42}(H_2O)_{16}](2,6\text{-}NDS)_6$, where 2,6-NDS = 2,6-napthalene disulfonate This cluster augments the very small library of Group 13 clusters that have been isolated from aqueous solution and closely resembles one other Ga(III) cluster with 32 metal centers that had been isolated using curcurbit ligands. These clusters have uncommon linked $Ga(O)_4$ centers and sets of both protonated and unprotonated μ_3-oxo.

Reprinted from *Inorganics*. Cite as: Casey, W.H.; Olmstead, M.M.; Hazlett, C.R.; Lamar, C.; Forbes, T.Z. A New Nanometer-Sized Ga(III)-Oxyhydroxide Cation. *Inorganics* **2015**, 3, 21–26.

1. Introduction

Large cations that form in a hydrolyzed solution of Group 13 trivalent metals [1] attract intense interest from a wide range of scientists. Geochemists use these clusters as experimental models to understand reaction dynamics for adsorbate uptake and isotope-exchange pathways affecting the metal-hydroxide solids [2] that make up soil. These clusters have also found a wide range of industrial uses in the semiconducting industry [3], in water treatment [4], in pharmaceutical products and in cosmetics [5]. However, unlike the hundreds of polyoxometalate ions that have been made using Group 5 and 6 metals, only a few dozen cation clusters have been isolated so far from hydrolyzed Group 13 metals, and reports of Ga^{III} clusters are particularly sparse. These clusters tend to fall into two categories: cation derivatives of the Baker-Figgis-Keggin structures, usually the ε-isomer, or a series of "flat" clusters [6–8] that are less symmetric and have no central $M(O)_4$ site.

Focusing on Ga^{III} oxyhydroxo clusters, there is the work of Johnson *et al.* [6–8], who developed the chemistry and applications for the 'flat' clusters [7], which had previously been made only with aminocarboxylate termination ligands to prevent condensation [9]. The existence of a $[GaO_4Ga_{12}(OH)_{24}(OH_2)_{12}]^{7+}$ ion having the structure of the ε isomer of the Keggin series was inferred from X-ray studies of solutions [10] and on pillared clays [9,11]. This Keggin structure of $[GaO_4Ga_{12}(OH)_{24}(OH_2)_{12}]^{7+}$ was predicted by Bradley [12] but has not yet been isolated in a crystal structure in spite of the relative ease with which the Al^{III} version can be crystallized. Fedin's group [12] produced the most noteworthy advance when they used a macrocyclic curcubit ligand to isolate a large Ga(III) polyoxocation with 32 metal centers (henceforth, **Ga₃₂**). This cluster had two sets of corner-shared tetrahedral sites and aspects of the molecule that resemble the "flat" clusters in that it contains sheets of five linked edge-shared $Ga(O)_6$ octahedra with two bridging $Ga(O)_6$ bonded to the sheets via corner-shared μ_2-OH.

2. Results and Discussion

Here we report a similar gallium cluster but with 30 metal centers (henceforth, **Ga₃₀**) that was crystallized from a simple aqueous solution and 2,6-napthalene disulfonate (NDS) as a charge-balancing anion. The crystallizing solution was made with the standard approach used to isolate Al^{III} polyoxocations—a 25 mL of 0.25 M GaCl₃ solution was heated to 80 °C and 60 mL of 0.25 M NaOH solution was added dropwise at rates of 2 mL/min. The resulting solutions were stirred at temperature until precipitate disappeared and then split into aliquots that were sealed into a Teflon-lined reactor and heated further overnight at 80 °C. After 16 h, the solutions were taken from the oven and 1 mL of 0.15 M 2,6-NDS solution was added. After 48 h of aging at room temperature, precipitate was filtered away and the solution left sealed in the dark. After several weeks, small clear crystals became apparent at the bottom of the growth vessels. The final solution had pH = 2.54, which is close to estimates of the first hydrolysis constant for $[Ga(OH_2)_6]^{3+}$ [13].

The structure of the crystal (Table 1) was determined by X-ray methods at the Advanced Light Source. Central to the crystal were **Ga₃₀(NDS)₆** clusters. These had a center of symmetry and a stoichiometry of: $[Ga_{30}(\mu_4\text{-}O)_{12}(\mu_3\text{-}O)_4(\mu_3\text{-}OH)_4(\mu_2\text{-}OH)_{42}(H_2O)_{16}]^{12+}$. Four of the galliums [Ga(1), Ga(2), and symmetry equivalents] have tetrahedral geometry with bound oxygen atoms. The rest are octahedral. Twelve of the oxygen atoms [O(2), O(3), O(4), O(5), O(6), O(7) and symmetry equivalents] are μ_4-oxo bridging four Ga^{III}. Another four oxygen atoms [O(1) and O(10) and symmetry equivalents] are μ_3-oxo bonded to three Ga^{III} and there are, in addition, four μ_3-OH [O(8) and O(9) and symmetry equivalents]. Furthermore there are 42 μ_2-OH [O(11)–O(31) and symmetry equivalents] bridging two metal centers. Finally, there are 16 terminal H₂O molecules [O(32)–O(39) and symmetry equivalents] that are terminally bound to Ga^{III}. Thus, there are a total of 78 protons bonded to oxygens in the cluster. These were located by difference Fourier methods, and also by examination of hydrogen-bonding interactions. There are six (2,6-NDS) dianions. Some of these are disordered with respect to centrosymmetry. A total of 47 hydrate molecules per unit cell were included in the refinement. However, some of the O···O distances are unreasonably short and there is substantial disorder, so the actual number can only be estimated at +/− 5 H₂O. No attempt was made to find the hydrogen atoms for these water molecules, which were not bonded to the **Ga₃₀** cluster.

The **Ga₃₂** and **Ga₃₀** structures (Figure 1) are very close to one another topologically and differ primarily in the existence of the two corner-shared Ga(O)₆ in the **Ga₃₂** that decorate a core similar to the **Ga₃₀** structure. The longest dimensions for the two clusters are 1.78 and 1.51 nm, respectively. Neither molecule closely resembles a Keggin structure, but there are μ_4-oxo linking the Ga(O)₄ tetrahedra to each other and to the outer Ga(O)₆ in the structures. These clusters are also distinct in that the tetrahedral Ga(O)₄ come in two paired sets (Figures 1 and 2). Key structural parameters are reported in the Supplemental Information, along with the structure file for the **Ga₃₀**.

Table 1. The distinction between the μ_3-oxo and μ_3-OH is possible because of the associated Ga-O distances.

μ_3-oxide distances	Å	Bond valence for oxygen
Ga(1)–O(1)	1.847(2)	0.729
Ga(2)–O(1)	1.853(2)	0.717
Ga(10)'–O(1)	1.934(2)	0.576
Ga(5)–O(10)	1.917(2)	0.603
Ga(7)–O(10)	1.887(2)	0.654
Ga(14)–O(10)	1.909(3)	0.616
μ_3-hydroxide distances	Å	Bond valence for oxygen
Ga(5)–O(8) *	2.042(3)	0.43
Ga(10)–O(8) *	2.078(3)	0.39
Ga(14)–O(8) *	2.026(2)	0.449
Ga(7)–O(9) *	2.078(2)	0.39
Ga(12)–O(9) *	2.019(2)	0.458
Ga(13) –O(9) *	2.026(3)	0.449

* The BVS for O(1) is 2.02 and for O(10) it is 1.87. Without the inclusion of hydrogen, the BVS for O(8) is 1.27 and for O(9) it is 1.30. With hydrogen included, the BVS for O(8) is 1.92 and for O(9) it is 1.96 [14].

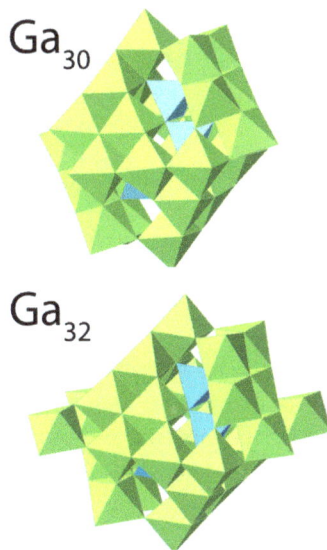

Figure 1. The topology of the **Ga₃₀** cluster shown in polyhedral representation and in a similar orientation as that of the **Ga₃₂** discovered by Gerasko *et al.* [12]. The central Ga(O)₄ sites in tetrahedral coordination are colored blue and the Ga(O)₆ are green.

4

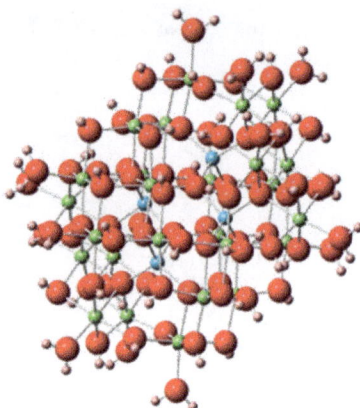

Figure 2. The Ga$_{30}$ cluster shown in ball-and-stick formalism and in a similar orientation as in the previous figure. Red spheres are oxygens, white are protons and Ga(III) are green if octahedrally coordinated and blue if tetrahedrally coordinated. The cluster stoichiometry is: $[Ga_{30}(\mu_4\text{-}O)_{12}(\mu_3\text{-}O)_4(\mu_3\text{-}OH)_4(\mu_2\text{-}OH)_{42}(H_2O)_{16}]^{12+}$ with all terminal oxygens as bound waters.

3. Conclusions

A new 30-metal-center oxy-hydroxide cation cluster was synthesized by hydrolysis of aqueous GaCl$_3$ and has a nominal stoichiometry: $[Ga_{30}(\mu_4\text{-}O)_{12}(\mu_3\text{-}O)_4(\mu_3\text{-}OH)_4(\mu_2\text{-}OH)_{42}(H_2O)_{16}]^{12+}$. This cluster augments the very, very small library of Ga^{III} polyoxocations. This is the second such cluster, both having similar structure, formed as a Ga(III)-hydroxide cation, suggesting that the reports of ε-Keggin-structured gallium molecules, like the $[GaO_4Ga_{12}(OH)_{24}(OH_2)_{12}]^{7+}$ may actually be this derivative structure, one of only three gallium-hydroxide cations that have been so isolated.

Acknowledgments

This work was supported by an NSF CCI grant through the Center for Sustainable Materials Chemistry, number CHE-1102637 and via NSF-CHE-1310368 to WHC. Support from the Advanced Light Source, supported by the Director, Office of Science, Office of Basic Energy Sciences, of the U.S. Department of Energy under Contract No. DE-AC02-05CH11231 is greatly appreciated. The authors thank Prof. Darren Johnson at the University of Oregon for encouragement. Images were generated using CrystalMaker®: a crystal and molecular structures program. CrystalMaker Software Ltd, Oxford, England (www.crystalmaker.com).

Author Contributions

The work was conceived by WHC and TF, with crystallization experiments undertaken by CL and CRH under direction of WHC and TF. Both Lamar and Hazlett were summer students visiting the Casey laboratory from their home institutions when these experiments were conducted. The X-ray data were collected and the structure solved by MMO. The manuscript was written with contributions from all authors. †These authors contributed equally. The structure is deposited in the Cambridge Structural Database as CCDC 1038690.

Conflicts of Interest

The authors declare no conflict of interest.

References

1. Casey, W.H. Large Aqueous Aluminum Hydroxide Molecules. *Chem. Rev.* **2006**, *106*, 1.
2. Casey, W.H.; Rustad, J.R. Reaction Dynamics, Molecular Clusters, and Aqueous Geochemistry. *Annu. Rev. Earth Sci.* **2007**, *35*, 21–27.
3. Johnson, D.W.; Rather, E.; Gatlin, J.T. Methods for producing gallium and other oxo/hydroxo-bridged metal aquo clusters. US Patent 20060199972 A1, 7 September 2006.
4. Stewart, T.A.; Trudell, D.E.; Alam, T.M.; Ohlin, C.A.; Lawler, C.; Casey, W.H.; Jett, S.; Nyman, M. Enhanced Water Purification: A Single Atom Makes a Difference. *Environ. Sci. Technol.* **2009**, *43*, 5416–5422.
5. Laden, K. *Antiperspirants and Deodorants*; Marcel Dekker: New York, NY, USA, 1999.
6. Rather, E.; Gatlin, J.T.; Nixon, P.G.; Tsukamoto, T.; Kravtsov, V.; Johnson, D.W. A Simple Organic Reaction Mediates the Crystallization of the Inorganic Nanocluster $[Ga_{13}(\mu_3\text{-}OH)_6 (\mu_2\text{-}OH)_{18}(H_2O)_{24}](NO_3)_{15}$. *J. Am. Chem. Soc.* **2005**, *127*, 3242–3243.
7. Mensinger, Z.L.; Wang, W.; Keszler, D.A.; Johnson, D.W. Oligomeric group 13 hydroxide compounds—a rare but varied class of molecules. *Chem. Soc. Rev.* **2012**, *41*, 1019–1030.
8. Oliveri, A.F.; Carnes, M.E.; Baseman, M.M.; Richman, E.K.; Hutchison, J.E.; Johnson, D.W. Single Nanoscale Cluster Species Revealed by ^{1}H NMR Diffusion-Ordered Spectroscopy and Small-Angle X-ray Scattering. *Angew. Chem. Int. Ed.* **2012**, *51*, 10992–10996.
9. Goodwin, J.C.; Teat, S.J.; Heath, S.L. How Do Clusters Grow? The Synthesis and Structure of Polynuclear Hydroxide Gallium(III) Clusters. *Angew. Chem. Int. Ed.* **2004**, *43*, 4037–4041.
10. Michot, L.J.; Montarges-Pelletier, E.; Lartiges, B.S.; d'Espinose de la Caillerie, J.-B.; Briois, V. Formation of the Ga_{13} Keggin ion: a combined EXAFS, and NMR study *J. Am. Chem. Soc.* **2000**, *122*, 6048–6056.
11. Bradley, S.M.; Kydd, R.A.; Yamdagni, R.; Fyfe, C.A. Gallium (Ga13), gallium-aluminum (GaAl12), and aluminum (Al13) polyoxocations and pillared clays. *Synth. Microporous Mater.* **1992**, 13–31.
12. Gerasko, O.A.; Mainicheva, E.A.; Naumov, D.Y.; Kuratieva, N.V.; Sokolov, M.N.; Fedin, V.P. Synthesis and Crystal Structure of Unprecedented Oxo/Hydroxo-Bridged Polynuclear Gallium(III) Aqua Complexes. *Inorg. Chem.* **2005**, *44*, 4133–4135.
13. Richens, D.T. *The Chemistry of Aqua Ions*; John Wiley: New York, NY, USA, 1997.
14. Brown, I.D. *The Chemical Bond in Inorganic Chemistry: The Bond Valence Model*; Oxford University Press: New York, NY, USA, 2002.

A Heteroaromatically Functionalized Hexamolybdate

Merinda R. Healey, Stephen P. Best, Lars Goerigk and Chris Ritchie

Abstract: A new heteroaromatic thiophene containing organoimido functionalized hexamolybdate has been synthesized and characterized in both solid form and solution. Structural analysis shows successful introduction of the organoimido ligand through replacement of one terminal oxo site on $[Mo_6O_{19}]^{2-}$ to yield the singly functionalized hexamolybdate. Spectroscopic and theoretical analysis indicates charge transfer between the inorganic and organic components, with a significantly red-shifted lowest lying transition of 399 nm *vs.* the parent Lindqvist ion of 325 nm. Additional characterization includes, thermal gravimetric analysis (TGA), infrared (IR), cyclic voltammetry (CV), nuclear magnetic resonance (NMR) and time-dependent density functional theory (TD-DFT) studies.

Reprinted from *Inorganics*. Cite as: Healey, M.R.; Best, S.P.; Goerigk, L.; Ritchie, C. A Heteroaromatically Functionalized Hexamolybdate. *Inorganics* **2015**, *3*, 82-100.

1. Introduction

Recent interest in the development of methodologies for the covalent grafting of organic molecules to the surface of polyoxometalates (POMs) has been driven by the utility of the resulting inorganic-organic hybrids in a range of applications [1–6]. One such methodology can be used to yield organoimido functionalized Lindqvist structures of general formula $[Mo_6O_{(19-x)}(NR)_x]^{2-}$ where x = 1–6 and (NR) can be aromatic or aliphatic in nature [7]. Since the first report of organoimido functionalized hexamolybdates, several groups have shown interest in the synthesis of this structural class, and have subsequently reported a variety of modified synthetic approaches and reaction conditions to achieve this goal [8–10]. That being said, drawing clear conclusions as to a rational synthetic approach from the literature is challenging when targeting multiple organoimido-substituted hexamolybdates. The use of aniline-based hydrochloride salts is a general synthetic route to the formation of mono-functionalized hexamolybdates and was first reported by Wei and Guo [11]. Both the aniline and its HCl salt were used for the synthesis of mono-functionalized complexes containing electron-withdrawing groups. This route involves the reaction of the α isomer of octamolybdate $[Mo_8O_{26}]^{4-}$ with the hydrochloride salt of the aromatic amine of choice in dry acetonitrile in the presence of the dehydrating and activating agent N,N'-Dicyclohexylcarbodiimide (DCC) [12].

Post-synthetic reactions involving organoimido hexamolybdates containing terminal halogen or ethynyl groups has now been well documented [9,13–17]. Both Sonogashira and Heck Pd-catalyzed carbon-carbon coupling reactions involving organoimido hexamolybdate complexes have been reported, showing their promise as building blocks in the design of elaborate molecular and polymeric compounds. Several key examples of post-synthetic Sonogashira and Heck couplings include the preparation of complexes bearing metal binding ligands [17] and the incorporation of the hexamolybdate in main chain polymers [16], as well as the preparation of polymers with POM containing pendant side chains [9]. From a synthetic standpoint it is significant to note the varying stabilities of $[Mo_6O_{(19-x)}(NR)_x]^{2-}$ complexes to the post-synthetic conditions used in these reactions, in particular the tolerance of the base. The complex reported herein is currently under investigation regarding its utility for inclusion in materials prepared using these methodologies.

2. Results and Discussion

Here, we report a novel mono-functionalized organoimido hexamolybdate $[Mo_6O_{18}(\textbf{L4})]^{2-}$ (**1-L4**) where **L4** = 4-(4-bromo-5-methylthiophen-2-yl)-2,6-dimethylaniline. **L4** was synthesized via the palladium catalysed cross coupling of 3,5-dibromo-2-methylthiophene (**L1**) with {4-[(diphenylmethylidene)amino]-3,5-dimethylphenyl}boronic acid (**L2**) to yield *N*-[4-(4-bromo-5-methylthiophen-2-yl)-2,6-dimethylphenyl]-1,1-diphenylmethanimine (**L3**) that was then subsequently deprotected and used as the HCl salt, as shown in (Scheme 1). The synthesis of (**1-L4**) involved the preparation of a solution of $(Bu_4N)_4[Mo_8O_{26}]$ (0.0561 mmol) in dry acetonitrile (2 mL) to which 1.34 equivalents of 4-(4-bromo-5-methylthiophen-2-yl)-2,6-dimethylaniline and 2 equivalents of DCC were added. This solution was then sealed in a 5 mL biotage microwave vial and reacted at 120 °C for 90 min at 4 bar in a biotage initiator microwave reactor. On cooling to room temperature, the reaction mixture was filtered and the solvent removed under vacuum. The resulting oil was washed with diethyl ether and subsequently recrystallized from acetonitrile via diethyl ether vapor diffusion. Fractional crystallization was then required, with the yellow tetrabutylammonium (TBA) salt of $[Mo_6O_{19}]^{2-}$ crystalizing initially. The remaining solution was decanted, concentrated and subjected to further diethyl ether diffusion resulting in the desired compound, in moderate yield, over 5–10 days.

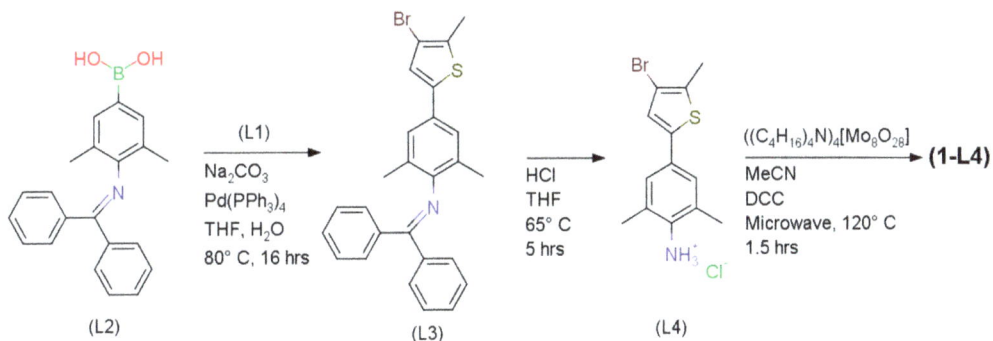

Scheme 1. Synthesis of the monofunctionalized organoimido hybrid (**1-L4**).

The molecular structure of (**1-L4**) (Figure 1) is characteristic of previously reported mono-imido hexamolybdates, with one of the terminal $[Mo\equiv O]^{4+}$ units being replaced by a more electron-rich $[Mo\equiv NR]^{4+}$ unit. The short Mo–N bond length (1.768(7) Å) and Mo–N–C bond angle (169.1(7)°) are in agreement with the assignment, and indicative of an Mo≡N triple bond character. Typical displacement of the central oxygen atom (O_c) within the hexamolybdate complex towards [O$_c$–Mo≡NR] (2.205(5) Å) *vs.* the average of the remaining non-substituted sites [O$_c$–Mo≡Ot] (2.34(1) Å) is also observed. Disorder of the thiophene portion of (**1-L4**) is observed due to free rotation around the single bond connecting the thiophene and aniline aromatic systems. The observed positions of carbon atoms C9, C12, and C13 were not affected by this disorder, whereas the positions of S1, C10, C11 the methyl substituent C14 and the bromine substituent Br1 were. Their positions were modeled using free variables, and by constraining the disordered atoms displacement parameters, atomic positions and bond lengths. The two observed positions refine well with occupancies of 58% and 42%. Furthermore, in the solid state the thiophene rings of two neighboring molecules are involved in pi-pi interactions resulting in the formation of a dimeric motif (Figure 1). The intermolecular separation was found to be 3.529(1) Å from ring center to ring center. Two charge balancing tetrabutyl-ammonium (TBA) cations are also required however only one could be successfully modeled, with the other showing significant disorder. A solvent mask was, therefore, implemented after the assignment of all locatable atoms from the difference map, with chemical and thermal analysis being consistent with the assignment of two TBA cations and the absence of any additional solvate in the bulk sample.

The electronic spectroscopy of (**1-L4**) (Figure A2) clearly shows a significant bathochromic shift and increase in intensity of the complexes lowest energy electronic absorption when compared to the parent plenary Linqvist polyanion [Mo$_6$O$_{19}$]$^{2-}$. Indeed this 72 nm shift to longer wavelengths (**1-L4** 399 nm *vs.* [Mo$_6$O$_{19}$]$^{2-}$ 325 nm) is among the largest observed for any mono-functionalized organoimido hexamolybdate with an $\varepsilon = 2.47 \times 10^4$ M^{-1} cm^{-1}. This absorption seems to be similar to other mono-functionalized compounds of extended conjugation presented in literature, such as that presented by Peng (382 nm, 392 nm, 403 nm) [15,17,18]. This significant bathochromic shift is indicative of extended conjugation between the organic and inorganic components, with the presence of the electron withdrawing –Br attached on the thiophene ring probably aiding electronic mobility within the hybrid material.

The nature of the absorption band for **1-L4** is further analyzed with linear-response time-dependent density functional theory (TD-DFT) calculations. Calculations at the TDA-CAMB3LYP/def2-TZVPP [19–21] //TPSS-D3/def2-TZVP [21–24] level of theory show qualitative agreement with experiment with a bright transition at 390 nm. The blue-shift of 9 nm compared to the experimental value is within the usual error margin for this level of theory [25] and expected for gas-phase calculations. Further analysis reveals that this bright electronic excitation is dominated by a transition from the highest occupied (HOMO) to the lowest unoccupied molecular orbital (LUMO). The HOMO is predominantly localized around the imido bridge and the adjacent six-membered ring, while the LUMO is additionally characterized by a strong contribution from the thiophene unit (Figure A9a,b). Inspection of the difference density (Figure A9c) further confirms that this excitation is of charge-transfer (CT) character with the imido bridge functioning as the electron donor and the thiophene as the electron acceptor; see also References [26] and [27] for related computational work.

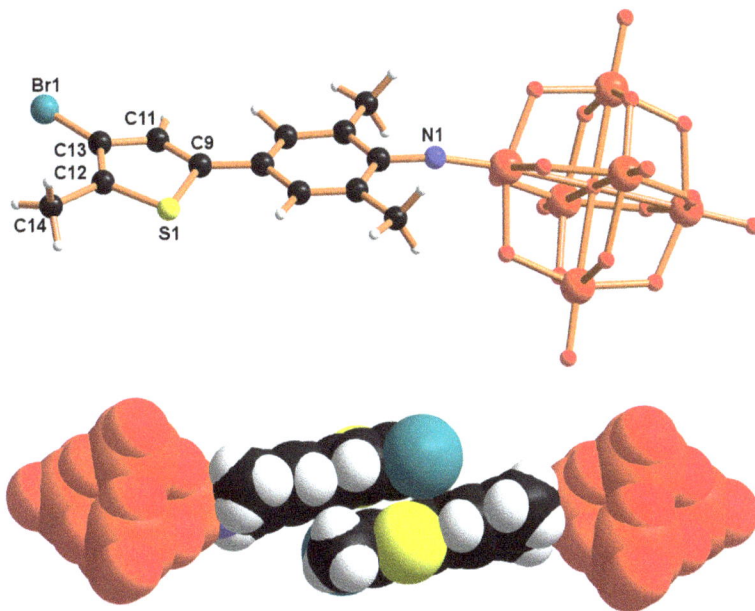

Figure 1. Graphical representation of (**1-L4**) (**top**) and space filling representation of the pi-stacked dimers observed in the solid state (**bottom**). Mo = Orange spheres; O = Red spheres; N = Blue sphere, C = Black spheres; S = Yellow sphere; Br = Teal sphere; H = White spheres [29].

The initial voltammetric response from solutions of (**1-L4**) reveal a reversible reduction with $E^{\circ\prime}$ = −1.07 V with an additional weak, reversible process with $E^{\circ\prime}$ = −0.92 V, due to contamination of the sample by $[Mo_6O_{19}]^{2-}$ (Figure 2). The presence of $[Mo_6O_{19}]^{2-}$ in the sample allows quantification of the shift in reduction potential with functionalization of the plenary Lindqvist polyanion. A scan to strongly reduce potentials lower than −1.65 V is accompanied by formation of daughter products which give rise to distinct anodic processes at potentials above 0.1 V (Figure A8), which are due to decomposition products. Similar anodic waves have previously been attributed to the oxidation of isopoly blues of unknown composition [28]. Consistent with earlier reports, there is evidence for surface adsorption of the reduced polyoxo species with subsequent voltammograms featuring increasingly current with broad and sharp features consistent with a mixture of surface immobilized and solute species.

FTIR spectroscopy of (**1-L4**) (Figure A4) shows the presence of several bands associated with the polyanion such as $v(Mo\text{-}O_t)$ and $v(Mo\text{-}O_b\text{-}Mo)$ stretches at 950 and 794 cm^{-1}, respectively. The characteristic $v(Mo\text{-}N)$ is observed as a sharp shoulder band at 975 cm^{-1}.

The solution stability of (**1-L4**) has been confirmed by ^1H NMR studies (Figure A7), with all protons being unambiguously assigned *versus* that observed for the (**L4**) starting material. Aryl aniline protons give a singlet at 7.30 ppm 2H, meanwhile the thiophene proton also gives a singlet at 7.23 ppm 1H. These peaks integrate well with the two sets of aniline methyl protons 2.62 ppm 6H and thiophene methyl protons 2.43 ppm 3H. Signals at 0.97, 1.35, 1.60, and 3.08 ppm are attributed to the protons in the tetrabutylammonium counterions. Higher integrations of the cation protons indicate the possibility of some residual hexamolybdate that could not be separated via fractional crystallization or during the isolation process. (1-L4) was also studied by LC-MS showing isotopic cluster anions centered at m/z 579.16, and 1159.39. The signals are thus assigned as the parent cluster M^{2-} (M = [$Mo_6O_{18}NC_{13}SH_{12}Br$]), calculated 579.17 and $M^{2-} + H^+$, calculated 1159.38 (Figure A1).

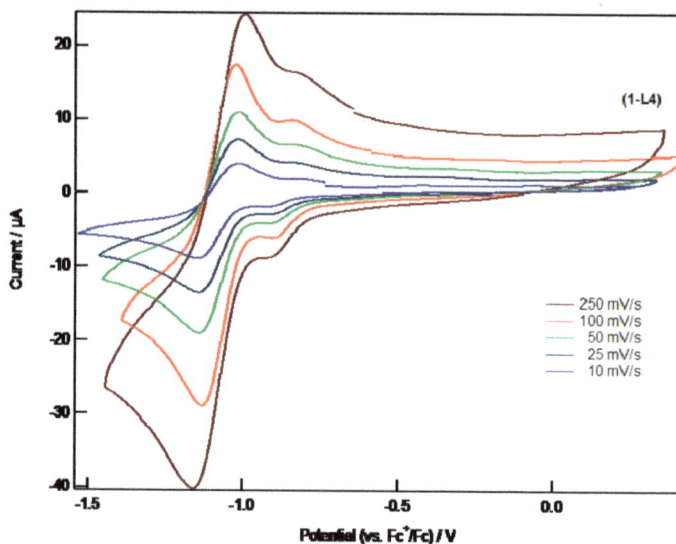

Figure 2. Cyclic voltammetry of a 2 mM solution of (1-L4) with (Bu_4N)PF_6 (0.1 M) in CH_3CN. The scans were recorded from a freshly polished electrode using Pt working (3 mm diameter) and counter electrodes and an Ag pseudo-reference electrode. Potentials were corrected *vs.* ferrocene.

3. Conclusions

A new heteroaromatically derivatized hexamolybdate has been synthesized and extensively characterized. Theoretical TD-DFT calculations qualitatively agree with experimental observations, with the lowest lying excitation having charge transfer character whereby the HOMO is localized around the organoimido bridge and the LUMO on the thiophene. The electrochemical response of (1-L4) is similar to that of the parent Lindqvist polyanion with the potential of the first reversible reduction shifted cathodically by 150 mV as a result of derivitization. The reversibility of the process indicates that the reduced compound does not undergo dissociation of the ligand or fragmentation over the timeframe of the cyclic voltammetric experiment. Further reduction results in rapid decomposition of the complex. The molecule represents a rare example of a halogenated organoimido functionalized hexamolybdate with potential for further post-synthetic modification as a result, which is currently under investigation.

Acknowledgments

CR and LG are recipients of Australian Research Council Discovery Early Career Researcher Awards (project numbers DE130100615 and DE140100550) LG also acknowledges funding from the Selby Scientific Foundation through the 2014 Selby Research Award and generous allocation of computing time from the National Computational Infrastructure (NCI) National Facility within the National Computational Merits Allocation Scheme (project fk5). MRH acknowledges funding from the Australian Government and the University of Melbourne in the form of an Australian Postgraduate Award (APA).

Author Contributions

CR conceived the work, with the synthetic work, crystallization and characterization completed by MRH. The X-ray data was collected by MRH and the structure solved and refined with the assistance of CR. SPB, with assistance of MRH conducted the electrochemistry and LG conducted the TD-DFT calculations. The manuscript was written with contributions from all authors.

Appendix

IR, ^{1}H NMR spectra, TGA and X-ray data collection and refinement details of **L3, L4** and **1-L4** are available where applicable. Crystal structures of **(1-L4)** and **(L3)** are deposited in the Cambridge Structural Database as CCDC 1057033 and 1057032.

A1. Instrumentation

A1.1. FT-IR Spectroscopy (KBr Disc)

FT-IR spectroscopy was performed on a Bruker Tensor 27 FT-IR spectrometer (University of Melbourne, Melbourne, Australia). Samples were prepared as KBr pellets. Signals are listed as wavenumbers (cm^{-1}) with the following abbreviations: vs = very strong, s = strong, m = medium, w = weak and b = broad.

A1.2. Elemental Analysis

Chemical analysis was performed on Carlo Erba Elemental Analyser EA 1108, (The Campbell Microanalytical Laboratory, Department of Chemistry, University of Otago, Otago, New Zealand).

A1.3. Microwave Reactor

Synthesis of **(1-L4)** was performed using a Biotage Initiator (University of Melbourne, Melbourne, Australia).

A1.4. UV-Vis Spectrometer

UV-Vis spectroscopy was performed on Agilent Technologies Cary 60 UV-Vis (University of Melbourne, Melbourne, Australia), using standard quartz cuvettes ($d = 1$ cm).

A1.5. Mass Spectroscopy (ESI)

Mass spectroscopy was performed on a high resolution Agilent QTOF LCMS 6520 (University of Melbourne, Melbourne, Australia).

A1.6. Thermogravimetric Analysis (TGA)

Thermal analysis was performed on a Mettler TGA/SDTA851^e Module (University of Melbourne, Melbourne, Australia). Method: gas flow: N_2 at 25 mL/min, temperature: 25 °C (5 min), 25–450 °C (rate: 5 °C/min)

A1.7. NMR Spectroscopy

NMR spectroscopy was performed on a Varian 400 MHz NMR Spectrometer (University of Melbourne, Melbourne, Australia).

A1.8. Column Chromatography

Chromatographic separations were performed using a Grace Technologies, Reveleris X2 (University of Melbourne, Melbourne, Australia).

A1.9. Electrochemistry

Electrochemical experiments were conducted using a purpose built cell previously described [30]. Experiments employed 3 mm diameter platinum working, silver pseudo-reference and platinum foil counter electrodes. Solutions for electrochemical analysis were prepared under strictly anaerobic conditions using a Vacuum Atmospheres glove box. The applied potential was controlled using a PAR model 362 potentiostat where waveforms were generated using EChem V1.5.2 software in conjunction with a Powerlab 4/20 interface (ADInstruments, University of Melbourne, Melbourne, Australia).

A2. Experimental Section

Chemicals were used as purchased without further purification. Solvents were degassed and dried over 3 Å molecular sieves using standard laboratory procedures. Previously reported compounds (**L1**) and (**L2**) were synthesised as described in the original paper [31].

Synthesis of (L3): *N*-[4-(4-bromo-5-methylthiophen-2-yl)-2,6-dimethylphenyl]-1,1-diphenylmethanimine

To a solution of *N*-(4-borate-2,6-dimethylphenyl)-1,1-diphenylmethanimine (820 mg, 2.25 mmol), Na_2CO_3 (360 mg, 3.40 mmol) and 3,5-dibromo-2-methylthiophene (563 mg, 2.25 mmol) in degassed THF (5 mL) and degassed water (3 mL), was added $Pd(PPh_3)_4$ (26 mg, 0.02 mmol) and the solution was purged with N_2 for 5 min. The reaction vessel was then sealed and heated to 80 °C for 16 h. Upon cooling, saturated aqueous NH_4Cl was added and the solution extracted with diethyl ether and dried with magnesium sulphate. The solution was then dried in vacuo and purified using column chromatography (silica, hexane : ethyl acetate, 0–5% ethyl acetate gradient flow over 10 min, R_f – 0.41 5% Ethyl acetate in hexane), to give a yellow solid. (1.036 g, 0.827 mmol, 37% yield). Recrystallization from acetone afforded large yellow blocks after 2 days via slow evaporation. ^{1}H NMR (CDCl$_3$, 400 MHz): δ 2.06 (s, 6H, Ar–CH$_3$), 2.38 (s, 3H, Ar–CH$_3$), 6.97 (s, 1H, ArH), 7.06 (s, 2H, ArH), 7.12(t, 2H, ArH), 7.23(t, 2H, ArH), 7.29 (t, 1H, ArH), 7.43 (t, 2H, ArH), 7.50 (t, 1H, ArH), 7.79 (d, 2H, ArH). Elemental analysis (%) calcd. For C$_{26}$H$_{22}$NSBr: C, 67.81; H, 4.82; N, 3.04. Found: C, 68.05; H, 4.96; N, 3.01. Selected IR data (KBr, cm^{-1}): 3053 (m), 3026 (w), 2966 (m), 2912 (m), 2853 (w), 1620 (vs), 1597 (s), 1576 (m), 1539 (m), 1490 (s), 1462 (s), 1445 (vs), 1377(w), 1329 (w), 1313 (s), 1286 (s), 1227 (s), 1180 (m), 1163 (m), 1137 (s), 1059 (s), 1028 (m), 999 (m), 976 (w), 949 (s), 912 (m), 885 (m), 864 (m), 849 (m), 820 (vs), 785 (s), 771 (s), 760 (m), 746 (m), 700 (vs), 648 (s), 617 (w), 590 (m), 557 (m), 505 (m).

Synthesis of (L4): (4-bromo-5-methylthiophen-2-yl)-2,6-dimethylaniline·HCl

To a solution of *N*-[4-(4-bromo-5-methylthiophen-2-yl)-2,6-dimethylphenyl]-1,1-diphenylmethanimine (1.00 g, 2.17 mmol) in THF (20 mL) was added HCl (2M, 10 mL) and heated at 65 °C for 5 h. The solution was then cooled, dried in vacuo and washed with ether to give an off white product (642.8 mg, 1.93 mmol, 89% yield). ^{1}H NMR (d6-DMSO, 400 MHz): δ 2.19 (s, 6H, Ar–CH$_3$), 2.34 (s, 3H, Ar–CH$_3$), 7.19 (s, 2H, ArH), 7.21 (s, 1H, ArH). Elemental analysis (%) calcd. For C$_{13}$H$_{15}$NSClBr: C, 46.93; H, 4.55; N, 4.21. Found: C, 46.37; H, 4.48; N, 4.19. Selected IR data (KBr, cm^{-1}): 3058(w), 2922 (w), 2852 (w), 2581 (w), 1621 (s), 1521(w), 1494 (m), 1468 (w), 1384 (m), 1318 (m), 1099 (w), 1063 (w), 818 (w), 786 (m).

Synthesis of (1-L4): Mo$_6$O$_{18}$(4-(4-bromo-5-methylthiophen-2-yl)-2,6-dimethylaniline)

A solution of (Bu$_4$N)$_4$[Mo$_8$O$_{26}$] (120.8 mg, 0.0561 mmol), dicyclohexylcarbodiimide (23.1 mg, 0.112 mmol), 4-(4-bromo-5-methylthiophen-2-yl)-2,6-dimethylaniline.HCl (25 mg, 0.0751 mmol) in dry acetonitrile was heated in a microwave reactor at 120 °C for 90 min. The solution was then cooled, filtered and the remaining solvent removed in vacuo. The oil was then washed with diethyl ether and dissolved in acetonitrile with vapor diffusion of diethyl ether resulting in the crystallization of hexamolybdate, which was removed by filtration. Further diffusion resulted in the crystallization of the desired red product (12 mg, 0.00730 mmol, 13% yield) after one week. ^{1}H NMR (CD$_3$CN, 400 MHz): δ 0.97 (t, 24H, –CH$_3$, [Bu$_4$N]$^+$), 1.35 (m, 16H, –CH$_2$–, [Bu$_4$N]$^+$), 1.60 (m, 16 H, –CH$_2$–, [Bu$_4$N]$^+$), 2.43 (s, 3H, Ar–CH$_3$), 2.62 (s, 6H, Ar–CH$_3$), 3.08 (m, 16H, –CH$_2$–, [Bu$_4$N]$^+$), 7.23 (s, 1H, ArH), 7.30 (s, 2H, ArH). Elemental analysis (%) calcd. For C$_{45}$H$_{84}$N$_3$SBrMo$_6$O$_{18}$: C, 32.90; H, 5.16; N, 2.56. Found: C, 32.98; H, 5.25; N, 2.53. Selected IR data (KBr, cm^{-1}): 2961(s), 2932 (w), 2873 (m), 1626 (m), 1595 (w), 1479 (m), 1382 (m), 1320 (m), 1161 (w), 975 (s, Mo–N), 950 (vs, Mo–O$_t$), 880 (w), 794 (vs, Mo–O–Mo), 590 (m), 448 (m).

A3. Characterization Section

A3.1. Mass Spectroscopy

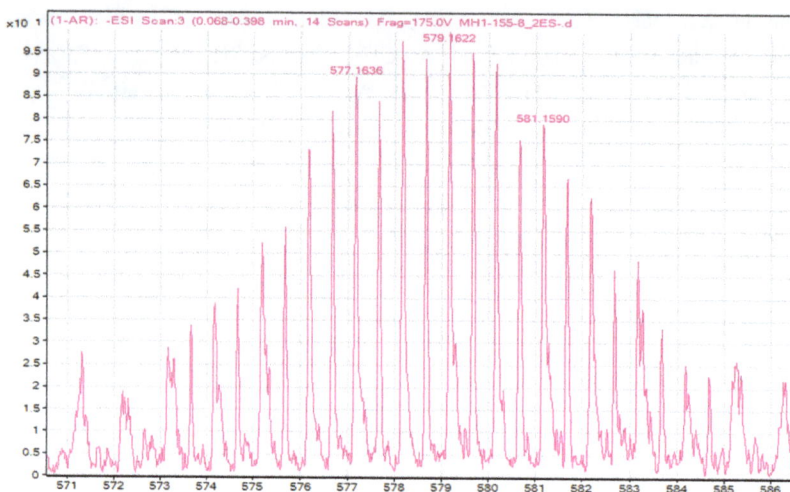

Figure A1. Mass Spectrum of Compound (**1-L4**) (M^{2-}, m/z = 579.16). ESI mass spectrometry was performed in acetonitrile in negative ion mode.

A3.2. Electronic Spectroscopy

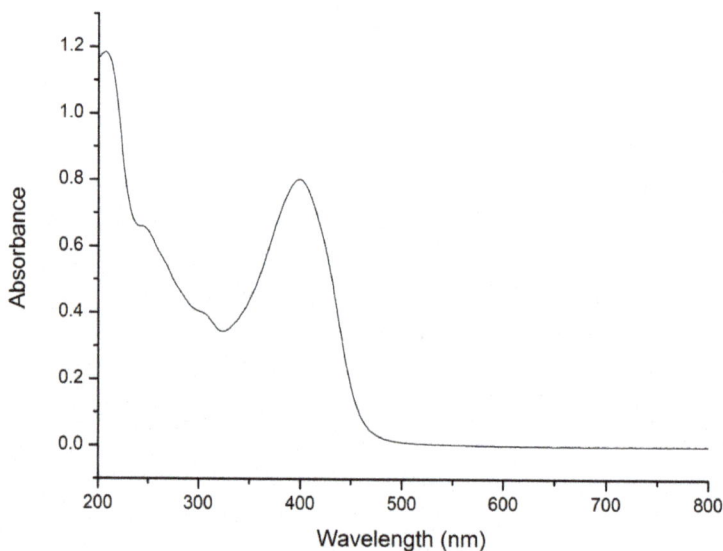

Figure A2. UV-Vis spectrum of (**1-L4**) in MeCN: λ_{max1} = 399 nm, ε = 2.47 × 10^4 M^{-1} cm^{-1}, λ_{max1} = 305 nm, ε = 1.23 × 10^4 M^{-1} cm^{-1}, λ_{max1} = 246 nm, ε = 2.03 × 10^4 M^{-1} cm^{-1}, λ_{max1} = 207 nm, ε = 3.65 × 10^4 M^{-1} cm^{-1}.

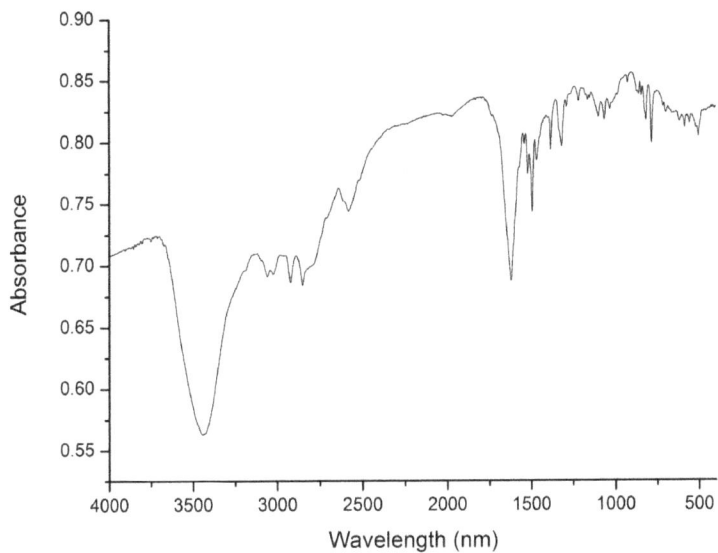

Figure A3. IR spectrum of (**L4**).

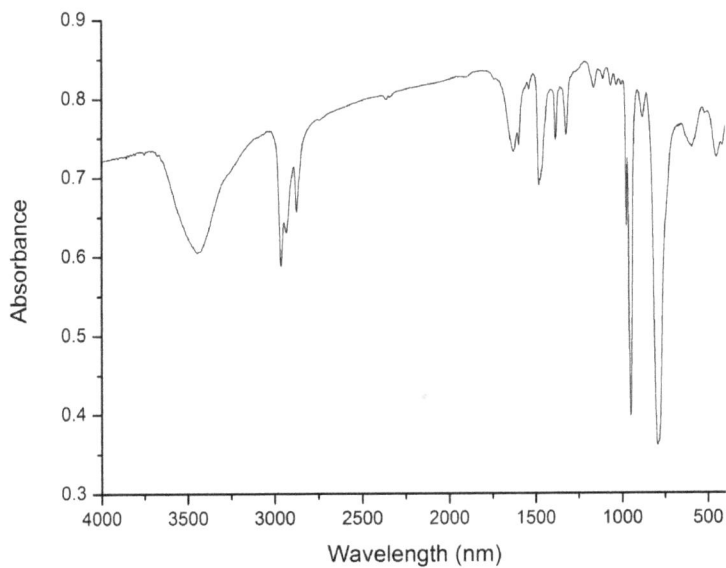

Figure A4. IR spectrum of (**1-L4**).

A3.3. Nuclear Magnetic Resonance Spectroscopy

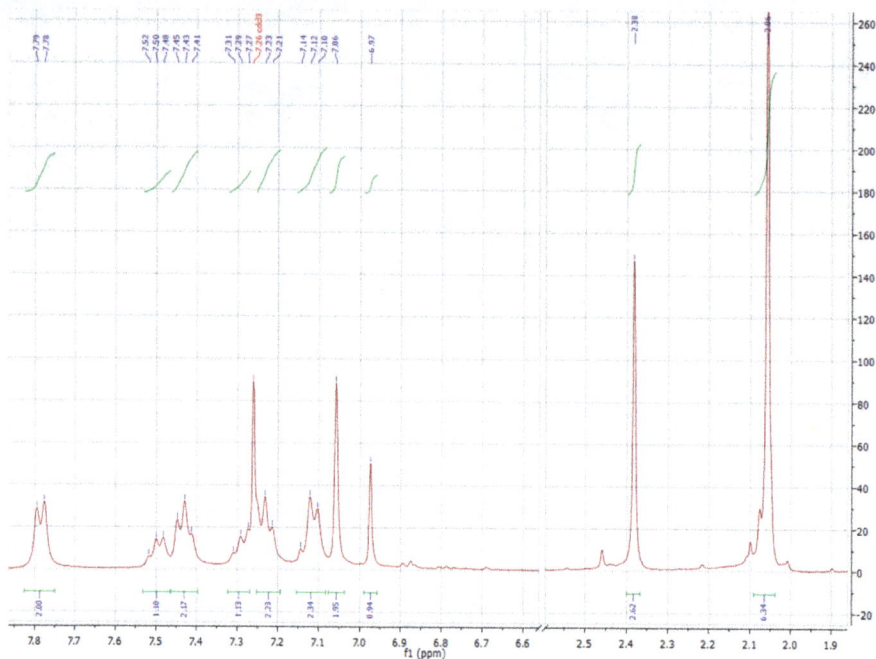

Figure A5. NMR Spectrum (400 MHz) of (**L3**) in CDCl₃.

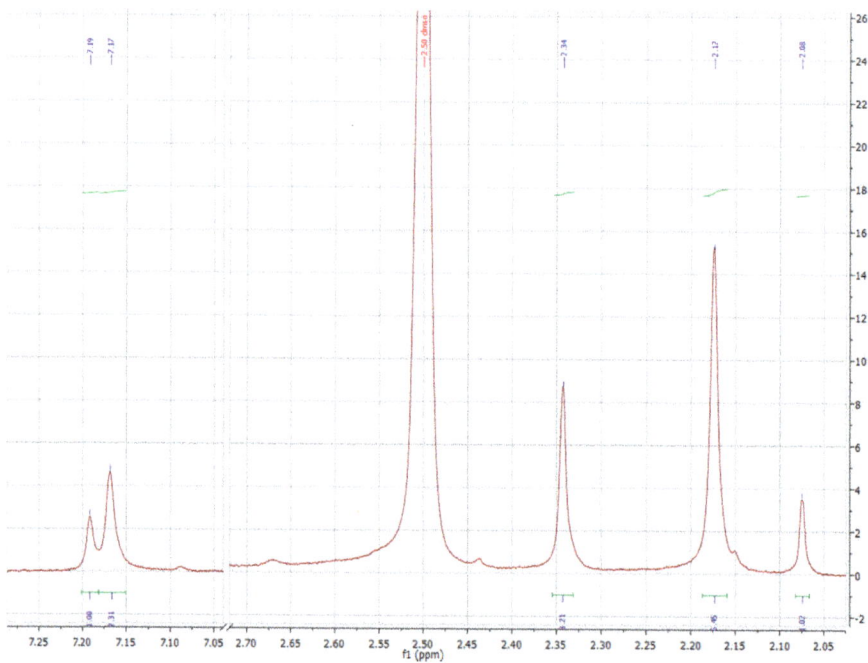

Figure A6. NMR Spectrum (400 MHz) of (**L4**) in d6-DMSO.

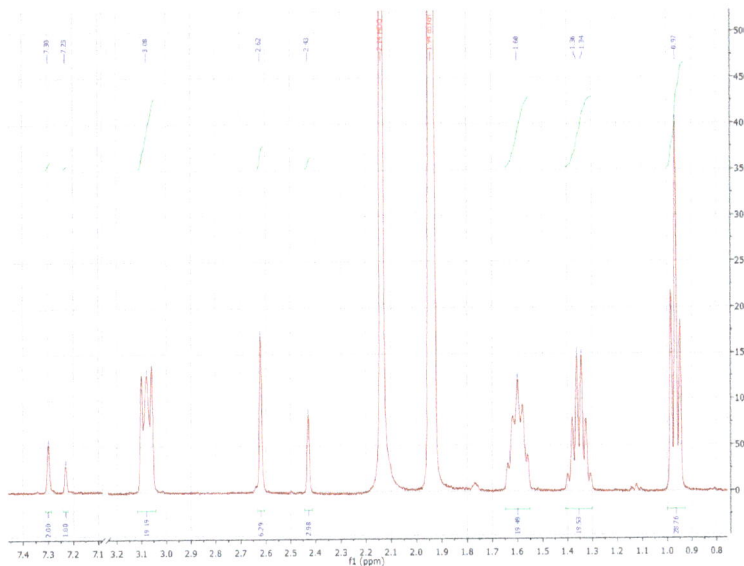

Figure A7. NMR Spectrum (400 MHz) of (**1-L4**) in d3-MeCN.

A3.4. Electrochemistry

Following the completion of voltammetric experiments a sample of ferrocene (Fc) was added to the solution and all potentials are referenced against the Fc^+/Fc couple.

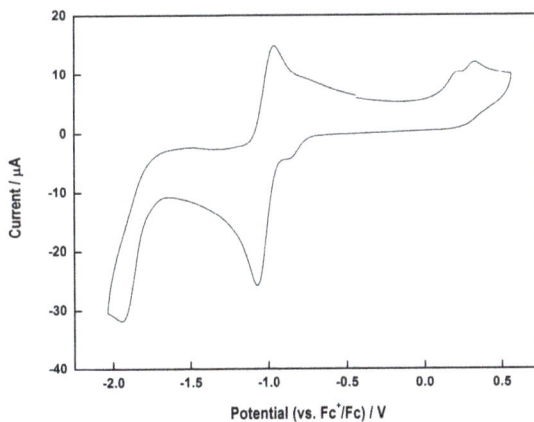

Figure A8. Cyclic voltammetry of (**1-L4**) showing the second reduction and accompanying anodic processes.

A3.5. Computational Details

The geometry optimization of one (**1-L4**) unit was carried out with TURBMOLE 6.4 [32], the TPSS [22] meta-generalized-gradient-approximation density functional, and Ahlrichs' triple-ζ atomic-orbital (AO) basis set with one set of polarization functions (def2-TZVP) [21]. The modified Stuttgart-Dresden def2-ECP [21] effective core potential was applied to the Mo atoms. Grimme's DFT-D3 [23] London-dispersion correction with Becke-Johnson damping [24] was applied during this step. Subsequent linear-response time-dependent density functional theory calculations were carried out with ORCA 3.0.2 [33] within the Tamm-Dancoff approximation (TDA) [19]. For all heavy atoms, the def2-TZVPP AO basis set was applied, while hydrogen atoms were treated with the def2-SV double-ζ basis to speed up the calculations [21]. The def2-ECP was again applied to all Mo atoms.

The results discussed in the manuscript were obtained with the range-separated CAMB3LYP density functional [20]. Additional calculations with the range-separated wB97X-D3 [34] functional confirmed the charge-transfer character of the first bright excitation and were in qualitative agreement with the CAMB3LYP calculations (absorption at 379 nm). The resulting highest occupied and lowest unoccupied molecular orbitals obtained at the CAMB3LYP level are shown in Figure A9, along with the difference density, and they clearly indicate the charge-transfer character of the excitation.

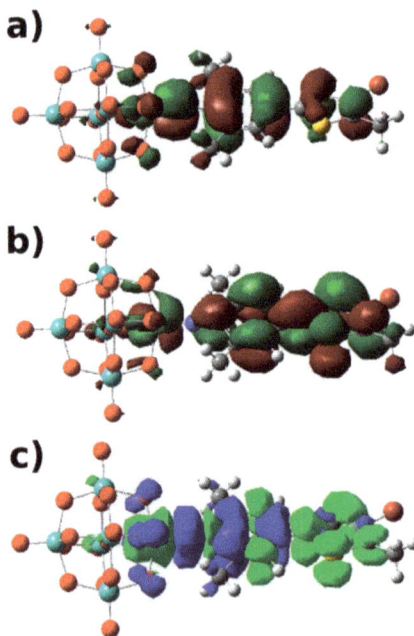

Figure A9. (a) Highest occupied molecular orbital obtained at the CAMB3LYP/def2-TZVPP level of theory displayed with an isovalue of 0.02 $e^-/Å^{33}$. **(b)** Lowest unoccupied molecular orbital obtained at the CAMB3LYP/def2-TZVPP level of theory displayed with an isovalue of 0.02 $e^-/Å^{33}$. **(c)** Difference density for the CT transition obtained at the TDA-CAMB3LYP/def2-TZVPP level of theory displayed with an isovalue of 0.0004 $e^-/Å^{33}$. Green lobes indicate an increase in electron density upon electronic excitation and blue indicates a decrease. All surface plots were generated with Gaussview 5.0 [34].

A3.6. Cartesian Coordinates

Table A1. Cartesian Coordinates (Å) of 1-L4 Optimized at the TPSS-D3/def2-TZVP Level of Theory.

	Cartesian Coordinates (Å)		
H	3.76704225	2.03847542	0.30065425
C	8.79961983	−2.01560166	0.41244729
H	8.87077619	−2.73103735	−0.41655383
H	9.69833234	−1.39231526	0.40394112
H	8.78913391	−2.59102756	1.34634501
C	0.96256278	0.13987740	−0.04620859
C	1.56254262	−1.14797725	−0.14917984
C	1.77530740	1.29901199	0.10936217
C	2.94484244	−1.24373548	−0.09804489
H	3.40180409	−2.22830515	−0.19445924
C	1.10163712	2.63869224	0.22587839
H	0.50481002	2.85061645	−0.66856868
H	1.83866424	3.43685064	0.36309774
H	0.39680786	2.64829174	1.06459060
C	7.59386438	−1.14230232	0.28740460
C	6.18839822	0.74582657	0.03186394
H	5.97589682	1.79675983	−0.11316036
C	3.15321125	1.15160124	0.15798309
C	3.77009862	−0.11209650	0.05490087
C	0.67156702	−2.34846316	−0.30992473
H	−0.01800377	−2.43524581	0.53732293
H	1.26335447	−3.26621562	−0.38903123
H	0.03677621	−2.24804774	−1.19713465
C	5.21655957	−0.23465473	0.11026329
C	7.50055059	0.21728381	0.13429621
N	−0.37806453	0.25874441	−0.08920770
O	−6.30236442	1.57923483	−1.41436027
O	−4.37228812	0.26702048	−0.09343012
O	−2.55751051	1.59712153	−1.44792530
O	−6.29892939	1.59573674	1.21226320
O	−4.41902666	2.91549342	−0.08453437
O	−2.55711831	1.61203372	1.24718283
O	−4.42922706	0.27669877	−2.74280726
O	−6.29344006	−1.04499882	1.23892385
O	−4.43233273	0.25667581	2.55667783
O	−6.29498094	−1.05586386	−1.40947046
O	−2.55700529	−1.08720380	−1.42217494
O	−4.46304402	3.15757998	−2.93033868
O	−2.55518400	−1.06732613	1.24923151
O	−4.45481996	3.10375150	2.79799406
O	−8.47713797	0.24262438	−0.08541032

Table A1. *Cont.*

	Cartesian Coordinates (Å)		
O	−4.44104229	−2.38607044	−0.10306189
O	−4.42260783	−2.62719372	2.74380416
O	−4.43177656	−2.57234737	−2.98509932
S	5.98186692	−1.79775572	0.31727666
BR	9.04012899	1.35225338	0.04645216
MO	−4.45217513	1.95216728	−1.72269566
MO	−6.77320756	0.24249817	−0.08640557
MO	−4.44116305	1.89427511	1.59437427
MO	−2.16185379	0.29537333	−0.10077312
MO	−4.40705584	−1.42246356	1.53532479
MO	−4.41479636	−1.36432115	−1.77989579

A3.7. Crystallography

Table A2. X-ray Data Collection. Single crystal X-ray data was collected using an Agilent Technologies SuperNova Dual Wavelength single crystal X-ray diffractometer at 130 K using Mo-Kα radiation (λ = 0.71073 Å) for (**L3**) and Cu-Kα radiation (λ = 1.5418 Å) for (**1-L4**) fitted with a mirror monochromator. Crystals were transferred directly from the mother liquor to the oil, to prevent solvent loss. The data was reduced using CrysAlisPro software (Version 1.171.36.28) (University of Melbourne, Melbourne, Australia) using a numerical absorption correction based on Gaussian integration over a multifaceted crystal model. Data was solved using direct methods by SHELXT and refined using a full-matrix least square procedure based upon F^2. All ordered non-H atoms were refined anisotropically.

Parameter	L3	1-L4
Formula	$C_{26}H_{22}NSBr$	$Mo_6O_{18}N_3C_{45}H_{84}SBr$
F_w	460.41	1642.76
Crystal Class	Monoclinic	Monoclinic
Space Group	$P2_1/n$	$P2_1/n$
a (Å)	12.5989(5)	11.03023(16)
b (Å)	11.5694(5)	16.9232(2)
c (Å)	15.2670(6)	32.4507(5)
α (°)	90	90
β (°)	101.954(4)	98.8238(13)
γ (°)	90	90
V (Å³)	2177.07(16)	5985.79(15)
Z	4	4
T (K)	130.01(10)	130.01(10)
μ (mm⁻¹)	1.996	11.658
reflns measd	13729	105913
unique reflns	5687	12538
R1 [$I > 2\sigma(I)$]	0.0497	0.0898
wR2(all data)	0.1727	0.2471

A3.8. Thermal Gravimetric Analysis

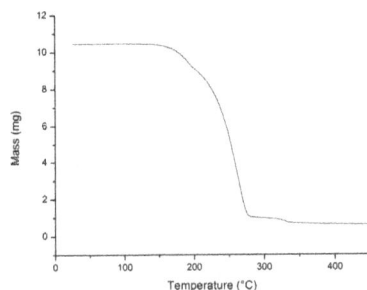

Figure A10. TGA of Compound (**L4**), showing that no solvent is present within the structure, which is in line with the elemental analysis.

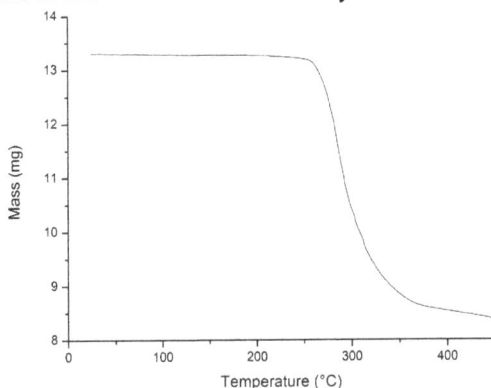

Figure A11. TGA of Compound (**1-L4**), showing that no solvent is present within the structure, which is in line with the crystal structure and elemental analysis.

Conflicts of Interest

The authors declare no conflict of interest.

References

1. An, H.Y.; Wang, E.B.; Xiao, D.R.; Li, Y.G.; Su, Z.M.; Xu, L. Chiral 3D architectures with helical channels constructed from polyoxometalate clusters and copper-amino acid complexes. *Angew. Chem. Int. Ed. Engl.* **2006**, *45*, 904–908.
2. Zhang, J.; Song, Y.; Cronin, L.; Liu, T. Self-assembly of organic-inorganic hybrid amphiphilic surfactants with large polyoxometalates as polar head groups. *J. Am. Chem. Soc.* **2008**, *130*, 14408–14409.
3. Song, Y.F.; McMillan, N.; Long, D.L.; Thiel, J.; Ding, Y.; Chen, H.; Gadegaard, N.; Cronin, L. Design of hydrophobic polyoxometalate hybrid assemblies beyond surfactant encapsulation. *Chemistry* **2008**, *14*, 2349–2354.

4. Song, Y.; McMillan, N.; Long, D.; Kane, S.; Malm, J.; Riehle, M.O.; Pradeep, C. P.; Gadegaard, N.; Cronin, L. Micropatterned surfaces with covalently grafted unsymmetrical polyoxometalate-hybrid clusters lead to selective cell adhesion. *J. Am. Chem. Soc.* **2009**, *131*, 1340–1341.

5. Aronica, C.; Chastanet, G.; Zueva, E.; Borshch, S.A.; Clemente-Juan, J.M.; Luneau, D. A mixed-valence polyoxovanadate(III,IV) cluster with a calixarene cap exhibiting ferromagnetic V(III)-V(IV) interactions. *J. Am. Chem. Soc.* **2008**, *130*, 2365–2371.

6. Pan, D.; Chen, J.; Tao, W.; Nie, L.; Yao, S. Polyoxometalate-modified carbon nanotubes: New catalyst support for methanol electro-oxidation. *Langmuir* **2006**, *22*, 5872–5876.

7. Strong, J.B.; Yap, G.P.A.; Ostrander, R.; Liable-Sands, L.M.; Rheingold, A.L.; Thouvenot, R.; Gouzerh, P.; Maatta, E.A. A New Class of Functionalized Polyoxometalates: Synthetic, Structural, Spectroscopic, and Electrochemical Studies of Organoimido Derivatives of $[Mo_6O_{19}]^{2-}$. *J. Am. Chem. Soc.* **2000**, *122*, 639–649.

8. Proust, A.; Thouvenot, R.; Chaussade, M.; Robert, F.; Gouzerh, P. Phenylimido derivatives of $[Mo_6O_{19}]^{2-}$: Syntheses, X-ray structures, vibrational, electrochemical, ^{95}Mo and ^{14}N NMR studies. *Inorg. Chim. Acta* **1994**, *224*, 81–95.

9. Moore, A.R.; Kwen, H.; Beatty, A.M.; Maatta, E.A. Organoimido-polyoxometalates as polymer pendants. *Chem. Commun.* **2000**, *996*, 1793–1794.

10. Wei, Y.; Xu, B.; Barnes, C.L.; Peng, Z. An Efficient and Convenient Reaction Protocol to Organoimido Derivatives of Polyoxometalates. *J. Am. Chem. Soc.* **2001**, *123*, 4083–4084.

11. Wu, P.; Li, Q.; Ge, N.; Wei, Y.; Wang, Y.; Wang, P.; Guo, H. An Easy Route to Monofunctionalized Organoimido Derivatives of the Lindqvist Hexamolybdate. *Eur. J. Inorg. Chem.* **2004**, *2004*, 2819–2822.

12. Li, Q.; Wu, P.; Xia, Y.; Wei, Y.; Guo, H. Synthesis, spectroscopic studies and crystal structure of a polyoxoanion cluster incorporating para-bromophenylimido ligand, $(Bu_4N)_2[Mo_6O_{18}(NC_6H_4Br-p)]$. *J. Organomet. Chem.* **2006**, *691*, 1223–1228.

13. Xu, B.; Wei, Y.; Barnes, C.L. Hybrid Molecular Materials Based on and Organic Conjugated Systems. *Angew. Chem. Int. Ed. Engl.* **2001**, *40*, 2290–2292.

14. Lu, M.; Wei, Y.; Xu, B.; Cheung, C.F.C.; Peng, Z.; Powell, D.R. Hybrid molecular dumbbells: bridging polyoxometalate clusters with an organic pi-conjugated rod. *Angew. Chem. Int. Ed. Engl.* **2002**, *41*, 1566–1568.

15. Kang, J.; Nelson, J.A.; Lu, M.; Xie, B.; Peng, Z.; Powell, D.R. Charge-transfer hybrids containing covalently bonded polyoxometalates and ferrocenyl units. *Inorg. Chem.* **2004**, *43*, 6408–6413.

16. Lu, M.; Xie, B.; Kang, J.; Chen, F.; Peng, Z. Synthesis of Main-Chain Polyoxometalate-Containing Hybrid Polymers and Their Applications in Photovoltaic Cells. *Chem. Mater.* **2005**, *17*, 402–408.

17. Zhu, Y.; Wang, L.; Hao, J.; Yin, P.; Zhang, J.; Li, Q.; Zhu, L.; Wei, Y. Palladium-catalyzed Heck reaction of polyoxometalate-functionalised aryl iodides and bromides with olefins. *Chem. Eur. J.* **2009**, *15*, 3076–3080.

18. Xu, B.; Peng, Z.; Wei, Y.; Powell, D.R. Polyoxometalates covalently bonded with terpyridine ligands. *Chem. Commun.* **2003**, *375*, 2562–2563.

19. Hirata, S.; Head-Gordon, M. Time-dependent density functional theory within the Tamm-Dancoff approximation. *Chem. Phys. Lett.* **1999**, *314*, 291–299.

20. Yanai, T.; Tew, D.P.; Handy, N.C. A new hybrid exchange-correlation functional using the Coulomb-attenuating method (CAM-B3LYP). *Chem. Phys. Lett.* **2004**, *393*, 51–57.

21. Weigend, F.; Ahlrichs, R. Balanced basis sets of split valence, triple zeta valence and quadruple zeta valence quality for H to Rn: Design and assessment of accuracy. *Phys. Chem. Chem. Phys.* **2005**, *7*, 3297–3305.

22. Tao, J.; Perdew, J.; Staroverov, V.; Scuseria, G. Climbing the Density Functional Ladder: Nonempirical Meta-Generalized Gradient Approximation Designed for Molecules and Solids. *Phys. Rev. Lett.* **2003**, *91*, 146401.

23. Grimme, S.; Antony, J.; Ehrlich, S.; Krieg, H. A consistent and accurate ab initio parametrization of density functional dispersion correction (DFT-D) for the 94 elements H-Pu. *J. Chem. Phys.* **2010**, *132*, 154104.

24. Grimme, S.; Ehrlich, S.; Goerigk, L. Effect of the damping function in dispersion corrected density functional theory. *J. Comput. Chem.* **2011**, *32*, 1456–1465.

25. Goerigk, L.; Grimme, S. Assessment of TD-DFT methods and of various spin scaled CIS(D) and CC2 versions for the treatment of low-lying valence excitations of large organic dyes. *J. Chem. Phys.* **2010**, *132*, 184103.

26. Janjua, M.R.S.A. Quantum mechanical design of efficient second-order nonlinear optical materials based on heteroaromatic imido-substituted hexamolybdates: First theoretical framework of POM-based heterocyclic aromatic rings. *Inorg. Chem.* **2012**, *51*, 11306–11314.

27. Melcamu, Y.Y.; Wen, S.; Yan, L.; Zhang, T.; Su, Z. Theoretical investigation of second-order nonlinear optical response by linking hexamolybdate with graphene in the donor-acceptor (D-A) framework. *Mol. Simul.* **2013**, *39*, 214–219.

28. Osakai, T.; Himeno, S.; Saito, A.; Hori, T. Electrochemical reduction of hexamolybdate(2−) ion in acidic aqueous-organic media. *J. Electroanal. Chem. Interfacial Electrochem.* **1990**, *285*, 209–221.

29. MarvinSketch. Available online: http://www.chemaxon.com (accessed on 27 April 2015).

30. Borg, S.J.; Best, S.P. Spectroelectrochemical cell for the study of interactions between redox-activated species and moderate pressures of gaseous substrates. *J. Electroanal. Chem.* **2002**, *535*, 57–64.

31. Lee, B.Y.; Kwon, H.Y.; Lee, S.Y.; Na, S.J.; Han, S.I.; Yun, H.; Lee, H.; Park, Y.W. Bimetallic anilido-aldimine zinc complexes for epoxide/CO_2 copolymerization. *J. Am. Chem. Soc.* **2005**, *127*, 3031–3037.

32. Furche, F.; Ahlrichs, R.; Hättig, C.; Klopper, W.; Sierka, M.; Weigend, F. Turbomole. *Wiley Interdiscip. Rev. Comput. Mol. Sci.* **2014**, *4*, 91–100.

33. Neese, F. The ORCA program system. *Wiley Interdiscip. Rev. Comput. Mol. Sci.* **2012**, *2*, 73–78.

34. Lin, Y.; Li, G.; Mao, S.; Chai, J. Long-Range Corrected Hybrid Density Functionals with Improved Dispersion Corrections. *J. Chem. Theory Comput.* **2013**, *9*, 263–272.

Biomolecules Electrochemical Sensing Properties of a PMo₁₁V@N-Doped Few Layer Graphene Nanocomposite

Diana M. Fernandes, Marta Nunes, Ricardo J. Carvalho, Revathi Bacsa,

Israel-Martyr Mbomekalle, Philippe Serp, Pedro de Oliveira and Cristina Freire

Abstract: A novel hybrid nanocomposite, PMo₁₁V@N-doped few layer graphene, was prepared by a one-step protocol through direct immobilization of the tetrabutylammonium salt of a vanadium-substituted phosphomolybdate (PMo₁₁V) onto N-doped few layer graphene (N-FLG). The nanocomposite characterization by FTIR and XPS confirmed its successful synthesis. Glassy carbon modified electrodes with PMo₁₁V and PMo₁₁V@N-FLG showed cyclic voltammograms consistent with surface-confined redox processes attributed to Mo-centred reductions ($Mo^{VI} \rightarrow Mo^{V}$) and a vanadium reduction ($V^{V} \rightarrow V^{IV}$). Furthermore, PMo₁₁V@N-FLG modified electrodes showed good stability and well-resolved redox peaks with high current intensities. The observed enhancement of PMo₁₁V electrochemical properties is a consequence of a strong electronic communication between the POM and the N-doped few layer graphene. Additionally, the electro-catalytic and sensing properties towards acetaminophen (AC) and theophylline (TP) were evaluated by voltammetric techniques using a glassy carbon electrode modified with PMo₁₁V@N-FLG. Under the conditions used, the square wave voltammetric peak current increased linearly with AC concentration in the presence of TP, but showing two linear ranges: 1.2×10^{-6} to 1.2×10^{-4} and 1.2×10^{-4} to 4.8×10^{-4} mol dm⁻³, with different AC sensitivity values, 0.022 A/mol dm⁻³ and 0.035 A/mol dm⁻³, respectively (detection limit, DL = 7.5×10^{-7} mol dm⁻³).

Reprinted from *Inorganics*. Cite as: Fernandes, D.M.; Nunes, M.; Carvalho, R.J.; Bacsa, R.; Mbomekalle, I.-M.; Serp, P.; de Oliveira, P.; Freire, C. Biomolecules Electrochemical Sensing Properties of a PMo₁₁V@N-Doped Few Layer Graphene Nanocomposite. *Inorganics* **2015**, *3*, 178-193.

1. Introduction

Taking advantage of the electroactivity of some drugs and biomolecules, the application of electrochemical sensors for biological analysis has been growing rapidly, mainly due to the simplicity, accuracy, precision, low cost and rapidity of the electrochemical techniques [1]. In order to develop electrochemical sensors with higher selectivity and sensitivity, the chemical modification of electrode surfaces has been a major focus of research. The modified electrodes present lower overpotential values and improved mass transfer kinetics, decreasing the effect of interferences and avoiding surface fouling [2]. Nanostructured materials, in particular, carbon-based nanomaterials such as carbon nanotubes and graphene, have attracted considerable interest in this field, owing to their unique physical, chemical and electrochemical properties. They present low residual current, readily renewable surfaces and wide potential windows, providing an important and feasible platform for electroanalysis [3,4].

Graphene (G), in particular, emerged as a "superstar" material in the last years, being characterized by a two-dimensional (2D), single-layer sheet of sp^2-hybridized carbon atoms that are closely packed into a hexagonal lattice structure [5]. Its properties, such as fast electron transportation, high thermal conductivity, excellent mechanical strength and high surface area, suggest its ability to detect analyte molecules and to promote a fast electron transfer between the electrode and the analyte, which make it a promising electrocatalyst [3,5]. In fact, several reports published show the good sensitivity and electrocatalytic activity of pristine graphene and graphene-based nanocomposites on the electrochemical sensing of biomolecules such as dopamine (DA) [6,7], uric (UA) and ascorbic (AA) acids [8], nucleic acids [9] and glucose [10].

More recently, advances in graphene research showed that the chemical doping with heteroatoms can be an effective strategy to modulate their electronic properties and surface chemistry. Among the several potential dopants, nitrogen is an excellent candidate due to its comparable size with carbon and five valence-electronic structure, which result in strong covalent bonds between the nitrogen and carbon atoms [11]. The resulting material—nitrogen-doped graphene (N-G)—presents higher electrical conductivity, much larger functional surface area, more biocompatibility and more chemically active sites for functionalization than pristine G [12], whereby it has been applied in N-G-based electrochemical sensors with enhanced performances to the detection of DA, AA and UA [13,14], glucose [15], H_2O_2 [16], antibiotics [17] and pesticides [12].

However, the properties of G-type materials may be yet more prized by the incorporation of these materials in nanocomposites, in order to couple their unique properties with interesting properties of other materials [4,10]. In this context, polyoxometalates (POMs), a kind of transition metal oxide nanoclusters [18], can be a good option owing to their rich chemical and structural variety and, mainly, due to their electronic properties. POMs have the ability to undergo fast, reversible and stepwise multi-electron reactions without decomposition, making them promising electro-catalysts [19]. The incorporation of POMs in nanocomposites with different materials has been reported [20,21]. Several POMs@carbon nanotubes (CNTs) composites have been prepared and applied in batteries [22], supercapacitors [23] and for the reduction of chemical species [24], while POM composites with metal nanoparticles, ionic liquids and conducting polymers have been applied for the reduction of H_2O_2, iodate, nitrite [18,25] and bromate [26], as well as in the sensing of small biomolecules (DA and AA) [19,27]. Previously, we also reported the preparation of several POM@MOFs nanocomposites and their application as electro-catalysts towards nitrite and iodate reductions and AA and DA oxidations [28,29].

Acetaminophen (AC), also known as paracetamol, is an analgesic and antipyretic drug widely used for the relief of mild to moderate pain associated with headache, backache, arthritis and postoperative pain. However, overdoses can lead to the accumulation of toxic metabolites, causing severe and sometimes fatal hepatotoxicity and nephrotoxicity [30]. Theophylline (TP) is a methyl-xanthine derivative which exists widely in nature and is one of the most commonly used medications for the treatment of the symptoms of chronic asthma [3]. Several techniques, including spectrophotometry and liquid chromatography (LC), have been applied for the determination of AC in pharmaceutical formulations and biological fluids, but these methods are time-consuming and expensive. Due to the advantages of low cost, fast response, simple instrumentation and high sensitivity, voltammetric methods are therefore a better solution for the determination of AC in the presence of TP.

This work reports on the preparation of a new nanocomposite, $PMo_{11}V@N$-FLG, by the immobilisation of the tetrabutylammonium salt of vanadium-substituted phosphomolybdate $[PMo_{11}VO_{40}]^{-5}$ ($PMo_{11}V$) into nitrogen-doped few-layer graphene (N-FLG). Investigations were focused on the nanocomposite characterization and its electrochemical properties upon immobilization on a GC electrode surface. The electro-catalytic properties towards AC and TP oxidation were evaluated. In addition, detection of AC in the presence of an excess of TP was also determined.

2. Results and Discussion

2.1. Nanocomposite Preparation and Characterization

The tetra-butylammonium salt of $PMo_{11}V$ was prepared according to the literature and characterized by several techniques [31]. The results obtained are in good agreement with the literature. Figure 1 shows the FTIR spectra of N-FLG, TBA-$PMo_{11}V$ and $PMo_{11}V@N$-FLG. The FTIR spectrum of TBA-$PMo_{11}V$ shows the vibration bands due to the C–H stretching vibration, at 2871 and 2960 cm^{-1}, and those due to C–H bending vibration at 1479 and 1374 cm^{-1} which are characteristic of the TBA salt, two bands in the interval 1080–1040 cm^{-1}, assigned to the splitting of the P–O stretching vibration, a band at 945 cm^{-1} assigned to the Mo–O_d vibration and the bands at 872 cm^{-1} and 799 cm^{-1} to the Mo–O_b–Mo and Mo–O_c–Mo stretching modes, respectively [31–33]. The FTIR spectrum of N-FLG (Figures 1 (black) and S1) presents very weak and poorly resolved vibrational bands: it shows a broad absorption at around 3455 cm^{-1} assigned to the stretching vibrations of the OH groups which may be due to hydroxyl groups and residual adsorbed water, and the vibration band at 1639 cm^{-1} may be due to C=N stretching vibrations [34]. The two bands at 2855 and 2926 cm^{-1} correspond to the aromatic sp^2 C–H stretching vibration. The sharp band at 1383 cm^{-1} is attributed to C–N stretching vibration [35]. The band at 1533 cm^{-1} is assigned to C=C stretching vibrations of the aromatic carbon, and the band at 1119 cm^{-1} can be assigned to stretching vibration of C–O [36,37].

The preparation of the nanocomposite $PMo_{11}V@N$-FLG was carried out by one-step protocol through the direct immobilization of the prepared TBA-$PMo_{11}V$ onto N-FLG. Accordingly, POM immobilization is based in hydrophobic interactions between the alkyl chains of the TBA cation in the sample with the N-FLG surface, which are also hydrophobic: the POM is linked electrostatically to the TBA cations and consequently adsorbed on the N-FLG surface. However, some interactions between Mo and/or V with the N groups present in the N-FLG cannot be excluded.

FTIR spectrum of the nanocomposite confirms the success of the POM immobilization since it exhibits the bands due to N-FLG in the range 1161 to 1631 cm^{-1}, as well as the vibrational bands due to POM in the range 798–1081 cm^{-1} and TBA at 2960, 2871, 1479 and 1374 cm^{-1}.

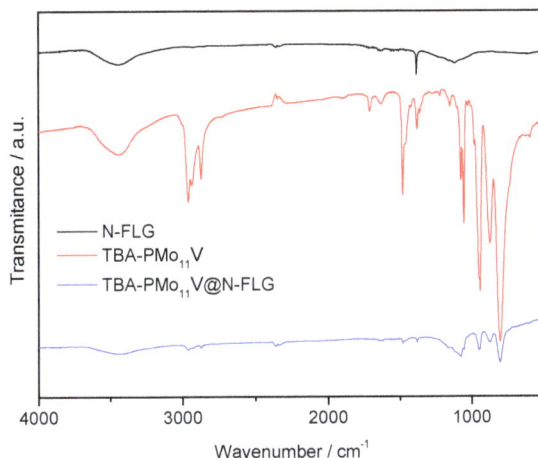

Figure 1. FTIR spectra in the range 4000–500 cm^{-1} for N-FLG (**black**), PMo$_{11}$V (**red**) and PMo$_{11}$V@N-FLG (**blue**).

The deconvoluted high resolution XPS spectra of N-FLG and PMo$_{11}$V@N-FLG are shown in Figures S2 and S3 as well as the obtained binding energies (Table S1); the surface atomic percentages of each component in both materials are summarized in Table 1. The presence in the nanocomposite of the target elements of each component, N-FLG (C, O, N) and POM (P, Mo, O, N and V) also confirms the successful fabrication of the nanocomposite. The atomic ratios of 1:11 and 1:1 for V:Mo and P:V, respectively, suggest that the POM structure is maintained in the nanocomposite. Furthermore, comparison between the XPS results obtained here with those obtained for the same POM immobilized onto commercial pristine graphene flakes (PMo$_{11}$V@GF) [38] shows that a higher amount of POM was immobilized onto N-FLG which is one advantage for applications in electrochemical sensing.

The C 1s high-resolution spectrum of pristine N-FLG (Figure S2a) was fitted with seven peaks: a main peak at 285.0 eV assigned to the graphitic structure (sp^2), a peak at 286.2 eV attributed to C–C and C–N, a peak at 287.2 eV assigned to C–O, a peak at 288.5 eV related to C=O, a peak at 289.9 eV due to –COO and the peaks at 291.2 and 292.4 eV are attributed to the characteristic shake-up satellite for the π–π* transition and to its extension, respectively [39,40]. For PMo$_{11}$V@N-FLG, the C 1s high-resolution spectrum (Figure S3a, Table S1) was also fitted with seven peaks: 285.0, 285.8, 286.6, 287.2, 288.2, 289.3 and 291.6 eV with the same assignment as for N-FLG. The slight differences in the peak positions may be due to some structure rearrangements.

The O 1s high resolution spectrum of N-FLG was fitted with two peaks at 531.3 and 533.0 eV ascribed to O=C from ketone, quinone moieties and to O–C from ether and phenol groups, respectively. The PMo$_{11}$V@N-FLG O 1s high resolution spectrum was deconvoluted into three peaks with binding energies at 530.9, 532.1 and 533.6 eV. The last two are assigned as described for N-FLG and the first is ascribed to O–Mo [38].

The N 1s high-resolution spectrum of pristine N-FLG was deconvoluted into four peaks: the peak at 399.1 eV is ascribed to pyridinic-N, the peak at 401.6 eV is attributed to pyrrolic-N, the peak at

404.5 eV is assigned to graphitic N and finally that at 407.3 eV may be due to some oxidized N [39,41,42]. The presence of N on PMo₁₁V@N-FLG and its atomic percentage could be assessed, but with low certainty due to the overlap of the N 1s and Mo 3p peaks in the N 1s high resolution spectrum. Nevertheless, the spectrum was fitted taking into consideration the existence of the Mo 3p ($3p_{1/2}$ and $3p_{3/2}$) and N 1s peaks. The area obtained through modulation of Mo 3p is similar to that obtained for Mo 3d, suggesting a correct fitting of peaks taking into consideration the N 1s peaks. The two peaks in the N 1s spectrum at 400.9 and 402.6 eV assigned to pyridinic and pyrrolic-N, respectively. However, two other very low intense peaks could be observed at 404.8 and 406.5 eV, but its low intensity prevent its inclusion in the fitting.

The Mo 3d high resolution spectrum was deconvoluted taking into account the $3d_{5/2}$ and $3d_{3/2}$ doublets caused by spin-orbital coupling. The Mo 3d high resolution spectrum of PMo₁₁V@N-FLG was fitted with two couples of peaks with binding energies, respectively, at 231.8 and 234.9 eV assigned to Mo^{5+} and at 233.1 and 236.2 eV assigned to Mo^{6+}. The presence of the Mo^{5+} species suggests that partial photoreduction of POMs occurred when the sample was exposed to the X-ray source or during the nanocomposites preparation.

Table 1. XPS surface atomic percentages for N-FLG and PMo₁₁V@N-FLG [a].

Sample	Atomic %					
	C 1s	O 1s	N 1s	P 2p	V 2p	Mo 3d
N-FLG	91.9	3.8	4.2			
PMo₁V@N-FLG	74.6	16.9	3.1	0.45	0.41	4.4

[a] Determined by the areas of the respective bands in the high-resolution XPS spectra.

2.2. Electrochemical Behaviour

The electroactive surface area of bare GCE, N-FLG/GCE and PMo₁₁V@N-FLG/GCE were determined from cyclic voltammograms of 1×10^{-3} mol dm^{-3} K₃[Fe(CN)₆] in KCl 1 mol dm^{-3} (Figure S4) using the Randles-Sevcik equation and assuming that electrode processes are controlled by diffusion:

$$i_{pa} = 2.69 \times 10^5 \, n^{3/2} \, A \, D_r^{1/2} \, R_\infty \, v^{1/2} \tag{1}$$

where n is the number of electrons involved in the process (1 in this case), A is the electrode surface area (cm²), D_r the diffusion coefficient (cm² s^{-1}), R the concentration of the species (mol cm^{-3}), v is the scan rate (V s^{-1}) and i_{pa} the intensity of the anodic peak current (A) [43]. The A values obtained were: 0.0443, 0.0374 and 0.0754 cm² for bare GCE, N-FLG/GCE and PMo₁₁V@N-FLG/GCE, respectively. These results indicate that the modification of GCE with N-FLG leads to a small decrease of the electroactive area (≈15%) while GCE modification with PMo₁₁V@N-FLG/GCE changes dramatically the electroactive area with an increase close to 70%, indicating that the introduction of the PMo₁₁V onto N-FLG provides a more conductive pathway for the electron-transfer of [Fe(CN)₆]$^{3-/4-}$. In addition, the anodic/cathodic peak-to-peak separations (ΔE_p) varies between 0.072 and 0.094 V for GCE, 0.378 and 0.781 V for N-FLG/GCE and 0.076 and 0.086 V for PMo₁₁V@N-FLG/GCE for the lowest (0.020 V s^{-1}) and highest (0.50 V s^{-1}) scan rates used.

The electrochemical behaviour of $PMo_{11}V$ and $PMo_{11}V@N$-FLG modified electrodes was studied in pH 2.5 H_2SO_4/Na_2SO_4 buffer solution (Figure 2). In the potential range +0.60 to −0.40 V, the $PMo_{11}V$-modified electrodes revealed five redox processes, denoted as V_1 and Mo_1 to Mo_4, Figure 3a: $E_{pcV1} = 0.311$ V, $E_{pcMo1} = 0.056$ V, $E_{pcMo2} = -0.058$ V, $E_{pcMo3} = -0.181$ V and $E_{pcMo4} = -0.376$ V $vs.$ Ag/AgCl. Peak V_1 is assigned to a vanadium redox process ($V^V{\rightarrow}V^{IV}$) and peaks Mo_1 to Mo_4 are attributed to molybdenum redox processes ($Mo^{VI}{\rightarrow}Mo^V$).

Figure 2b depicts the cyclic voltammograms obtained with a $PMo_{11}V@N$-FLG modified electrode in pH 2.5 H_2SO_4/Na_2SO_4 buffer solution and in the potential range +0.90 to −0.50 V. The nanocomposite also shows five redox processes: $E_{pcV1} = 0.317$ V, $E_{pcMo1} = 0.079$ V, $E_{pcMo2} = -0.032$ V, $E_{pcMo3} = -0.227$ V and $E_{pcMo4} = -0.370$ V $vs.$ Ag/AgCl, corresponding to one-electron oxidation process of vanadium and four two-electron redox processes assignable to the molybdenum centres. All the peaks are much better resolved and have higher current intensities ($\approx$10 times higher) compared with the POM-modified electrode, suggesting faster electron-transfer kinetics, which is associated with the exceptional electronic properties of N-FLG.

In the experimental timescale employed (scan rates in the range 0.02–0.5 V s^{-1}) both cathodic (E_{pc}) and anodic (E_{pa}) peak potentials varied less than 0.011 V for $PMo_{11}V$ and 0.010 V for $PMo_{11}V@N$-FLG. Figure 3 depicts the plots of log i_p $versus$ log v for the Mo_1 and Mo_3 waves of $PMo_{11}V$ and $PMo_{11}V@N$-FLG modified electrodes. Both i_{pc} and i_{pa} are directly proportional to the scan rate for all peaks, with r^2 between 0.998 and 0.991 for $PMo_{11}V$ and 0.998 and 0.992 for $PMo_{11}V@N$-FLG, which indicate surface-confined processes [38]. In addition, the anodic to cathodic peak-to-peak separations (ΔE_p) of all redox couples vary between 0.052 (0.02 V s^{-1}) and 0.019 V (0.5 V s^{-1}) and the ratios i_{pa}/i_{pc} are close to one (0.96 $\pm$ 0.05).

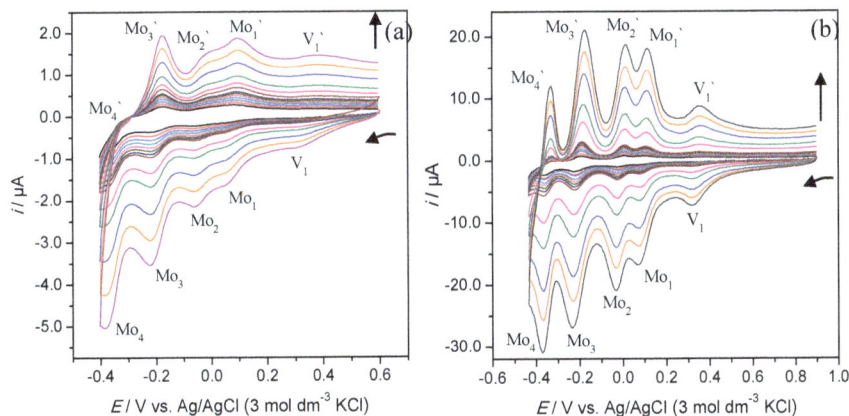

Figure 2. Cyclic voltammograms in pH 2.5 H_2SO_4/Na_2SO_4 buffer solution at different scan rates from 0.02–0.5 V s^{-1} of $PMo_{11}V$ (**a**) and $PMo_{11}V@N$-FLG (**b**).

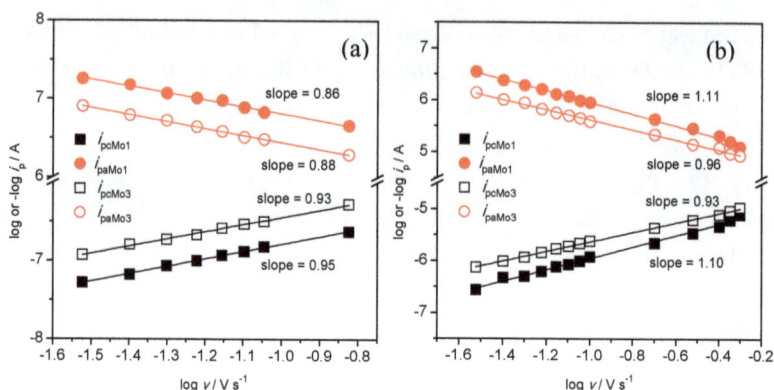

Figure 3. Plots of log i_{pc} and i_{pa} vs. log v of PMo₁₁V (**a**) and PMo₁₁V@N-FLG (**b**).

Electrochemical surface coverages of modified electrodes with PMo₁₁V and PMo₁₁V@N-FLG were calculated from cyclic voltammetry using the Mo₁ reduction process according to the equation:

$$\Gamma = (4i_p RT)/(n^2 F^2 vA),\qquad(2)$$

where i_p is the peak current (A), n is the number of electrons transferred (2 in this case), v is the scan rate (V s⁻¹), A is the geometric area of the electrode (0.07065 cm²), R is the gas constant, T is the temperature (298 K) and F is Faraday's constant. Peak currents (Mo₁) were plotted against scan rate (0.02 to 0.5 V s⁻¹) and the value of i_p/v obtained was used to calculate the surface coverage. This led to $\Gamma = 0.006$ nmol cm⁻² for PMo₁₁V and $\Gamma = 0.077$ nmol cm⁻² for PMo₁₁V@N-FLG. These results show that electrochemical surface coverage for the nanocomposite modified electrodes is significantly higher (almost 13 times) than that of the POM-modified electrodes. This constitutes an outstanding advantage when developing modified electrodes with carbon nanomaterials for applications in electrocatalysis and electroanalytical determinations. The N-FLG allowed the immobilization of a much larger quantity of the electroactive PMo₁₁V on the electrode surface due to its high surface area, and simultaneously improved their electrochemical responses.

2.3. Electro-Catalytic Performance of Nanocomposite Modified Electrodes

The electrochemical oxidation of acetaminophen (AC) and theophylline (TP) was initially investigated individually by cyclic voltammetry at the three different electrodes. Figure 4 shows the cyclic voltammograms of 5.0×10^{-4} mol dm⁻³ solutions of AC and TP at a bare GCE and at N-FLG/GCE and PMo₁₁V@N-FLG/GCE modified electrodes. With a bare GCE electrode and under the experimental conditions used, AC and TP are irreversibly oxidized at ≈0.762 V and at ≈1.352 V vs. Ag/AgCl, respectively.

At a N-FLG modified electrode, the oxidation of TP occurs at almost the same potential as with a bare GCE (1.360 V), and no significant changes are observed in the peak current, whereas for the PMo₁₁V@N-FLG modified electrode the peak current increases ≈31% when compared with that obtained with a bare GCE (Figure 4a) and the TP cathodic peak shifts 0.055 V to less positive potentials.

For AC, the modification of a GCE with N-FLG also leads to small differences in the peak currents, but E_{pa} shifts to less positive potentials ($E_{pa} = 0.676$ V). For PMo$_{11}$V@N-FLG/GCE, the AC oxidation peak is observed at even less positive potentials, c.a. $E_{pa} = 0.622$ V vs. Ag/AgCl (E_{pa} cathodic shift of ≈0.140 V when compared to the bare GCE) and the peak current increases by ≈57%, similarly to TP. Additionally, a cathodic peak is observed at $E_{pc} \approx 0.559$ V which is attributed to the partial reduction of AC, suggesting that the oxidation of AC in the PMo$_{11}$V@N-FLG modified electrode becomes a quasi-reversible process [36]; the new set of peaks in the potential range 0.45–0.0 V corresponds to the PMo$_{11}$V redox processes.

These results show that the modification of a GCE with PMo$_{11}$V@N-FLG leads to an improvement in the overall electrochemical performance towards TP and AC oxidations, which is due to the electro-catalytic properties of PMo$_{11}$V.

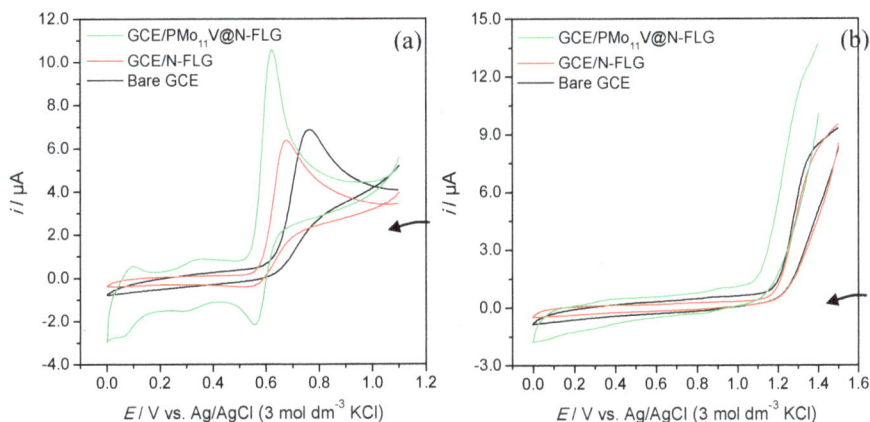

Figure 4. Cyclic voltammograms of 5.0×10^{-4} mol dm^{-3} AC (**a**) and TP (**b**) at a bare GCE, N-FLG/GCE and PMo$_{11}$V@N-FLG/GCE in pH 2.5 H$_2$SO$_4$/Na$_2$SO$_4$ buffer solution, scan rate 0.050 V s^{-1}.

The electrochemical oxidation of AC and TP was also investigated individually by square wave voltammetry. Figure 5 shows the SWV curves of 5.0×10^{-4} mol dm^{-3} solutions of AC and TP at a bare GCE and at N-FLG/GCE and PMo$_{11}$V@N-FLG/GCE modified electrodes. As observed for cyclic voltammetric results, the immobilization of N-FLG and PMo$_{11}$V@N-FLG on the GCE surface lead to some changes in the square-wave curves of each species when compared to the GCE. At the bare electrode, the oxidation peaks of AC and TP are observed at 0.725 and 1.309 V vs. Ag/AgCl.

The modification of a GCE with N-FLG leads to small changes in the oxidation potential of TP, but there is a decrease in peak currents of ≈24%. However, for the oxidation of AC there is an increase of 19% in the peak current and the peak is observed at c.a. 0.645 V vs. Ag/AgCl (E_{pa} cathodic shift of ≈0.080 V). These changes suggest that N-FLG has itself a strong interaction with the AC species leading to better detection. Still, much more significant results were obtained for GCE modification with the PMo$_{11}$V@N-FLG nanocomposite. For the oxidation of TP there is an increase of the peak currents of 47% and a E_{pa} cathodic shift of 0.045 V, while for the AC oxidation the E_{pa}

cathodic shift is 0.126 V and there was an increase of 234% in the peak current, suggesting that this modified electrode is more suitable for the determination of AC rather than TP. These effects are crucial for the application of modified electrodes in electro-catalysis. In a $PMo_{11}V@N$-FLG modified electrode, the high electrical conductivity and large surface area of N-FLG is associated with the unique electro-catalytic properties of $PMo_{11}V$—its ability to accept and release a high number of electrons without decomposition. Consequently, both components contribute to the overall improvement of the electrocatalytic performance. The sensitivity of the $PMo_{11}V@N$-FLG modified electrode towards the oxidation of AC and TP was found to be 0.040 and 0.011 A/mol dm^{-3}, respectively.

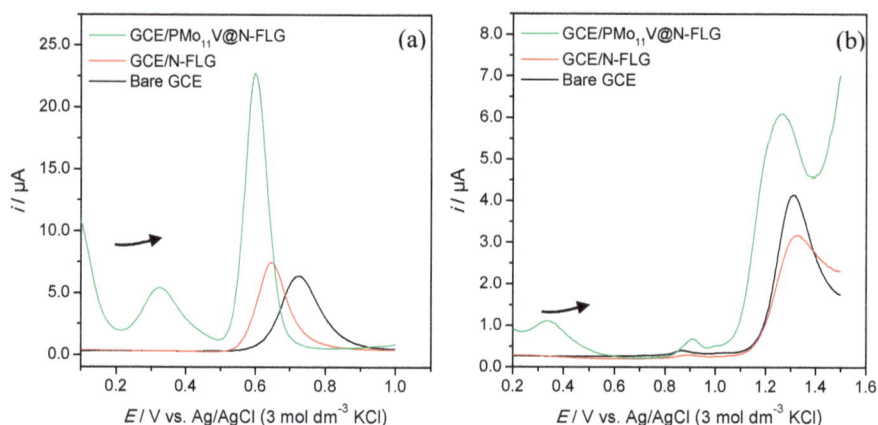

Figure 5. Square wave voltammograms of 5.0×10^{-4} mol dm^{-3} AC (**a**) and TP (**b**) at a bare GCE, N-FLG/GCE and $PMo_{11}V@N$-FLG/GCE in pH 2.5 H$_2$SO$_4$/Na$_2$SO$_4$ buffer solution, scan rate 0.050 V s^{-1}.

2.4. Determination of AC in the Presence of TP at a $PMo_{11}V@N$-FLG/GCE

Figure 6 shows selected square wave voltammograms recorded for the addition of increasing concentrations of AC. The oxidation peak current of AC increases linearly with the increasing concentration of AC from 1.2×10^{-6} to 4.8×10^{-4} mol dm^{-3} in the presence of TP: inset in Figure 6. Two different linear regions can be observed: one from 1.2×10^{-6} to 1.2×10^{-4} mol dm^{-3} ($i = 0.022\ c_{AC} + 0.018$, $r^2 = 0.998$), leading to a detection limit 7.5×10^{-7} mol dm^{-3} (detection limit calculated through the equation 3 σ/slope, where σ is the standard deviation of the blank), and a second linear region from 1.2×10^{-4} to 4.8×10^{-4} mol dm^{-3} ($i = 0.035\ c_{AC} - 1.58$, $r^2 = 0.995$). Taking into account that the 'breaking point' in the plot in the inset of Figure 6 corresponds to an AC concentration of 1.2×10^{-4} mol dm^{-3}, which is very similar to the concentration of TP in solution (1.0×10^{-4} mol dm^{-3}), these results may suggest some competition effects between TP and AC: when TP is in larger quantities than AC, a lower AC sensitivity (0.022 A/mol dm^{-3}) is observed when compared to the sensitivity value when AC concentration exceeds that of TP (0.035 A/mol dm^{-3}).

To the best of our knowledge, there are no reports in the literature concerning the use of modified electrodes with a $PMo_{11}V@N$-FLG nanocomposite for the determination of AC in the presence of

TP. Furthermore, the obtained detection limit is better than that (8.1×10^{-6} mol dm^{-3}) reported for a similar study at a ferrocene-derivative modified-graphene paste electrode [3].

Finally, the stability and robustness of the PMo$_{11}$V@N-FLG/GCE was evaluated by measuring and comparing the i_{pa} of 0.5 mM AC in a pH 2.5 H$_2$SO$_4$/Na$_2$SO$_4$ buffer solution at the modified electrode at the begging and at the end of all the experiments (data not shown). No significant change in peak position was observed and the peak current only decreased by 11%, confirming the electrochemical stability of the PMo$_{11}$V@N-FLG modified electrode. This loss might be due to the partial depletion of nanocomposite from the electrode surface due to consecutive use.

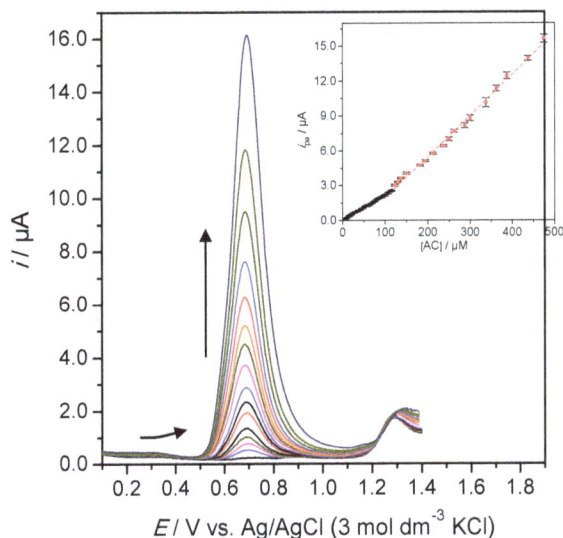

Figure 6. Square-wave voltammetric response at a PMo$_{11}$V@N-FLG/GCE in pH 2.5 buffer solution containing 1.0×10^{-4} mol dm^{-3} TP and different concentrations of AC from 1.2×10^{-6} to 4.8×10^{-4} mol dm^{-3}. Inset: plots of the anodic peak current as a function of the AC concentration.

3. Experimental Section

3.1. Materials and Methods

The N-doped few layer graphene (N-FLG) was prepared by fluidized bed chemical vapor deposition process by the decomposition of a mixture of ethylene and ammonia in the presence of a ternary oxide powder catalyst at 650 °C [44]. The N-FLG powder was recovered after washing the powder in 35% HCl at 25 °C. The tetra-butylammonium salt of the vanadium-phosphomolybdate, PMo$_{11}$V was prepared according to the literature and characterized by several techniques [31] and the PMo$_{11}$V@N-FLG was prepared using an adapted procedure [38]. Briefly, an acetonitrile solution (2.5 mL) of the compound (20 mg) was added to a toluene dispersion (25 mL) of the N-FLG (20 mg). The resulting yellowish solutions were vigorously stirred at room temperature for 10 min. The solutions were left to rest for 1 h and became colourless (the absence of POM in solution was

34

confirmed by UV-Vis spectroscopy), suggesting a 100% efficiency POM immobilization onto the surfaces of N-FLG, leading to: 1.82 mmol of POM/g nanocomposite. Then, the toluene was removed and the resulting nanocomposite was dried under vacuum at 40 °C.

Theophylline (Fluka), acetaminophen (Sigma-Aldrich, Sintra, Portugal, >99%) and acetonitrile (Romil, Cambridge, UK) were used as received.

The Fourier transform infrared (FTIR) spectra were performed in a Jasco FT/IR-460 Plus spectrophotometer (Jasco, Easton, USA) in the range 400–4000 cm^{-1}, using a resolution of 4 cm^{-1} and 32 scans; the spectra were obtained in KBr pellets (Merck, spectroscopic grade) containing 0.2% weight of PMo$_{11}$V@N-FLG.

The X-ray photoelectron spectroscopy (XPS) measurements were performed at CEMUP (Porto, Portugal), in a VG Scientific ESCALAB 200A spectrometer (VG Scientific, UK) using non-monochromatized Al Kα radiation (1486.6 eV). To correct possible deviations caused by electric charge of the samples, the C 1s band at 285.0 eV was taken as the internal standard. The XPS spectra were deconvoluted with the XPSPEAK 4.1 software (VG Scientific, UK), using non-linear least squares fitting routine after a Shirley-type background subtraction. The surface atomic percentages were calculated from the corresponding peak areas and using the sensitivity factors provided by the manufacturer.

3.2. Electrochemical Studies

An Autolab PGSTAT 30 potentiostat/galvanostat controlled by a GPES software (EcoChimie B.V, Utrecht, The Netherlands) was used for cyclic voltammetry (CV) and square-wave voltammetry (SWV). A conventional three-electrode compartment cell was used. The working electrode was a glassy carbon electrode, GCE, (3 mm diameter, BAS, MF-2012), the auxiliary and reference electrodes were a platinum wire (7.5 cm, BAS, MW-1032) and Ag/AgCl (sat. KCl) (BAS, MF-2052), respectively. The cell was enclosed in a grounded Faraday cage and all studies were carried out at room temperature and kept under an argon flow. A combined glass electrode (Crison) connected to a pH meter Basic 20$^+$ (Crison) was used for the pH measurements.

Electrolyte solutions for cyclic voltammetry were prepared using ultra-pure water (resistivity 18.2 MΩ cm at 25 °C, Millipore, Porto, Portugal). The H$_2$SO$_4$/Na$_2$SO$_4$ buffer solution with pH = 2.5 used for electrochemical studies was prepared by mixing appropriate amounts of a 0.2 mol dm^{-3} H$_2$SO$_4$ solution with a 0.5 mol dm^{-3} Na$_2$SO$_4$ solution.

The dispersion used to produce the modified electrodes was prepared as follows: a N,N-dimethylformamide (DMF) dispersion (3 mL) of N-FLG or PMo$_{11}$V@N-FLG (1 mg) was sonicated for 10 min.

Prior to modification, the GCE electrode was conditioned by a polishing/cleaning procedure using diamond pastes of 6, 3 and 1 µm (MetaDi II, Buehler, Düsseldorf, Germany) followed by aluminium oxide of particle size 0.3 µm (Buehler, Düsseldorf, Germany) on a microcloth polishing pad (BAS Bioanalytical Systems Inc., Warsaw, Poland), and then the electrode was rinsed with ultra-pure water and finally sonicated for 5 min in ultra-pure water. Electrode modification consisted in depositing a 3 µL drop of the selected material in dimethylformamide solution onto the surface of the GCE and the solvent was allowed to evaporate for about 30 min at room temperature.

4. Conclusions

The vanadium-substituted phosphomolybdate $PMo_{11}V$ was successfully immobilized on N-doped few layer graphene. X-ray photoelectron spectroscopy revealed that the POM structure was kept in the $PMo_{11}V@N$-FLG nanocomposite. The preparation of modified electrodes with $PMo_{11}V@N$-FLG was easy and quick to perform, leading to reproducible and stable modified electrodes. $PMo_{11}V@N$-FLG/GCE revealed five well defined, surface-confined redox processes that were attributed to four molybdenum centres ($Mo^{VI} \rightarrow Mo^{V}$) and a vanadium centre ($V^{V} \rightarrow V^{IV}$). The $PMo_{11}V@N$-FLG nanocomposite immobilization on a GCE allowed a higher $PMo_{11}V$ surface coverage compared to the direct immobilization of $PMo_{11}V$. This constitutes an outstanding advantage for electrocatalytic and sensing applications: $PMo_{11}V@N$-FLG showed excellent electrocatalytic properties towards AC and TP oxidations and allowed the detection of AC in the presence of an excess of TP.

Acknowledgments

The authors thank Fundação para a Ciência e a Tecnologia (FCT, Portugal) for financial support through projects PEst-C/EQB/LA0006/2013, FCOMP-01-0124-FEDER-037285 and Operation NORTE-07-0124-FEDER-000067–Nanochemistry. DF (SFRH/BPD/74877/2010) and MN (SFRH/BD/79171/2011) also thank FCT for their grants. Thanks are also due to the COST Action CM-1203 PoCheMoN.

Author Contributions

The synthesis of N-FLG was performed by Revathi Bacsa and Philippe Serp and the synthesis of $PMo_{11}V$ and $PMo_{11}V@N$-FLG by Diana M. Fernandes. Marta Nunes contributed to experiments involving the XPS analysis. The electrochemistry experiments were conducted by Diana M. Fernandes and Ricardo J. Carvalho. Cristina Freire, Israel-Martyr Mbomekalle and Pedro de Oliveira helped in the analysis of data and discussion. Cristina Freire and Diana M. Fernandes supervised the research and were responsible for writing of the manuscript.

Conflicts of Interest

The authors declare no conflict of interest.

References

1. Beitollahi, H.; Taher, M.A.; Hosseini, A. Fabrication of a nanostructure-based electrochemical sensor for simultaneous determination of epinephrine and tryptophan. *Measurement* **2014**, *51*, 156–163.
2. Karimi-Maleh, H.; Moazampour, M.; Ahmar, H.; Beitollahi, H.; Ensafi, A.A. A sensitive nanocomposite-based electrochemical sensor for voltammetric simultaneous determination of isoproterenol, acetaminophen and tryptophan. *Measurement* **2014**, *51*, 91–99.

3. Tajik, S.; Taher, M.A.; Beitollahi, H. Application of a new ferrocene-derivative modified-graphene paste electrode for simultaneous determination of isoproterenol, acetaminophen and theophylline. *Sens. Actuators B Chem.* **2014**, *197*, 228–236.

4. Lawal, A.T. Synthesis and utilisation of graphene for fabrication of electrochemical sensors. *Talanta* **2015**, *131*, 424–443.

5. Wu, S.X.; He, Q.Y.; Tan, C.L.; Wang, Y.D.; Zhang, H. Graphene-Based Electrochemical Sensors. *Small* **2013**, *9*, 1160–1172.

6. Kim, Y.R.; Bong, S.; Kang, Y.J.; Yang, Y.; Mahajan, R.K.; Kim, J.S.; Kim, H. Electrochemical detection of dopamine in the presence of ascorbic acid using graphene modified electrodes. *Biosens. Bioelectron.* **2010**, *25*, 2366–2369.

7. Wang, Y.; Li, Y.M.; Tang, L.H.; Lu, J.; Li, J.H. Application of graphene-modified electrode for selective detection of dopamine. *Electrochem. Commun.* **2009**, *11*, 889–892.

8. Ping, J.F.; Wu, J.; Wang, Y.X.; Ying, Y.B. Simultaneous determination of ascorbic acid, dopamine and uric acid using high-performance screen-printed graphene electrode. *Biosens. Bioelectron.* **2012**, *34*, 70–76.

9. Lim, C.X.; Hoh, H.Y.; Ang, P.K.; Loh, K.P. Direct Voltammetric Detection of DNA and pH Sensing on Epitaxial Graphene: An Insight into the Role of Oxygenated Defects. *Anal. Chem.* **2010**, *82*, 7387–7393.

10. Ruiyi, L.; Juanjuan, Z.; Zhouping, W.; Zaijun, L.; Junkang, L.; Zhiguo, G.; Guangli, W. Novel graphene-gold nanohybrid with excellent electrocatalytic performance for the electrochemical detection of glucose. *Sens. Actuators B Chem.* **2015**, *208*, 421–428.

11. Barsan, M.M.; Prathish, K.P.; Sun, X.; Brett, C.M.A. Nitrogen doped graphene and its derivatives as sensors and efficient direct electron transfer platform for enzyme biosensors. *Sens. Actuators B Chem.* **2014**, *203*, 579–587.

12. Dong, X.; Jiang, D.; Liu, Q.; Han, E.; Zhang, X.; Guan, X.; Wang, K.; Qiu, B. Enhanced amperometric sensing for direct detection of nitenpyram via synergistic effect of copper nanoparticles and nitrogen-doped graphene. *J. Electroanal. Chem.* **2014**, *734*, 25–30.

13. Sheng, Z.H.; Zheng, X.Q.; Xu, J.Y.; Bao, W.J.; Wang, F.B.; Xia, X.H. Electrochemical sensor based on nitrogen doped graphene: Simultaneous determination of ascorbic acid, dopamine and uric acid. *Biosens. Bioelectron.* **2012**, *34*, 125–131.

14. Li, S.M.; Yang, S.Y.; Wang, Y.S.; Lien, C.H.; Tien, H.W.; Hsiao, S.T.; Liao, W.H.; Tsai, H.P.; Chang, C.L.; Ma, C.C.M.; *et al.* Controllable synthesis of nitrogen-doped graphene and its effect on the simultaneous electrochemical determination of ascorbic acid, dopamine, and uric acid. *Carbon* **2013**, *59*, 418–429.

15. Luo, S.P.; Chen, Y.; Xie, A.J.; Kong, Y.; Wang, B.; Yao, C. Nitrogen Doped Graphene Supported Ag Nanoparticles as Electrocatalysts for Oxidation of Glucose. *ECS Electrochem. Lett.* **2014**, *3*, B20–B22.

16. Tian, Y.; Wang, F.L.; Liu, Y.X.; Pang, F.; Zhang, X. Green synthesis of silver nanoparticles on nitrogen-doped graphene for hydrogen peroxide detection. *Electrochim. Acta* **2014**, *146*, 646–653.

17. Borowiec, J.; Wang, R.; Zhu, L.H.; Zhang, J.D. Synthesis of nitrogen-doped graphene nanosheets decorated with gold nanoparticles as an improved sensor for electrochemical determination of chloramphenicol. *Electrochim. Acta* **2013**, *99*, 138–144.

18. Wang, R.Y.; Jia, D.Z.; Cao, Y.L. Facile synthesis and enhanced electrocatalytic activities of organic-inorganic hybrid ionic liquid polyoxometalate nanomaterials by solid-state chemical reaction. *Electrochim. Acta* **2012**, *72*, 101–107.

19. Zhou, C.L.; Li, S.; Zhu, W.; Pang, H.J.; Ma, H.Y. A sensor of a polyoxometalate and Au-Pd alloy for simultaneously detection of dopamine and ascorbic acid. *Electrochim. Acta* **2013**, *113*, 454–463.

20. Ji, Y.; Huang, L.; Hu, J.; Streb, C.; Song, Y.F. Polyoxometalate-functionalized nanocarbon materials for energy conversion, energy storage and sensor systems. *Energy Environ. Sci.* **2015**, *8*, 776–789.

21. Herrmann, S.; Ritchie, C.; Streb, C. Polyoxometalate—conductive polymer composites for energy conversion, energy storage and nanostructured sensors. *Dalton Trans.* **2015**, *44*, 7092–7104.

22. Kawasaki, N.; Wang, H.; Nakanishi, R.; Hamanaka, S.; Kitaura, R.; Shinohara, H.; Yokoyama, T.; Yoshikawa, H.; Awaga, K. Nanohybridization of Polyoxometalate Clusters and Single-Wall Carbon Nanotubes: Applications in Molecular Cluster Batteries. *Angew. Chem. Int. Ed.* **2011**, *50*, 3471–3474.

23. Cuentas-Gallegos, A.; Martinez-Rosales, R.; Baibarac, M.; Gomez-Romero, P.; Rincon, M.E. Electrochemical supercapacitors based on novel hybrid materials made of carbon nanotubes and polyoxometalates. *Electrochem. Commun.* **2007**, *9*, 2088–2092.

24. Guo, W.H.; Xu, L.; Xu, B.B.; Yang, Y.Y.; Sun, Z.X.; Liu, S.P. A modified composite film electrode of polyoxometalate/carbon nanotubes and its electrocatalytic reduction. *J. Appl. Electrochem.* **2009**, *39*, 647–652.

25. Ma, H.Y.; Gu, Y.; Zhang, Z.J.; Pang, H.J.; Li, S.; Kang, L. Enhanced electrocatalytic activity of a polyoxometalates-based film decorated by gold nanoparticles. *Electrochim. Acta* **2011**, *56*, 7428–7432.

26. Papagianni, G.G.; Stergiou, D.V.; Armatas, G.S.; Kanatzidis, M.G.; Prodromidis, M.I. Synthesis, characterization and performance of polyaniline-polyoxometalates (XM12, X = P, Si and M = Mo, W) composites as electrocatalysts of bromates. *Sens. Actuators B Chem.* **2012**, *173*, 346–353.

27. Zhu, W.; Zhang, W.J.; Li, S.; Ma, H.Y.; Chen, W.; Pang, H.J. Fabrication and electrochemical sensing performance of a composite film containing a phosphovanadomolybdate and cobalt(II) tetrasulfonate phthalocyanine. *Sens. Actuators B Chem.* **2013**, *181*, 773–781.

28. Fernandes, D.M.; Granadeiro, C.M.; de Sousa, P.M.P.; Grazina, R.; Moura, J.J.G.; Silva, P.; Paz, F.A.A.; Cunha-Silva, L.; Balula, S.S.; Freire, C. SiW11Fe@ MIL-101(Cr) Composite: A Novel and Versatile Electrocatalyst. *ChemElectroChem* **2014**, *1*, 1293–1300.

29. Fernandes, D.M.; Barbosa, A.D.S.; Pires, J.; Balula, S.S.; Cunha-Silva, L.; Freire, C. Novel Composite Material Polyoxovanadate@MIL-101(Cr): A Highly Efficient Electrocatalyst for Ascorbic Acid Oxidation. *ACS Appl. Mater. Interfaces* **2013**, *5*, 13382–13390.

30. Wang, S.F.; Xie, F.; Hu,R.F. Carbon-coated nickel magnetic nanoparticles modified electrodes as a sensor for determination of acetaminophen. *Sens. Actuators B Chem.* **2007**, *123*, 495–500.

31. Himeno, S.; Ishio, N. A voltammetric study on the formation of V(V)- and V(IV)-substituted molybdophosphate(V) complexes in aqueous solution. *J. Electroanal. Chem.* **1998**, *451*, 203–209.

32. Gaunt, A.J.; May, I.; Sarsfield, M.J.; Collison, D.; Helliwell, M.; Denniss, I.S. A rare structural characterisation of the phosphomolybdate lacunary anion, $[PMo_{11}O_{39}]^{7-}$. Crystal structures of the Ln(III) complexes, $(NH_4)_{11}$ $[Ln(PMo_{11}O_{39})_2]\cdot16H_2O$ (Ln = Ce^{III}, Sm^{III}, Dy^{III} or Lu^{III}). *Dalton Trans.* **2003**, doi:10.1039/B301995K.

33. Copping, R.; Gaunt, A.J.; May, I.; Sarsfield, M.J.; Collison, D.; Helliwell, M.; Denniss, I.S.; Apperley, D.C. Trivalent lanthanide lacunary phosphomolybdate complexes: a structural and spectroscopic study across the series $[Ln(PMo_{11}O_{39})_2]^{11-}$. *Dalton Trans.* **2005**, doi:10.1039/B500408J.

34. Olalde, B.; Aizpurua, J.M.; Garcia, A.; Bustero, I.; Obieta, I.; Jurado, M.J. Single-walled carbon nanotubes and multiwalled carbon nanotubes functionalized with poly(L-lactic acid): a comparative study. *J. Phys. Chem. C* **2008**, *112*, 10663–10667.

35. Kong, X.K.; Sun, Z.Y.; Chen, M.; Chen, C.L.; Chen, Q.W. Metal-free catalytic reduction of 4-nitrophenol to 4-aminophenol by N-doped graphene. *Energy Environ. Sci.* **2013**, *6*, 3260–3266.

36. Kumarasinghe, A.R.; Samaranayake, L.; Bondino, F.; Magnano, E.; Kottegoda, N.; Carlino, E.; Ratnayake, U.N.; de Alwis, A.A.P.; Karunaratne, V.; Amaratunga, G.A.J. Self-Assembled Multilayer Graphene Oxide Membrane and Carbon Nanotubes Synthesized Using a Rare Form of Natural Graphite. *J. Phys. Chem. C* **2013**, *117*, 9507–9519.

37. Lee, D.W.; de los Santos, L.; Seo, J.W.; Felix, L.L.; Bustamante, A.; Cole, J.M.; Barnes, C.H.W. The Structure of Graphite Oxide: Investigation of Its Surface Chemical Groups. *J. Phys. Chem. B* **2010**, *114*, 5723–5728.

38. Fernandes, D.M.; Freire, C. Carbon Nanomaterial-Phosphomolybdate Composites for Oxidative Electrocatalysis. *ChemElectroChem* **2015**, *2*, 269–279.

39. Wang, Y.; Shao, Y.Y.; Matson, D.W.; Li, J.H.; Lin, Y.H. Nitrogen-Doped Graphene and Its Application in Electrochemical Biosensing. *ACS Nano* **2010**, *4*, 1790–1798.

40. Lipinska, M.E.; Rebelo, S.L.H.; Pereira, M.F.R.; Gomes, J.; Freire, C.; Figueiredo, J.L. New insights into the functionalization of multi-walled carbon nanotubes with aniline derivatives. *Carbon* **2012**, *50*, 3280–3294.

41. Xu, H.Y.; Xiao, J.J.; Liu, B.H.; Griveau, S.; Bedioui, F. Enhanced electrochemical sensing of thiols based on cobalt phthalocyanine immobilized on nitrogen-doped graphene. *Biosens. Bioelectron.* **2015**, *66*, 438–444.

42. Liang, X.Q.; Zhong, J.; Shi, Y.L.; Guo, J.; Huang, G.L.; Hong, C.H.; Zhao, Y.D. Hydrothermal synthesis of highly nitrogen-doped few-layer graphene via solid-gas reaction. *Mater. Res. Bull.* **2015**, *61*, 252–258.

43. Bard, A.J.; Faulkner, L.R. *Electrochemical Methods, Fundamentals and Applications*; Wiley: New York, NY, USA, 2001.
44. Bacsa, R.R.; Cameán, I.; Ramos, A.; Garcia, A.B.; Tishkova, V.; Bacsa, W.S.; Gallagher, J.R.; Miller, J.T.; Navas, H.; Jourdain, V.; *et al.* Few layer graphene synthesis on transition metal ferrite catalysts. *Carbon* **2015**, *89*, 350–360.

Polyoxoniobates and Polyoxotantalates as Ligands—Revisited

Pavel A. Abramov, Maxim N. Sokolov and Cristian Vicent

Abstract: This short review summarizes our contribution to the coordination chemistry of noble metals (organometallic fragments of Rh, Ir, Ru and hydroxo Pt(IV)) and polyoxocomplexes of niobium and tantalum.

Reprinted from *Inorganics.* Cite as: Abramov, P.A.; Sokolov, M.N.; Vicent, C. Polyoxoniobates and Polyoxotantalates as Ligands—Revisited. *Inorganics* **2015**, *3*, 160-177.

1. Introduction

The group 5 (V, Nb, Ta) polyoxometalates (commonly defined as anionic polynuclear oxocomplexes) form two large but apparently distinct families, which, despite certain resemblance, show almost no overlap [1–9]. Vanadium(V) prefers aggregation into decaniobate $[V_{10}O_{28}]^{6-}$ and rearrangement of this very stable structure (search in CBSD yields about 150 structurally characterized salts of this anion with various organic cations) can be realized only in the presence of PO_4^{3-} as template at low pH, producing a "bicapped" Keggin-type $[PV_{14}O_{42}]^{9-}$ anion [10–15]. Vanadium (IV) (but not Nb(IV) or Ta(IV)) can aggregate around differently-shaped anionic templates (like Cl^-, NCS^- etc.) producing a family of reduced polyoxovanadates built from tetragonal pyramids, with fascinating magnetic properties [16,17]. The chemistry of polyoxoniobates (PONb) and tantalates is much less studied and for a long time it was confined to preparations of alkali metal salts of hexaniobate and hexatantalate, $[Nb_6O_{19}]^{8-}$ and $[Ta_6O_{19}]^{8-}$. The beginning of the modern era in this chemistry can be dated back to the works of Spinner, Pope and Stucky in the late 1960s–1980s [18–24], who obtained evidence for condensation of $[Nb_6O_{19}]^{8-}$ into larger entities, and discovered its ability to coordinate certain transition metals. This chemistry is quite different from polyvanadates and is dominated by Lindqvist hexametalates $[M_6O_{19}]^{8-}$, in spite of the existence of the decaniobate, $[Nb_{10}O_{28}]^{6-}$ [25,26] and recently reported decatantalate $[Ta_{10}O_{28}]^{6-}$ [27]. By contrary, free $[V_6O_{19}]^{8-}$ is unknown, and the reason for this is perhaps that, if existed, it would have an extraordinarily high charge density (defined as the anionic charge of the POM divided by the number of non-hydrogen atoms of the POM). Among the known POM, the highest charge density (0.32) is achieved precisely for $[M_6O_{19}]^{8-}$ [9]. As V is smaller than Nb, $[V_6O_{19}]^{8-}$ would have the highest charge density per volume ratio. It can be stabilized only by coordination of four positively charged $\{Cp^*Rh\}^{2+}$ units [28,29].

The high charge density means that $[M_6O_{19}]^{8-}$ anions are expected to act as efficient ligands towards various metals. This is indeed so, and a few examples of coordination of $[Nb_6O_{19}]^{8-}$ to Mn(IV) and Ni(IV) were reported, as early as in 1967–1969 [18–24], and Co(III), reported only in 2011 [30], giving complexes of the type $[M(Nb_6O_{19})_2]^{n-}$, where two hexaniobate units (each acting as tridentate ligand) sandwich an octahedrally coordinated central metal ion. There is also a strong association of hexaniobate with alkali metal cations, even in water, where the existence of neutral $[A_8Nb_6O_{19}]$ (A = Rb, Cs) species was established by Small Angle X-ray Scattering

(SAXS) [31]. Curiously enough, no tantalate analogues of $[M(Nb_6O_{19})_2]^{n-}$ could be obtained [32]. It is doubtless highly desirable to expand this still rather modest list of existing complexes. However, strongly basic conditions, necessary to prevent hydrolysis of hexaniobate (and precipitation of niobium oxide), severely limit the choice of suitable metal precursors (they would simply give insoluble hydroxides). This drawback can be obviated in two different ways. First, in order to limit the hydrolysis and ensuing condensation, a part of the coordinating positions of a transition metal can be occupied by a polydentate (or polyhaptic) ligand (or simply by a strongly coordinated unidentate ligand, such as CO). This can be achieved for Co(III), Cr(III), Cu(II) and Ni(II) in the presence of polyamines [33–37]. Particularly attractive for this purpose are the isolobal fragments d^6-*fac*-{ML$_3$}, where ML$_3$ stands for $\{M(CO)_3\}^+$ (M = Mn, Re), $\{(arene)Ru\}^{2+}$, $\{Cp^*Rh\}^{2+}$, $\{Cp^*Ir\}^{2+}$. In fact, a decade ago Pope *et al.* used $[M_6O_{19}]^{8-}$ (M = Nb, Ta) as robust inorganic tridentate ligands in reactions with $[Mn(CO)_3(CH_3CN)_3]ClO_4$ or $[Re(CO)_5Br]$, which gave rise to *trans*-$[(M'(CO)_3)_2M_6O_{19}]^{6-}$ (M' = Mn, Re; M = Nb, Ta) [38]. Proust *et al.* reported the syntheses of a series of complexes combining $\{(arene)Ru\}^{2+}$ and $[Nb_6O_{19}]^{8-}$ with Ru/POM stoichiometries ranging from 1:1–4:1 [39]. Coordination of organometallic fragments, such as $\{(p\text{-cymene})Ru\}^{2+}$ or $\{Cp^*Rh\}^{2+}$, to POM adds extra possibilities for the synthesis of hybrid (*i.e.*, combining chemically very different moieties within the same structure) complexes. This development reflects not only academic interest in such compounds, being a logical development of POM chemistry in general, but also is driven by the well-known catalytic properties of many POM complexes of transition metals, in particular, with noble metals [40].

The second way is to use a metal which forms stable well-defined and water soluble hydroxocomplexes, such as Pt(IV), for which both $[Pt(OH)_4(H_2O)_2]$ and $[Pt(OH)_6]^{2-}$ exist as individual species [41]. In this case, a particular attraction for coordinating Pt(IV) to $[Nb_6O_{19}]^{8-}$ comes from the reports about use of PONb in photocatalysis. In this case, a photoactivator is necessary, which role is typically fulfilled by $[PtCl_6]^{2-}$ [42]. From this point of view, direct coordination of Pt(IV) to PONb is very desirable. Pt(IV), being octahedral and sufficiently oxophilic, could either substitute a Nb atom in the hexaniobate structure, similar to $[H_2Pt^{IV}V_9O_{28}]^{5-}$ which is simply a Pt(IV)-monosubstituted classical decavanadate [43,44], or be "grafted" as a capping atom. As we will see, the latter is the case.

In this short review, we focus on the recent advances in the field of group V (in particular, Nb and Ta) POM complexes, concentrating on the synthesis of Lindqvist-based POM with incorporated noble metal centers and their potential applications as electrocatalysts for water oxidation. The second part addresses related studies with heteropolyniobate homologues coordinating noble metal fragments. The third part deals with the current use of analytical tools to unravel the integrity and reaction dynamics in the solution and selected examples are discussed.

2. Reactivity of $[M_6O_{19}]^{8-}$ towards Noble Metal Complexes

In this section, studies of reactivity of the Lindqvist $[M_6O_{19}]^{8-}$ (M = Ta and Nb) POM towards a noble metal complex are discussed, with an emphasis on finding suitable experimental conditions for noble metal grafting, the main structural features of the resulting solids, as well as their behavior in aqueous solution.

2.1. Reaction [M₆O₁₉]⁸⁻ with {Cp*Rh}²⁺ and {Cp*Ir}²⁺

2.1. Reaction $[M_6O_{19}]^{8-}$ with $\{Cp^*Rh\}^{2+}$ and $\{Cp^*Ir\}^{2+}$

Heating of $[Cp^*RhCl_2]_2$ with $K_7H[Nb_6O_{19}]\cdot13H_2O$ or $Cs_8[Ta_6O_{19}]\cdot14H_2O$ in water at 80–90 °C in 2:1 molar ratio yields yellow-orange solutions, from which $K_4[\{Cp^*Rh\}_2Nb_6O_{19}]\cdot20H_2O$ **(1)** and $Cs_4[\{Cp^*Rh\}_2Ta_6O_{19}]\cdot18H_2O$ **(2)** can be readily isolated in high yields (80% for Nb and 70% for Ta) [45]. Crystal structures of both complexes contain hybrid bi-capped organometallic-POM anions *trans*-$[\{Cp^*Rh\}_2M_6O_{19}]^{4-}$ (Figure 1) with the typical Lindqvist-type oxide metal core (metal octahedron with one μ_6-, twelve μ_2- and six terminal oxygen ligands) decorated at an opposite pair of the triangular $\{M_3O_3\}$ faces with two capping organometallic fragments $\{Cp^*Rh\}^{2+}$. In both cases, exclusively *trans*-isomers crystallize. No *cis*-isomers were observed in solution or in solid. Both **1** and **2** produce basic solutions in water; 3 mM solutions have the pH value of 8.8, which is much lower in comparison to that produced by alkali metal salts of parent $[M_6O_{19}]^{8-}$ of comparable concentration (10.5–11.0). The ESI(-) mass spectra of aqueous solutions of **1** and **2** display prominent peaks assigned to the $[\{Cp^*Rh\}_2M_6O_{19} + 2H]^{2-}$ dianions, accompanied by minor signals from the singly-charged $[\{Cp^*Rh\}_2M_6O_{19} + 3H]^-$ anions. However, the protonation degree observed by ESI-MS does not necessarily mirrors that in the solution because of the influence of the ESI process itself on the protonation [46]. For example, the most efficient ionization mechanism observed by ESI-MS for the decaniobate $[Nb_{10}O_{28}]^{6-}$ ion was the uptake of up to three protons, whereas solution studies indicated no protonation [26,47].

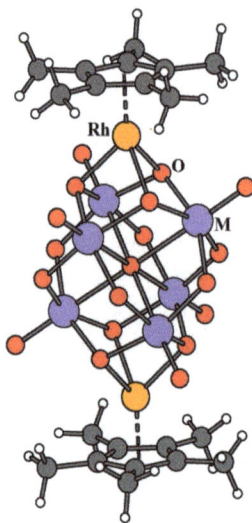

Figure 1. View of *trans*-$[\{Cp^*Rh\}_2M_6O_{19}]^{4-}$ in the crystal structure of $K_4[(Cp^*Rh)_2Nb_6O_{19}]\cdot20H_2O$ **(1)** and $Cs_4[(Cp^*Rh)_2Ta_6O_{19}]\cdot18H_2O$ **(2)** (ball and stick model).

The reactions of $[M_6O_{19}]^{8-}$ with $[Cp*RhCl_2]_2$ in 1:1 molar ratio under the same conditions gave solutions, from which only poorly diffracting yellow solids could be isolated. These solids can be extracted with methanol yielding yellow solutions. The ESI mass spectra of the methanolic extracts were consistent with the presence of POM/Rh complexes featuring 1:1 stoichiometry of general formula ($[\{Cp*Rh\}M_6O_{19}]^{6-}$ (**3**, M = Nb; **4**, M = Ta).

Grafting of $\{Cp*Ir\}^{2+}$ fragments onto the $[M_6O_{19}]^{8-}$ (M = Nb, Ta) proceeds essentially in the same manner with $[Cp*IrCl_2]_2$ as the source of the $\{Cp*Ir\}^{2+}$ fragment. Both 2:1 $[\{Cp*Ir\}_2Nb_6O_{19}]^{4-}$ (**5**, M = Nb; **6**, M = Ta) and 1:1 $[\{Cp*Ir\}Nb_6O_{19}]^{6-}$ (**7**, M = Nb; **8**, M = Ta) complexes can be produced in high yields by varying the reagent ratio. In this case, we succeeded in isolation and X-ray characterization of potassium (M = Nb) and sodium (M = Nb, Ta) salts of 2:1 *trans*-$[\{Cp*Ir\}_2M_6O_{19}]^{4-}$, and sodium salts of the single-capped $[\{Cp*Ir\}M_6O_{19}]^{6-}$ complexes (both for Nb and Ta) (Figure 2).

Compound **5**$^{4-}$ was tested as electrocatalyst for water oxidation. The cyclic voltammogram of the *trans*-$[\{Cp*Ir\}_2Nb_6O_{19}]^{4-}$ (**5**$^{4-}$) in 1 M Na_2SO_4 exhibited a strong anodic current due to water oxidation already at 0.8 V *versus* Ag/AgCl. The current (296 µA) at 1.6 V in the presence of the complex was 5.1 times higher than that (57.7 µA) for the blank solution.

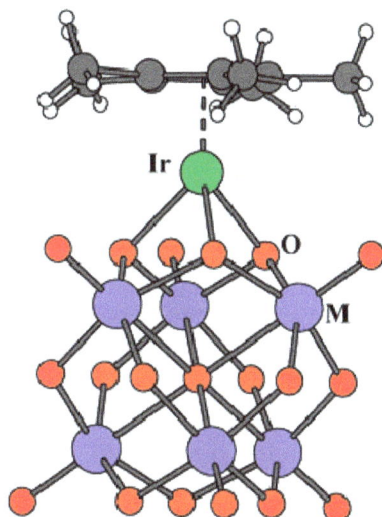

Figure 2. Ball and stick model of single-capped $[\{Cp*Ir\}M_6O_{19}]^{6-}$ hybrid anions (M = Nb, Ta).

2.2. Complexation of $[Ta_6O_{19}]^{8-}$ with $\{(C_6H_6)Ru\}^{2+}$

$Na_8[Ta_6O_{19}]\cdot24.5H_2O$ reacts with $[(C_6H_6)RuCl_2]_2$ (POM/Ru ratio 1:1) in water, yielding a yellow solution which contains single-capped $[(C_6H_6)RuTa_6O_{19}]^{6-}$ (**9**$^{6-}$) complex as the main product [48]. The progress of the reaction was monitored by ESI-MS as shown in Figure 3.

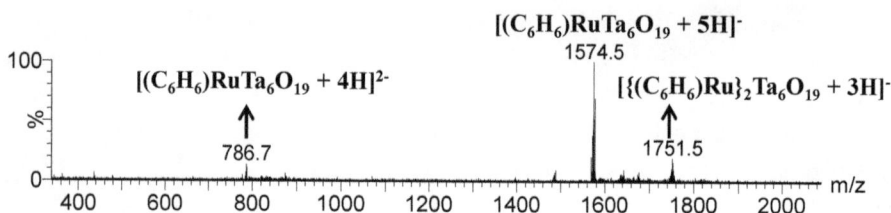

Figure 3. ESI(−) mass spectrum of aqueous solutions recorded at Uc = 10 V of the products of reaction between $Na_8[Ta_6O_{19}]\cdot24.5H_2O$ and $[(C_6H_6)RuCl_2]_2$ with POM / Ru ratio 1:1.

The ESI(−) mass spectrum of this reaction features the $[(C_6H_6)RuTa_6O_{19} + 5H]^-$ anion (*m/z* 1574.5) as the base peak, accompanied by a minor signal due to the doubly charged $[(C_6H_6)RuTa_6O_{19} + 4H]^{2-}$ species (*m/z* 786.7). A minor peak at *m/z* 1751.5 was attributed to the double-capped $[\{(C_6H_6)Ru\}_2Ta_6O_{19} + 3H]^-$ anion with two coordinated $\{(C_6H_6)Ru\}^{2+}$ units. Evaporation of the reaction solution gave yellow plates of $Na_{10}[\{(C_6H_6)RuTa_6O_{18}\}_2(\mu\text{-}O)]\cdot39.4H_2O$ (**10**) in 80% yield, with unexpected $[\{(C_6H_6)RuTa_6O_{18}\}_2(\mu\text{-}O)]^{10-}$ (**10**$^{10-}$) anion (see scheme 1) built from two fragments $\{(C_6H_6)RuTa_6O_{18}\}$ joined via a common linear μ_2-O bridge. The formation of $[\{(C_6H_6)RuTa_6O_{18}\}_2(\mu\text{-}O)]^{10-}$ can be viewed as coordination-induced condensation of two hexatantalates according to the equilibrium:

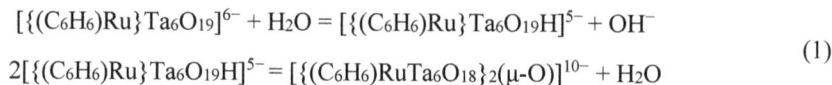

$$[\{(C_6H_6)Ru\}Ta_6O_{19}]^{6-} + H_2O = [\{(C_6H_6)Ru\}Ta_6O_{19}H]^{5-} + OH^-$$

$$2[\{(C_6H_6)Ru\}Ta_6O_{19}H]^{5-} = [\{(C_6H_6)RuTa_6O_{18}\}_2(\mu\text{-}O)]^{10-} + H_2O$$

(1)

It is worth stressing that such condensation is unknown for free, non-coordinated hexatantalate. Perhaps, the reduced charge of the 1:1 complex $[\{(C_6H_6)Ru\}Ta_6O_{19}]^{6-}$ facilitates interaction between two negatively charged species; coordination of electron-accepting fragment $\{(C_6H_6)Ru\}^{2+}$ may affect electronic density on the oxygen atoms of the POM part as well.

Decreasing the POM/Ru ratio to 1:2 leads to the bis-capped complex *trans*-$[\{(C_6H_6)Ru\}_2Ta_6O_{19}]^{4-}$ (**11**), isolated as $Na_4[\{(C_6H_6)Ru\}_2Ta_6O_{19}]\cdot20H_2O$ ($Na_4([11]\cdot20H_2O)$ in high yield. The ESI(−) mass spectrum displays the signal from the triply protonated $[\{(C_6H_6)Ru\}_2Ta_6O_{19} + 3H]^-$ anion (*m/z* 1751.5) as the main peak, accompanied by a minor signal from the doubly protonated $[\{(C_6H_6)Ru\}_2Ta_6O_{19} + 2H]^{2-}$ dianion (*m/z* 875.3). A small amount of monocapped $[\{(C_6H_6)Ru\}Ta_6O_{19} + 5H]^-$ species coexists with the $[\{(C_6H_6)Ru\}_2Ta_6O_{19}]^{4-}$ anion, thus suggesting a plausible dynamic equilibrium between mono- and bis-capped species. The relationship between the Ru/Ta species is summarized in Scheme 1:

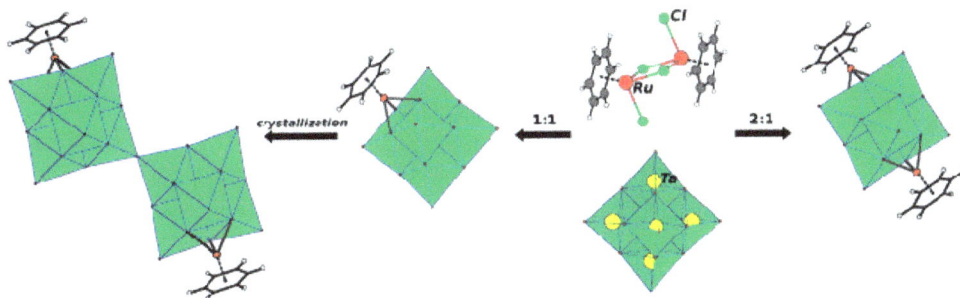

Scheme 1. Formation of 1:1 and 1:2 complexes from $[Ta_6O_{19}]^{8-}$ and $[(C_6H_6)RuCl_2]_2$.

2.3. Reaction $[Nb_6O_{19}]^{8-}$ and Pt^{4+}

Hydrothermal reaction (150 °C) of $[Pt(OH)_4(H_2O)_2]$ with $K_7H[Nb_6O_{19}]\cdot13H_2O$ in 1:1 molar ratio gives a yellow solution, which shows only one signal in ^{195}Pt NMR spectrum at 3189 ppm. Evaporation of this solution yields yellow crystals of $Cs_2K_{10}[Nb_6O_{19}\{Pt(OH)_2\}]_2\cdot13H_2O$ (**Cs_2K_{10}-12**). Compound **Cs_2K_{10}-12** was characterized by X-ray structural analysis (see Figure 4 left) [49].

Platinum niobate **12** is a unique dimeric complex, unprecedented in the PONb chemistry, and consists of two Pt(IV) centers coordinated each by three oxygen atoms from a $\{Nb_3O_3\}$ face of a hexaniobate (d(Pt-O)$_{av}$ 2.028(13), 2.010(14), and 2.124(14) Å), a terminal oxygen of the adjacent $[Nb_6O_{19}]^{8-}$ anion (d(Pt-O) 1.998(11) Å), and bearing two terminal OH groups (d(Pt-O) = 2.015(16) and 1.994(14) Å). Freshly prepared aqueous solution of **12** gives ^{195}Pt NMR spectrum with only one signal at 3189 ppm, the same as is observed in the mother liquor. The negative ESI mass spectrum also strongly suggests that the integrity of the dimeric $[Nb_6O_{19}\{Pt(OH)_2\}]_2^{12-}$ polyanion is preserved in solution.

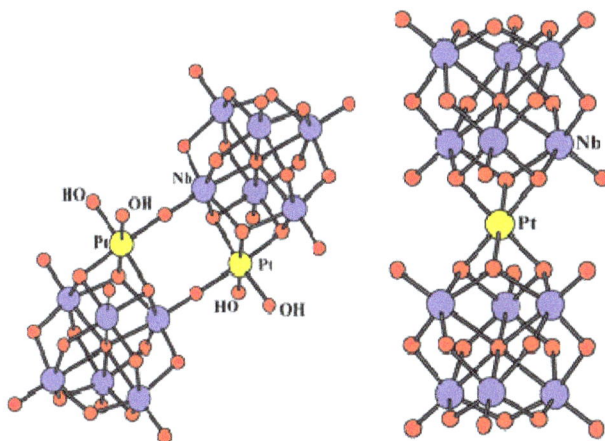

Figure 4. Structure of $[Nb_6O_{19}\{Pt(OH)_2\}]_2^{12-}$ (**12^{12-}**) (left) and structure of $[Pt(Nb_6O_{19})_2]^{12-}$ (**13^{12-}**) (right), ball and stick models.

Reaction of [Pt(OH)$_4$(H$_2$O)$_2$] and K$_7$H[Nb$_6$O$_{19}$]·13H$_2$O in the 1:2 molar ratio at 160–190 °C gives a yellow solution, which has two ^{195}Pt NMR signals at 3189 and 3422 ppm with relative intensities 0.75/0.25. Slow evaporation of this solution in a glass beaker produces light-yellow crystals of Na$_2$K$_{10}$[Pt(Nb$_6$O$_{19}$)$_2$]·18H$_2$O (**Na$_2$K$_{10}$-13**) in 20% yield (Na$^+$ comes from glass leaching). The sandwich-type anion [Pt(Nb$_6$O$_{19}$)$_2$]$^{12-}$ (**13^{12-}**) is a new (fourth) member of the well-known [M(Nb$_6$O$_{19}$)$_2$]$^{n+}$ (M = Mn, Ni, Co) family (Figure 4 right). All the Pt-O distances are equivalent, being 2.042(1) Å. The salt **Na$_2$K$_{10}$-13** is isotypic with K$_6$Na$_2$[Co^{III}H$_5$(Nb$_6$O$_{19}$)$_2$]·26.5H$_2$O [30].

Further evaporation of the solution gives **Cs$_2$K$_{10}$-12**. A freshly prepared solution of **Na$_2$K$_{10}$-13** demonstrates only one, slightly broadened, signal at 3422 ppm. Aged solutions give rise to increasingly intense signal at 3189 ppm, corresponding to the formation of [Nb$_6$O$_{19}${Pt(OH)$_2$}]$_2^{12-}$ (**12**). This observation implies a dynamics in solution, where [Pt(Nb$_6$O$_{19}$)$_2$]$^{12-}$ (**13**) slowly converts into [Nb$_6$O$_{19}${Pt(OH)$_2$}]$_2^{12-}$ (**12**). Taking advantage of the versatility and high sensitivity of the ESI-MS technique for tracing solution speciation in the POM chemistry [47,50], we recorded ESI-MS spectrum of aqueous solution of **Cs$_2$K$_{10}$-12** equilibrated for one week. ESI-MS results suggest coexistence of at least four different species in solution. Beside **12** and **13**, the ESI-MS data suggest the presence of free Lindqvist [Nb$_6$O$_{19}$]$^{8-}$ polyanion and a species that corresponds to the [Nb$_6$O$_{19}$Pt(OH)$_3$]$^{7-}$ anion whose structure might putatively correspond to capping of [Nb$_6$O$_{19}$]$^{8-}$ with a {Pt(OH)$_3$}$^+$ group. Remarkably, the platinum niobate **Cs$_2$K$_{10}$-12** exhibits electrocatalytic activity for water oxidation in 0.1 M Na$_2$SO$_4$. Oxidation process generates a significant rise in the current for a catalytic process mediated by the complex. The current in the presence of the platinum niobate is several times higher than for the blank solution. This electrochemical behavior is similar to the reported for the nonatungstate complex with a tetrarhodium-oxo core [51].

3. Reactivity of Heteropolyniobates towards Noble Metal Complexes

In 1979, S.K. Ray reported that reaction of K$_7$H[Nb$_6$O$_{19}$]·13H$_2$O with H$_6$TeO$_6$ gave a heteropolyniobate formulated as K$_8$H$_2$[TeNb$_{12}$O$_{38}$]·13H$_2$O [52]. This formula was deduced uniquely from elemental analysis. Such stoichiometry could imply that this compound would be analogous with the known heteropolyniobates [(Nb$_6$O$_{19}$)$_2$M]$^{n-}$ (M = Co, Ni, Mn, Pt) with octahedral Te^{VI} sandwiched between two hexaniobates. We were curious if it were really so and repeated these experiments, only to find out that the purported "dodecaniobate" was in fact [(OH)TeNb$_5$O$_{18}$]$^{6-}$, which can be regarded simply as a substituted hexaniobate with one {NbO}$^{3+}$ vertex replaced with a {TeOH}$^{5+}$ vertex. In fact, this complex had been already identified by W. Casey and co-workers, who prepared it from Nb$_2$O$_5$·xH$_2$O, Me$_4$NOH and H$_6$TeO$_6$ in methanolic solution [53]. Perhaps, the authors of [52] dealt with an approximately stoichiometric mixture of [(OH)TeNb$_5$O$_{18}$]$^{6-}$ and [Nb$_6$O$_{19}$]$^{8-}$, which led them to the wrong conclusion about the composition of the product. Surface charge density of telluroniobate is lower than in the hexaniobate, which means that their reactivity might be different. In order to check the significance of this difference, reactivity studies towards noble metal complexes have been extended in our group to the telluroniobate [(OH)TeNb$_5$O$_{19}$]$^{6-}$.

3.1. Reaction [(OH)TeNb₅O₁₉]⁶⁻ with {Cp*Rh}²⁺ and {Cp*Ir}²⁺

3.1. Reaction [(OH)TeNb$_5$O$_{19}$]$^{6-}$ with {Cp*Rh}$^{2+}$ and {Cp*Ir}$^{2+}$

We investigated reaction of $[(OH)TeNb_5O_{18}]^{6-}$ with $[Cp*MCl_2]_2$ as sources of $\{Cp*M\}^{2+}$ (M = Rh, Ir). Despite the difference in the surface charge density between $[(OH)TeNb_5O_{18}]^{6-}$ and $[Nb_6O_{19}]^{8-}$, formation of double-capped complexes takes place as easily as with hexaniobate. Crystal structure of $Na_2[\{Cp*Rh\}_2Te(OH)Nb_5O_{18}]\cdot24H_2O$ **(14)** shows coordination of two $\{Cp*Rh\}^{2+}$ fragments, with the Te atom localized in the central M_4 plane of Lindqvist-type anion as illustrated in Figure 5. Two other metal positions are fully occupied by niobium. However, this disorder corresponds to four different orientations of a *single* isomer which is the only possible structure for the *trans* complex in the solid state.

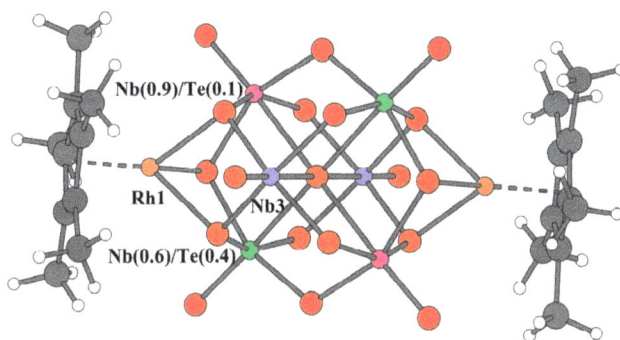

Figure 5. Structure of $[\{Cp*Rh\}_2Te(OH)Nb_5O_{19}]^{2-}$ anion **(14)**, positions of Te atom with different occupancies are colored in green and purple (ball and stick model).

3.2. Reaction [(OH)TeNb₅O₁₉]⁶⁻ with {(C₆H₆)Ru}²⁺

3.2. Reaction [(OH)TeNb$_5$O$_{19}$]$^{6-}$ with {(C$_6$H$_6$)Ru}$^{2+}$

Reactivity studies on $[(OH)TeNb_5O_{18}]^{6-}$ were also extended to $[(C_6H_6)RuCl_2]_2$ as source of the $\{(C_6H_6)Ru\}^{2+}$. ESI-MS can be conveniently used to monitor the advance of the reaction. Grafting of one and two $\{(C_6H_6)Ru\}^{2+}$ fragments can be immediately inferred from the characteristic m/z shifts to higher values, as illustrated in Figure 6. Moreover, the uptake of one or two Ru-containing fragments confers a new characteristic isotopic pattern due to multiple isotopes of the Ru at the natural abundance. Formation of single and double capped complexes $[(OH)TeNb_5O_{18}\{(C_6H_6)Ru\}]^{4-}$ **(15)** and $[(OH)TeNb_5O_{18}\{(C_6H_6)Ru\}_2]^{2-}$ **(16)** takes place as smoothly as with hexaniobate. Characteristic peaks at m/z 539.1 and 628.1 correspond to formulations $[(OH)TeNb_5O_{18}\{(C_6H_6)Ru\} + 2H]^{2-}$ and $[(OH)TeNb_5O_{18}\{(C_6H_6)Ru\}_2]^{2-}$, respectively. Other species based upon **15** and **16** associated Na^+ or K^+ adducts are also evident.

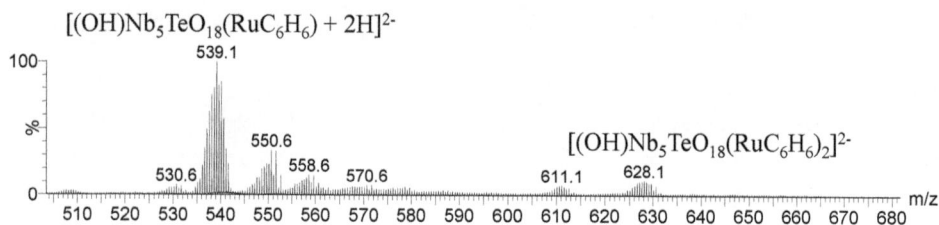

Figure 6. ESI mass spectrum of the reaction mixture between $[(OH)Nb_5O_{18}]^{6-}$ and $[(C_6H_6)RuCl_2]_2$ with POM / Ru ratio 1:1.

3.3. Reaction $[SiNb_{12}O_{40}]^{16-}$ with $\{(C_6H_6)Ru\}^{2+}$

Studies on the complexation with the siliconiobate $[SiNb_{12}O_{40}]^{16-}$ [54] are particularly relevant. Because of its high negative charge, it is expected to be much more "sticky" than well-known Group 6 Keggin complexes $[EM_{12}O_{40}]^{n-}$ (n = 2–6) (M = Mo, W) which are often regarded as non-coordinating [55]. It can be anticipated that quite a large number of metal fragments could be grafted to $[SiNb_{12}O_{40}]^{16-}$. Following this rationale, we undertook preliminary exploration of the reaction of $[SiNb_{12}O_{40}]^{16-}$ with $[(C_6H_6)RuCl_2]_2$, and ESI-MS results indicate that up to four $\{(C_6H_6)Ru\}^{2+}$ fragments can be grafted to $[SiNb_{12}O_{40}]^{16-}$, leading to the species formulated as $[\{(C_6H_6)Ru\}_4SiNb_{12}O_{40}]^{8-}$ (**17**).

4. Analytical Tools to Study Solution Speciation in Nb and Ta POM Chemistry

One of the general features of the reactivity of POMs is their lability in aqueous solutions up to the point that such solutions can be regarded as virtual libraries of different building blocks, which equilibrate together and can form different structures depending on counter cations, pH and *etc.* The simplest case is protonation. The hexametalate anions $[M_6O_{19}]^{8-}$ (M = Nb, Ta) are very basic. For example, hexaniobate in water exists as $[H_3Nb_6O_{19}]^{5-}$ (pH = 8), $[H_2Nb_6O_{19}]^{6-}$ (pH = 10), $[HNb_6O_{19}]^{7-}$ (pH = 12), and $[Nb_6O_{19}]^{8-}$ (pH = 14) [56]. Similarly, for hexatantalate three protonated species, $[HTa_6O_{19}]^{7-}$, $[H_2Ta_6O_{19}]^{6-}$, and $[H_3Ta_6O_{19}]^{5-}$ were found [57–59]. Even in aqueous NBu$_4$OH hexatantalate exists as $[H_2Ta_6O_{19}]^{6-}$ [60]. Small-angle X-ray scattering (SAXS) studies of $[Ta_6O_{19}]^{8-}$ and $[Nb_6O_{19}]^{8-}$ in water have been reported and important differences in the solution behavior have been observed due to the distinctive propensity to aggregation with counteractions for M = Ta and Nb [61]. Uv-vis spectroscopy complemented with DFT calculations have been also reported [62]. More complicated processes connected with rearrangement of POM backbone can take place [63–65]. It is obvious that proving the identity between species existing in solid state and solution is a very important part of this chemistry. However, the study of solution behavior of polyniobates and polytantalates is a challenging task. These species are colorless, diamagnetic, show no redox behavior and are, under most conditions, in a dynamic equilibrium. In this section, useful analytical tools that can provide complementary signatures for the Nb and Ta POM characterization in solution are discussed. In particular, the usefulness of ^{1}H NMR, ^{1}H NMR Diffusion Ordered Spectroscopy (DOSY), capillary electrophoresis and ESI-MS in concert is underlined. An illustrative example is given to demonstrate the complementarily of all techniques to determine stoichiometry and solution dynamics.

The possibilities of NMR are restricted to the use of ^{17}O, which unfortunately has very low natural abundance and sensitivity. It was used, for example, to follow the conversion of several niobates in water [66,67]. However, coordination of organometallic fragments to niobates and tantalates provides new opportunities for use of NMR for solution studies. In particular, structural information can be gained from ^1H NMR signals of the organic ligands attached to the grafted heterometal, which would tell us about the overall symmetry of the compound. Moreover, it is possible to estimate diffusion coefficients from ^1H NMR DOSY of the organic part that can be subsequently used to calculate molecular weight of a particular species. This ^1H DOSY NMR technique is becoming a powerful tool for the determination of the size (hydrodynamic radius) and the molecular weight of inorganic clusters [68–70]. This technique has not been virtually used in the chemistry of Nb and Ta POM prior to our work.

Another useful tool for solution studies is capillary electrophoresis (CE). It is rarely used in the chemistry of POMs [71,72]. The possibilities opened up by the use of CE for POMs mixtures were studied in a few cases [73]. Although in these works CE revealed itself as a promising analytical tool, its potential for POM chemistry is far from being fully exploited. Combination of CE, HPLC and other methods (in particular ESI-MS and NMR) can provide sufficient information about quantitative and qualitative composition of solutions, numbers of POM species, and their stability and evolution with time.

Accurate determination of the formulae and structure of new compounds remains an important issue in POM chemistry. In this respect, use of soft ionization mass spectrometric techniques, and in particular, ESI techniques, is becoming widespread. Excellent reviews on the use of soft ionization MS techniques can be found elsewhere; for example, Cronin *et al.* illustrated the versatility of ESI and cryospray ionization techniques not only for characterization of final POM assemblies, but also to prove mechanisms of self-assembly of supramolecular POM-based architectures in solution [47,50]. Ohlin *et al.* reported an important work on this topic highlighting different approaches to obtain reliable information on the dynamic solution of POM on the basis of ESI-MS [47]. However, there are several potential limitations to using ESI-MS alone for the investigation of POM. This is the case when dynamic speciation involving hydroxy-bridged POM species, the pH value, ionic strength or the counteraction plays an important role. In addition, during the ESI process itself, the samples undergo dramatic changes in pH and concentration, which may perturb the speciation equilibrium. Moreover, due to the high charge state and molecular weight, the use of ESI-MS spectrometers equipped with high resolution analysers is important to facilitate unambiguous molecular compositions' determination. In this part, we will discuss a case study, where we used ESI-MS, ^1H DOSY NMR and CE techniques, combined together, in order to elucidate the solution behavior of the $\{(C_6H_6)Ru\}^{2+}/[Ta_6O_{19}]^{8-}$ system to afford compounds **9** and **10** [48].

4.1. $\{(C_6H_6)Ru\}^{2+}/[Ta_6O_{19}]^{8-}$ System—Case Study

The ^1H DOSY NMR experiment performed on a sample of $Na_{10}[10]\cdot39.4H_2O$ in D_2O (see Figure 7) yielded the self-diffusion coefficients $D = 315 \pm 10$ $\mu m^2\cdot s^{-1}$ and $D = 325 \pm 10$ $\mu m^2\cdot s^{-1}$, respectively, for the ^{1}H NMR signals located at 5.85 ppm and 5.94 ppm. The Stokes-Einstein relation $D=\dfrac{kT}{6\pi\eta R_H}$ (which describes the diffusion coefficient D for spherical objects, where k is the Boltzmann constant, T the temperature taken equal to 300 K in this study, η of the solvent viscosity taken equal to that of water at 300 K (1.002 Pa.s) and R_H the hydrodynamic radius) gave hydrodynamic radii of 6.75–6.96 Å which are compatible with the radii expected for *solvated monomeric* species [{(C$_6$H$_6$)Ru}Ta$_6$O$_{19}$]$^{6-}$ and [{(C$_6$H$_6$)Ru}$_2$Ta$_6$O$_{19}$]$^{4-}$ but *not with the dimeric* [{(C$_6$H$_6$)RuTa$_6$O$_{18}$}$_2$(μ-O)]$^{10-}$.

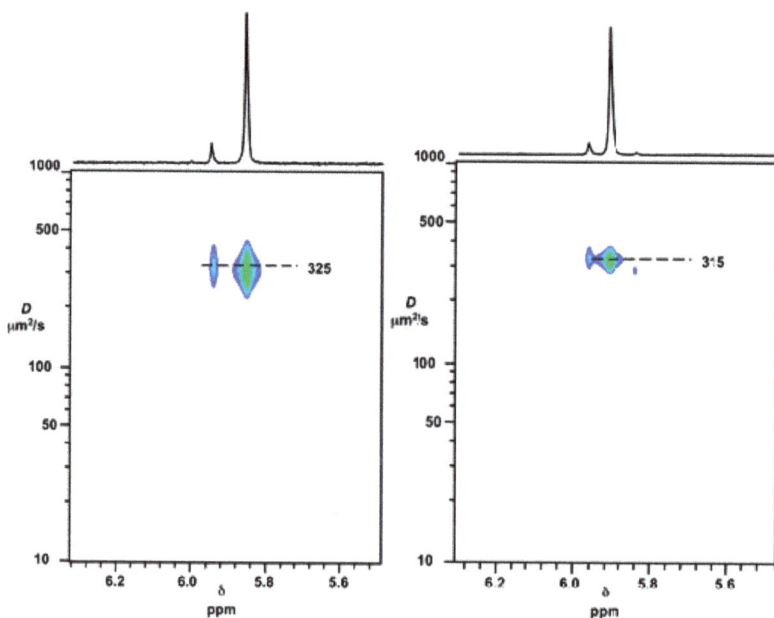

Figure 7. ^{1}H DOSY NMR spectra of **Na$_{10}$-10** in D$_2$O (**left**) and ^{1}H DOSY NMR spectra of **Na$_4$-9** in D$_2$O (**right**).

The ^{13}C NMR spectrum of Na$_4$[{(C$_6$H$_6$)Ru}$_2$Ta$_6$O$_{19}$]·20H$_2$O (**Na$_4$-9**) in D$_2$O exhibits two signals at 81.95 ppm and 81.82 ppm, with the complementary set of signals in the ^{1}H NMR spectrum at 5.97 (6%) ppm and 5.92 (93%) ppm and 5.85 (≤1%). This pattern corresponds to a mixture of *cis* (6%) and *trans* (93%) isomers of [{(C$_6$H$_6$)Ru}$_2$Ta$_6$O$_{19}$]$^{4-}$, similarly to what was observed for [{(p-cymene)Ru}$_2$Nb$_6$O$_{19}$]$^{4-6}$, while the minor peak at 5.85 ppm is to be attributed to the 1:1 complex [{(C$_6$H$_6$)Ru}Ta$_6$O$_{19}$]$^{6-}$. The ^{1}H DOSY NMR experiment gives the self-diffusion coefficients D = 315 ± 10 μm^2 s^{-1} and D = 325 ± 10 μm^2 s^{-1} for the signals at 5.92 and 5.97 ppm, respectively, whereas the low intensity of the minor peak has prevented accurate determination of the D value. These values correspond to hydrodynamic radii in the 0.675–0.696 nm range. This is identical to that obtained for the solutions of Na$_{10}$[{(C$_6$H$_6$)RuTa$_6$O$_{18}$}$_2$(μ-O)]·39.4H$_2$O and agree well with the attribution of ^{1}H NMR peaks to *cis*-[{(C$_6$H$_6$)Ru}$_2$Ta$_6$O$_{19}$]$^{4-}$ (5.97 ppm), *trans*-[{(C$_6$H$_6$)Ru}$_2$Ta$_6$O$_{19}$]$^{4-}$ (5.92 ppm), and [{(C$_6$H$_6$)Ru}Ta$_6$O$_{19}$]$^{6-}$ (5.85 ppm). These results necessarily imply two equilibria. The first corresponds to *cis-trans* isomerization process, strongly displaced towards the *trans* isomer. The latter corresponds to the de-coordination of a $\{(C_6H_6)Ru\}^{2+}$ moiety. This hypothesis is supported by the observation of a small broad peak at 5.4 ppm, attributed to $[(C_6H_6)_2Ru_2(OH)_3]^+$ (see Equation 2).

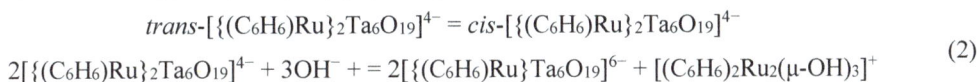

$$trans\text{-}[\{(C_6H_6)Ru\}_2Ta_6O_{19}]^{4-} = cis\text{-}[\{(C_6H_6)Ru\}_2Ta_6O_{19}]^{4-}$$

$$2[\{(C_6H_6)Ru\}_2Ta_6O_{19}]^{4-} + 3OH^- = 2[\{(C_6H_6)Ru\}Ta_6O_{19}]^{6-} + [(C_6H_6)_2Ru_2(\mu\text{-}OH)_3]^+ \qquad (2)$$

ESI(−) mass analysis of Na_{10}-**10** in water reveals only the presence of the *monomeric* **9**$^{6-}$ species, thus indicating that the dimerization depicted in Eq. 1 is a reversible process ongoing from the solution to the solid state and *vice versa*. The electrophorograms recorded for aqueous solutions of Na_{10}-**10** and Na_4-**9** are presented in Figure 8. The solutions of Na_{10}-**10**, to which some amount of Na_4-**9** has been added as reference, give two well-resolved peaks (Figure 7). The second peak corresponds to **9**, which was confirmed by UV-spectroscopy and by the standard addition method. The first peak must therefore correspond to the *monomeric* $[\{(C_6H_6)Ru\}Ta_6O_{19}]^{6-}$ species (or its dominant protonated form), because the signal from the dimer with its larger size and charge, should appear after the signal of **10**, and with a longer retention time. Thus, combination of ESI-MS and CE gives coherent evidence that in aqueous solutions $[\{(C_6H_6)RuTa_6O_{18}\}_2(\mu\text{-}O)]^{10-}$ *dissociates into the monomeric* $[\{(C_6H_6)Ru\}Ta_6O_{19}]^{6-}$.

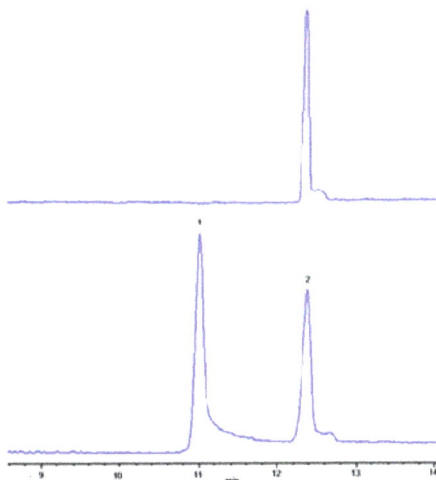

Figure 8. The electrophorogram of complex Na_4-**9** (up). The electrophorogram of complex Na_{10}-**10** solution with addition of complex Na_4-**9** as reference (down). Fused-silica capillary 50μm i.d. × 75 sm; sample buffer: $7.5 \cdot 10^{-3}$M borate (pH 9.18); running buffer: $7.5 \cdot 10^{-2}$ M borate (pH 9.18).

5. Conclusions

The chemistry of niobium and tantalum polyoxoanions still belongs to a little researched area of inorganic chemistry. Many unusual complexes still remain hidden, and isolation of principally new

$[(Ta_6O_{18})_2(\mu\text{-}O)]^{14-}$ backbone is a tantalyzing hint at this potential richness. In this short review, we summarized our recent research focused on the coordination of noble metals to hexametalates. Grafting of $\{Cp^*M\}^{2+}$ (M = Rh, Ir) on the surface of Lindqvist-type hexametalates (including $[(OH)TeNb_5O_{19}]^{6-}$) was studied with full characterization of the hybrid complexes in solid state and solution. Coordination of $\{(C_6H_6)Ru\}^{2+}$ to $[Ta_6O_{19}]^{8-}$ gives unprecedented $[(\{(C_6H_6)Ru\}Ta_6O_{18})_2(\mu\text{-}O)]^{10-}$. Complexation of Pt(IV) with $[Nb_6O_{19}]^{8-}$ was studied with different techniques. New dimeric complex $[(Nb_6O_{19}\{Pt(OH)_2\})_2]^{12-}$ is obtained when hexaniobate reacts with Pt(IV) in 1:1 molar ratio, while increasing the Pt/Nb$_6$ ratio to 2:1 gives sandwich-type complex $[Pt(Nb_6O_{19})_2]^{12-}$. We used different, both traditional and non-traditional, analytical tools for studying the solution behavior of individual reaction products and reaction mixtures, a challenging task for the systems where few analytical methods are readily available. The ^{1}H, ^{13}C and DOSY NMR techniques combined with ESI-MS and CE highlight are adequate tools for characterization of complex solutions. An illustrative example dealing with the $\{(C_6H_6)Ru\}^{2+}/[Ta_6O_{19}]^{8-}$ system demonstrates complementarity of all these techniques for determination of stoichiometry and solution dynamics, an approach both promising and highly recommendable to trace unambiguously the identity of large inorganic systems [74].

Acknowledgments

Authors thanks SCIC of the UJI and all members of our Russian-Spanish-French team for very fruitful collaboration. This work was supported by Russian Scientific Foundation (RScF 14-13-00645).

Author Contributions

The manuscript was written with equal contributions from all authors. MNS owns the general idea and plan of the publication, and he wrote parts 1 and 3, as well as Conclusions. PAA authored part 2, and CV was the main contributor to part 4.

Conflicts of Interest

The authors declare no conflict of interest.

References

1. Livage, J. Synthesis of polyoxovanadates via "chimie douce". *Coord. Chem. Rev.* **1998**, *178–180*, 999–1018.
2. Hayashi, Y. Hetero and lacunary polyoxovanadate chemistry: Synthesis, reactivity and structural aspects. *Coord. Chem. Rev.* **2011**, *255*, 2270–2280.

3. Müller, A.; Sessoli, R.; Krickemeyer, E.; Bögge, H.; Meyer, J.; Gatteschi, D.; Pardi, L.; Westphal, J.; Hovemeier, K.; Rohlfing, R.; *et al.* Polyoxovanadates: High-Nuclearity Spin Clusters with Interesting Host–Guest Systems and Different Electron Populations. Synthesis, Spin Organization, Magnetochemistry, and Spectroscopic Studies. *Inorg. Chem.* **1997**, *36*, 5239–5250.

4. Breen, J.M.; Zhang, L.; Clement, R.; Schmitt, W. Hybrid Polyoxovanadates: Anion-Influenced Formation of Nanoscopic Cages and Supramolecular Assemblies of Asymmetric Clusters. *Inorg. Chem.* **2012**, *51*, 19–21.

5. Zhang, L.; Schmitt, W. From Platonic Templates to Archimedean Solids: Successive Construction of Nanoscopic {$V_{16}As_8$}, {$V_{16}As_{10}$}, {$V_{20}As_8$}, and {$V_{24}As_8$} Polyoxovanadate Cages. *J. Am. Chem. Soc.* **2011**, *133*, 11240–11248.

6. Gatteschi, D.; Sessoli, R.; Müller, A.; Kögerler, P. Polyoxometalate chemistry: A source for unusual spin topologies. In *Polyoxometalate Chemistry*; Pope, M.T., Müller, A., Eds.; Kluwer: Dordrecht, The Netherlands, 2001; p. 319;

7. Choi, J.; Sanderson, L.A.W.; Musfeldt, J.L.; Ellern, A.; Kögerler, P. Optical properties of the molecule-based magnet $K_6[V_{15}As_6O_{42}(H_2O)]\cdot 8H_2O$. *Phys. Rev. B* **2003**, *68*, 064412;

8. Gatteschi, D.; Pardi, L.; Barra, A.L.; Müller, A. Polyoxovanadates: The Missing Link between Simple Paramagnets and Bulk Magnets? *Mol. Eng.* **1993**, *3*, 157–169.

9. Nymann, M. Polyoxoniobate chemistry in the 21st century. *Dalton Trans.* **2011**, *40*, 8049–8058.

10. Kato, R.; Kobayashi, A.; Sasaki, Yu. 1:14 Heteropolyvanadate of phosphorus: preparation and structure. *J. Am. Chem. Soc.* **1980**, *102*, 6571–6572.

11. Kato, R.; Kobayashi A.; Sasaki, Yu. The heteropolyvanadate of phosphorus. Crystallographic and NMR studies. *Inorg. Chem.* **1982**, *21*, 240–246.

12. Nomiya, K.; Kato, K.; Miwa, M. Preparation and spectrochemical properties of soluble vanadophosphate polyanions with bicapped-Keggin structure. *Polyhedron* **1986**, *5*, 811–813.

13. Khan, M.I.; Zubieta, J.; Toscano, P. Protonation sites in a heteropolyvanadate of phosphorus: X-ray crystal structure of $(Me_3NH)_4(NH_4)[H_4PV_{14}O_{42}]$. *Inorg. Chim. Acta* **1992**, *193*, 17–20.

14. Nakamura, S.; Yamawaki, T.; Kusaka, K.; Otsuka, T.; Ozeki, T. Hydrogen-bond Networks Involving Protonated Bicapped-Keggin Tetradecavanadophosphate Anions. *J. Clust. Sci.* **2006**, *17*, 245–256.

15. Grabau, M.; Forster, J.; Heussner, K.; Streb, C. Synthesis and Theoretical Hirshfeld Analysis of a Supramolecular Heteropolyoxovanadate Architecture. *Eur. J. Inorg. Chem.* **2011**, 1719–1724.

16. Fang, X.; Kögerler, P.; Speldrich, M.; Schilder, H.; Luban, M. A polyoxometalate-based single-molecule magnet with an $S = 21/2$ ground state. *Chem. Commun.* **2012**, *48*, 1218–1220.

17. Monakhov, K.Yu.; Linnenberg, O.; Kozłowski, P.; van Leusen, J.; Besson, C.; Secker, T.; Ellern, A.; López, X.; Poblet, J.M.; Kögerler, P. Supramolecular Recognition Influences Magnetism in $[X@HV^{IV}_8V^{V}_{14}O_{54}]^{6-}$ Self-Assemblies with Symmetry-Breaking Guest Anions. *Chem. Eur. J.* **2015**, *21*, 2387–2397.

18. Dale, B.W.; Pope, M.T. The heteropoly-12-niobomanganate(IV) anion. *J. Chem. Soc. Chem. Commun.* **1967**, 792–792.

19. Flynn, C.M., Jr.; Stucky, G.D. Heteropolyniobate complexes of manganese(IV) and nickel(IV). *Inorg. Chem.* **1969**, *8*, 332–334.

20. Flynn, C.M., Jr.; Stucky, G.D. Crystal structure of sodium 12-niobomanganate(IV),$Na_{12}MnNb_{12}O_{38} \cdot 50H_2O$. *Inorg. Chem.* **1969**, *8*, 335–344.

21. Kheddar, N.; Spinner, B. Constitution et domaines de stabilite des oxaloniobates en solution aqueuse. *Bull. Soc. Chim. Fr.* **1972**, *2*, 502–506 (In French).

22. Marty, A.; Abdmeziem, K.; Spinner, B. Une nouvelle condensation pour des sels du niobium V: les nonaniobates de tetramethyl et de tetraethylammonium. *Comptes Rendus C.* **1976**, *283*, 285–288.

23. Marty, A.; Abdmeziem, K.; Spinner, B. De noveaux isopolyanions du niobium V: comportement en solution aqueuse des nonaniobates de tetramethyl et tetraethylammonium. *Bull. Soc. Chim. Fr.* **1977**, *3–4*: 231–238 (In French).

24. Goiffon, A.; Philippot, E.; Maurin, M. Structure cristalline du niobate 7/6 de sodium $(Na_7)(H_3O)Nb_6O_{19} \cdot 14H_2O$. *Rev. Chim. Miner.* **1980**, *17*, 466–476

25. Graeber, E.J.; Morosin, B. The molecular configuration of the decaniobate ion ($Nb_{10}O_{28}^{6-}$) *Acta Cryst.* **1977**, *33*, 2137–2143.

26. Villa, E.M.; Ohlin, C.A.; Balogh, E.; Anderson, T.M.; Nyman, M.D.; Casey, W.H. Reaction Dynamics of the Decaniobate Ion $[H_xNb_{10}O_{28}]^{(6-x)-}$ in Water. *Angew. Chem. Int. Ed.* **2008**, *47*, 4844–4846.

27. Matsumoto, M.; Ozawa, Y.; Yagasaki, A.; Zhe, Y. Decatantalate—The Last Member of the Group 5 Decametalate Family. *Inorg. Chem.* **2013**, *52*, 7825–7827.

28. Hayashi, Y.; Ozawa, Y.; Isobe, K. The first vanadate hexamer capped by 4 cyclopentadienyl-rhodium or cyclopentadienyl-iridium groups. *Chem. Lett.* **1989**, 425–428.

29. Chae, H.K.; Klemperer, W.G.; Day, V.W. Organometal hydroxide route to $[(C_5Me_5)Rh]_4(V_6O_{19})$ *Inorg. Chem.* **1989**, *28*, 1423–1424.

30. Ma, P.T.; Chen, G.; Wang, G.; Wang, J.P. Cobalt—Sandwiched lindqvist hexaniobate dimer $[Co(III)H_5(Nb_6O_{19})_2]^{8-}$. *Rus. J. Coord. Chem.* **2011**, *37*, 772–775.

31. Antonio, M.R.; Nyman, M.; Anderson, T. M. Direct Observation of Contact Ion-Pair Formation in Aqueous Solution. *Angew. Chem. Int. Ed.* **2009**, *48*, 6136–6140.

32. Dale, B.W.; Buckley, J.M.; Pope, M.T. Heteropoly-niobates and -tantalates containing manganese(IV). *J. Chem. Soc. A.* **1969**, 301–304.

33. Flynn, C.M., Jr.; Stucky, G.D. Sodium 6-niobo(ethylenediamine)cobaltate(III) and its chromate(III) analog. *Inorg. Chem.* **1969**, *8*, 178–180.

34. Bontchev, R.P.; Nyman, M. Evolution of Polyoxoniobate Cluster Anions. *Angew. Chem. Int. Ed.* **2006**, *45*, 6670–6672.

35. Niu, J.Y.; Ma, P.T.; Niu, H.Y.; Li, J.; Zhao, J.W.; Song, Y.; Wang, J.P. Giant Polyniobate Clusters Based on $[Nb_7O_{22}]^{9-}$ Units Derived from a Nb_6O_{19} Precursor. *Chem.–Eur. J.*, **2007**, *13*, 8739–8748.

36. Chen, G.; Ma, P.T.; Wang, J.P.; Niu, J.Y. A new organic–inorganic hybrid polyoxoniobate based on Lindqvist-type anion and nickel complex. *J. Coord. Chem.* **2010**, *63*, 3753–3763.

37. Hegetschweiler, K.; Finn, R.C.; Rarig, R.S.; Sander, J.; Steinhauser, S.; Worle, M.; Zubieta, J. Surface complexation of $[Nb_6O_{19}]^{8-}$ with NiII: solvothermal synthesis and X-ray structural characterization of two novel heterometallic Ni-Nb-polyoxometalates. *J. Inorg. Chim. Acta* **2002**, *337*, 39–47.

38. Dickman, M.H.; Pope, M.T. Robust, Alkali-Stable, Triscarbonyl Metal Derivatives of Hexametalate Anions, $[M_6O_{19}\{M'(CO)_3\}_n]^{(8-n)-}$ (M = Nb, Ta; M' = Mn, Re; n = 1, 2). *Inorg. Chem.* **2001**, *40*, 2582–2586.

39. Laurencin, D.; Thouvenot, R.; Boubekeur, K.; Proust, A. Synthesis and reactivity of {Ru(*p*-cymene)}$^{2+}$ derivatives of $[Nb_6O_{19}]^{8-}$: A rational approach towards fluxional organometallic derivatives of polyoxometalates. *Dalton Trans.* **2007**, 1334–1345.

40. Hill, C.L.; Prosser-McCartha, C.M. Homogeneous catalysis by transition metal oxygen anion clusters. *Coord. Chem. Rev.* **1995**, *143*, 407–455.

41. Vasilchenko, D.; Tkachev, S.; Baidina, I.; Korenev, S. Speciation of Platinum(IV) in Nitric Acid Solutions. *Inorg. Chem.* **2013**, *52*, 10532–10541.

42. Huang, P.; Qin, C.; Su, Z.-M.; Xing, Y.; Lang, X.-L.; Shao, K.-Z.; Lan, Y.-Q.; Wang, E.-B. Self-Assembly and Photocatalytic Properties of Polyoxoniobates: $\{Nb_{24}O_{72}\}$, $\{Nb_{32}O_{96}\}$, and $\{K_{12}Nb_{96}O_{288}\}$ Clusters. *J. Am. Chem. Soc.* **2012**, *134*, 14004–14010.

43. Lee, U.; Joo, H.C.; Park, K.M.; Mal, S.S.; Kortz, U.; Keita, B.; Nadjo, V. Facile Incorporation of Platinum(IV) into Polyoxometalate Frameworks: Preparation of $[H_2Pt^{IV}V_9O_{28}]^{5-}$ and Characterization by 195Pt NMR Spectroscopy. *Angew. Chem. Int. Ed.* **2008**, *47*, 793–796.

44. Kortz, U.; Lee, U.; Joo, H.C.; Park, K.M.; Mal, S.S.; Dickman, M.H.; Jameson, G.B. Platinum-Containing Polyoxometalates. *Angew. Chem. Int. Ed.* **2008**, *47*, 9383–9384.

45. Abramov, P.A.; Sokolov, M.N.; Virovets, A.V.; Floquet, S.; Haouas, M.; Taulelle, F.; Cadot, E.; Vicent, C.; Fedin, V. Grafting $\{Cp^*Rh\}^{2+}$ on the surface of Nb and Ta Lindqvist-type POM. *Dalton Trans.* **2015**, *44*, 2234–2239.

46. Di Marco, V.B.; Bombi, G.G. Electrospray mass spectrometry (ESI-MS) in the study of metal-ligand solution equilibria. *Mass Spectrom. Rev.* **2006**, *25*, 347–379.

47. Ohlin, C.A. Reaction Dynamics and Solution Chemistry of Polyoxometalates by Electrospray Ionization Mass Spectrometry. *Chem.-Asian J.* **2012**, *7*, 262–270.

48. Abramov, P.A.; Sokolov, M.N.; Floquet, S.; Haouas, V.; Taulelle, V.; Cadot, V.; Peresypkina, E.V.; Virovets, A.V.; Vicent, C.; Kompankov, N.B.; Zhdanov, A.A.; Shuvaeva, O.V.; Fedin, V.P. Coordination-Induced Condensation of $[Ta_6O_{19}]^{8-}$: Synthesis and Structure of $[\{(C_6H_6)Ru\}_2Ta_6O_{19}]^{4-}$ and $[\{(C_6H_6)RuTa_6O_{18}\}_2(\mu\text{-}O)]^{10-}$. *Inorg. Chem.* **2014**, *53*, 12791–12798.

49. Abramov, P.A.; Vicent, C.; Kompankov, N.B.; Gushchin, A.L.; Sokolov, M.N. Platinum polyoxoniobates. *Chem. Commun.* **2015**, *51*, 4021–4023.

50. Miras, H.N.; Wilson, E.F.; Cronin, L. Unravelling the complexities of inorganic and supramolecular self-assembly in solution with electrospray and cryospray mass spectrometry. *Chem. Commun.* **2009**, 1297–1311.

51. Sokolov, M.N.; Adonin, S.A.; Abramov, P.A.; Mainichev, D.A.; Zakharchuk, N.F.; Fedin, V.P. Self-assembly of polyoxotungstate with tetrarhodium-core: Synthesis, structure and ^{183}W NMR studies. *Chem. Commun.* **2012**, *48*, 6666–6668.

52. Ray, S.K. Synthesis of a Te dodecaniobate. *J. Ind. Chem. Soc.* **1976**, *53*, 1238–1239.

53. Son, J.-H.; Wang, J.; Osterloh, F.E.; Yu, P.; Casey, W.H. A tellurium-substituted Lindqvist-type polyoxoniobate showing high H_2 evolution catalyzed by tellurium nanowires via photodecomposition. *Chem. Commun.* **2014**, *50*, 836–838.

54. Nyman, M.; Bonhomme, F.; Alam, T.M.; Parise, J.B.; Vaughan, G.M.B. $[SiNb_{12}O_{40}]^{16-}$ and $[GeNb_{12}O_{40}]^{16-}$: Highly Charged Keggin Ions with Sticky Surfaces. *Angew. Chem., Int. Ed.*, **2004**, *43*, 2787–2792.

55. Anyushin, A.V.; Smolentsev, A.I.; Mainichev, D.A.; Vicent, C.; Gushchin, A.L.; Sokolov, M.N.; Fedin, V.P. Synthesis and characterization of a new Keggin anion: $[BeW_{12}O_{40}]^{6-}$. *Chem. Commun.* **2014**, *50*, 9083.

56. Black, J.R.; Nyman, M.; Casey, W.H. Kinetics of ^{17}O-exchange reactions in aqueous metal-oxo nanoclusters. *Geochim. Cosmochim. Acta* **2006**, *70*, A53-A53.

57. Nelson, W.H.; Tobias, R.S. Polyanions of the Transition Metals. II. Ultracentrifugation of Alkaline Tantalum(V) Solutions; Comparison with Light Scattering. *Inorg. Chem.* **1964**, *3*, 653–658.

58. Spinner, B.; Kheddar, N. Nouveaux isopolyanions du tantale V. *Comp. Rend.* **1969**, *268C*, 1108–1111.

59. Arana, G.; Etxebarria, N.; Fernandez, L.A.; Madariaga, J.M. Hydrolysis of Nb(V) and Ta(V) in aqueous KCl at 25 °C. Part II: Construction of a thermodynamic model for Ta(V). *J. Solution Chem.* **1995**, *24*, 611–622.

60. Matsumoto, M.; Ozawa, Y.; Yagasaki, A. Which is the most basic oxygen in $[Ta_6O_{19}]^{8-}$?— Synthesis and structural characterization of $[H_2Ta_6O_{19}]^{6-}$. *Inorg. Chem. Commun,* **2011**, *14*, 115–117.

61. Fullmer, L.B.; Molina, P.I.; Antonio, M.R.; Nyman, M. Contrasting ion-association behaviour of Ta and Nb polyoxometalates. *Dalton Trans.* **2014**, *43*, 15295–15299.

62. Deblonde, G.J.P.; Moncomble, A.; Cote, G.; Belair, S.; Chagnes, A. Experimental and computational exploration of the UV-visible properties of hexaniobate and hexatantalate ions. *RSC Adv.* **2015**, *5*, 7619–7627.

63. Maekawa, M.; Ozawa, Y.; Yagasaki, A. Icosaniobate: A New Member of the Isoniobate Family *Inorg. Chem.* **2006**, *45*, 9608–9609.

64. Matsumoto, M.; Ozawa, Y.; Yagasaki, A. Reversible dimerization of decaniobate. *Polyhedron* **2010**, *29*, 2196–2201.

65. Tsunashima, R.; Long, D.L.; Miras, H.N.; Gabb, D.; Pradeep, C.P.; Cronin, L. The Construction of High-Nuclearity Isopolyoxoniobates with Pentagonal Building Blocks: $[HNb_{27}O_{76}]^{16-}$ and $[H_{10}Nb_{31}O_{93}(CO_3)]^{23-}$. *Angew. Chem. Int. Ed.* **2010**, *49*, 113–116.

66. Villa, E.M.; Ohlin, C.A.; Rustad, J.R.; Casey, W.H. Isotope-Exchange Dynamics in Isostructural Decametalates with Profound Differences in Reactivity *J. Am. Chem. Soc.* **2009**, *131*, 16488–16492.

67. Villa, E.M.; Ohlin, C.A.; Casey, W.H. Oxygen-Isotope Exchange Rates for Three Isostructural Polyoxometalate Ions. *J. Am. Chem. Soc.* **2010**, *132*, 5264–5272.

68. Bannani, F.; Floquet, S.; Leclerc-Laronze, N.; Haouas, M.; Taulelle, F.; Marrot, J.; Kögerler, P.; Cadot, E. Cubic Box *versus* Spheroidal Capsule Built from Defect and Intact Pentagonal Units. *J. Am. Chem. Soc.* **2012**, *134*, 19342–19345.

69. Van Lokeren, L.; Cartuyvels, E.; Absillis, G.; Willema, R.; Parac-Vogt, T.N. Phosphoesterase activity of polyoxomolybdates: as a tool for obtaining insights into the reactivity of polyoxometalate clusters. *Chem. Commun.* **2008**, 2774–2776.

70. Lemonnier, J.-F.; Floquet, S.; Kachmar, A.; Rohmer, M.-M.; Bénard, M.; Marrot, J.; Terazzi, E.; Piguet, C.; Cadot, E. Host–guest adaptability within oxothiomolybdenum wheels: structures, studies in solution and DFT calculations. *Dalton Trans.* **2007**, 3043–3054.

71. Sakurai, N.; Kadohata, K.; Ichinose, N. Application of high-speed liquid chromatography using solvent extraction of the molybdoheteropoly yellow to the determination of microamounts of phosphorus in waste waters. *Fresenius' Z. Anal. Chem.* **1983**, *314*, 634–637.

72. Kirk, A.D.; Riske, W.; Lyon, D.K.; Rapko, B.; Finke, R.G. Rapid, high-resolution, reversed-phase HPLC separation of highly charged polyoxometalates using ion-interaction reagents and competing ions. *Inorg. Chem.* **1989**, *28*, 792–797.

73. Hettiarachichi, K.; Ha, Y.; Tran, T.; Cheung, A.P. Application of HPLC and CZE to the analysis of polyoxometalates. *J. Pharm. Biomed. An.* **1995**, *13*, 515–523.

74. Oliveri, A.F.; Elliott, E.W.; Carnes, M.E.; Hutchison, J.E.; Johnson, D.W. Elucidating Inorganic Nanoscale Species in Solution: Complementary and Corroborative Approaches. *ChemPhysChem* **2013**, *14*, 2655–2661.

Single-Crystal to Single-Crystal Reversible Transformations Induced by Thermal Dehydration in Keggin-Type Polyoxometalates Decorated with Copper(II)-Picolinate Complexes: The Structure Directing Role of Guanidinium

Aroa Pache, Santiago Reinoso, Leire San Felices, Amaia Iturrospe, Luis Lezama and Juan M. Gutiérrez-Zorrilla

Abstract: Three new hybrid inorganic-metalorganic compounds containing Keggin-type polyoxometalates, neutral copper(II)-picolinate complexes and guanidinium cations have been synthesized in bench conditions and characterized by elemental analysis, infrared spectroscopy and single-crystal X-ray diffraction: the isostructural $[C(NH_2)_3]_4[\{XW_{12}O_{40}\}\{Cu_2(pic)_4\}]\cdot[Cu_2(pic)_4(H_2O)]_2\cdot6H_2O$ [X = Si (**1**), Ge (**3**)] and $[C(NH_2)_3]_8[\{SiW_{12}O_{40}\}_2\{Cu(pic)_2\}_3\{Cu_2(pic)_4(H_2O)\}_2]\cdot8H_2O$ (**2**). The three compounds show a pronounced two-dimensional character owing to the structure-directing role of guanidinium. In **1** and **3**, layers of $[\{XW_{12}O_{40}\}\{Cu_2(pic)_4\}]_n^{4n-}$ hybrid POM chains and layers of $[Cu_2(pic)_4(H_2O)]$ complexes and $[C(NH_2)_3]^+$ cations pack alternately along the z axis. The hydrogen-bonding network established by guanidinium leads to a trihexagonal tiling arrangement of all copper(II)-picolinate species. In contrast, layers of $[C(NH_2)_3]^+$-linked $[\{SiW_{12}O_{40}\}_2\{Cu(pic)_2\}_3]_n^{8n-}$ double chains where each Keggin cluster displays a $\{Cu_2(pic)_4(H_2O)\}$ moiety pointing at the intralamellar space are observed in **2**. The thermal stability of **1–3** has been studied by thermogravimetric analyses and variable temperature powder X-ray diffraction. Compounds **1** and **3** undergo single-crystal to single-crystal transformations promoted by reversible dehydration processes and the structures of the corresponding anhydrous phases **1a** and **3a** have been established. Despite the fact that the $[Cu_2(pic)_4(H_2O)]$ dimeric complexes split into $[Cu(pic)_2]$ monomers upon dehydration, the packing remains almost unaltered thanks to the preservation of the hydrogen-bonding network established by guanidinium and its associated Kagome-type lattice. Splitting of the dimeric complexes has been correlated with the electron paramagnetic resonance spectra.

Reprinted from *Inorganics*. Cite as: Pache, A.; Reinoso, S.; Felices, L.S.; Iturrospe, A.; Lezama, L.; Gutiérrez-Zorrilla, J.M. Single-Crystal to Single-Crystal Reversible Transformations Induced by Thermal Dehydration in Keggin-Type Polyoxometalates Decorated with Copper(II)-Picolinate Complexes: The Structure Directing Role of Guanidinium. *Inorganics* **2015**, *3*, 194-218.

1. Introduction

Over the past several years, the large family of anionic metal-oxygen clusters known as polyoxometalates (POMs) has been thoroughly employed as building blocks to construct a variety of inorganic-organic hybrid compounds [1–7]. The assembly of POMs with transition metal complexes bearing organic ligands (TMCs) is an effective strategy for designing such type of compounds. The POMs may adopt a variety of roles in these types of hybrid systems: (1) charge compensating anions; (2) ligands directly bonded to TMCs; (3) templates inducing the self-assembly of

MOFs [8–13]. The clusters can act as peculiar inorganic ligands able to bind several TMCs through terminal or bridging oxygen atoms [14–16], and this often results in assemblies with extended structures. Thus, many high-dimensional POM-based hybrids have been successfully synthesized to date [17–19].

A critical factor for the construction of such architectures rests on the choice of appropriate organic ligands. For example, carboxylate derivatives of heterocyclic amines with mixed N,O-donor atoms are likely to afford polymeric structures with high dimensionalities among the vast library of polydentate ligands [20–22]. One way of better controlling the structure of the hybrid compound is the use directing agents able to form extensive networks of weak intermolecular interactions. A great deal of attention has been paid to the structure-directing role of several organic species and a surprising variety of organically-templated inorganic frameworks are found in the literature [23–25]. Guanidinium has shown up as an excellent template because it can establish massive hydrogen-bonding networks due to its high molecular symmetry and extremely weak acid character [26]. This cation has been successfully applied in POM chemistry not only as a template of high-dimensional frameworks but also as a selective crystallizing agent for minor POM species in mixed solutions [27–29].

We have recently reported a series of hybrid compounds based on $[XW_{12}O_{40}]^{4-}$ Keggin-type anions (X = Si, Ge) and copper(II) complexes of tetradentate bis(aminopyridil) ligands that can reversibly undergo thermal desorption of water via single-crystal to single-crystal (SCSC) transformations with significant modifications in the bonding and coordination geometry around the Cu^{II} centers [30,31]. To date, full studies on SCSC transformations are still scarce for POM-based compounds [32–38] and those involving the temperature as the external stimulus inducing the solid-state phase transition are limited to the low-temperature polymorphs of $[Tm_2(H_2O)_{14}(H_6CrMo_6O_{24})][H_6CrMo_6O_{24}] \cdot 16H_2O$ and $[C(NH_2)_3]_6[Mo_7O_{24}] \cdot H_2O$, to the monitoring of the dehydration in the $H_5PV_2Mo_{10}O_{40} \cdot 36H_2O$ acid and in the porous $[Co_2(ppca)_2(H_2O)(V_4O_{12})_{0.5}]$ (ppca = 4-(pyridin-4-yl)pyridine-2-carboxylic acid) hybrid material [39–42]. This scarcity is certainly remarkable because the study of solid-state phase transitions induced by external stimuli such as the temperature, redox processes, or the interaction with guest molecules is at the forefront of the crystal engineering [43]. For example, several reports on SCSC transformations triggered by the removal, incorporation and/or exchange of solvent guest molecules can be found in the literature for related systems like metalorganic framework (MOF) materials [44–47]. These processes are often referred to as dynamic structural changes associated to compounds classified as third generation materials with potential applications in gas storage and separation, chemical sensing or magnetic switching [48,49].

We now intend to explore the thermostructural behavior of other hybrid systems related to our previous Keggin/bis(aminopyridyl) compounds to evaluate the role of the organic component in facilitating such SCSC transitions. We have first focused our studies on N,O-polydentate heterocyclic ligands, which represent a great first choice for the preparation of extended structures [50–52], and their concerted action with templating cations like $[C(NH_2)_3]^+$. Keggin-type anions have been kept as the inorganic building blocks in our systems because (i) these clusters and their numerous derivatives represent the most archetypal class of heteropolyoxometalates [53,54] and (ii) they are widely known to give rise to highly intricate hybrid structures by coordinating a large number of TMCs simultaneously [6,55].

In this work, we report the synthesis, crystal structure, thermal behavior and electron paramagnetic resonance (EPR) spectra of a series of guanidinium-templated compounds based on Keggin-typeanions and copper(II)-picolinate complexes: $[C(NH_2)_3]_4[\{XW_{12}O_{40}\}\{Cu_2(pic)_4\}]\cdot[Cu_2(pic)_4(H_2O)]_2\cdot6H_2O$ [X = Si (**1**), Ge (**3**)] and $[C(NH_2)_3]_8[\{SiW_{12}O_{40}\}_2\{Cu(pic)_2\}_3\{Cu_2(pic)_4(H_2O)\}_2]\cdot8H_2O$ (**2**). Compounds **1** and **3** undergo SCSC transformations promoted by thermally induced, reversible dehydration processes and the structures of the anhydrous phases $[C(NH_2)_3]_4[\{XW_{12}O_{40}\}\{Cu_2(pic)_4\}]\cdot[Cu(pic)_2]_4$ [X = Si (**1a**), Ge (**3a**)] have also been determined by single crystal X-ray diffraction.

2. Results and Discussion

2.1. Synthesis and Infrared Spectroscopy

Compounds **1–3** were prepared under mild bench conditions from the $[C(NH_2)_3]^+$-directed self-assembly of $[XW_{12}O_{40}]^{4-}$ (X = Si, Ge) and $[Cu(pic)_2]$ building blocks in acidic aqueous medium (pH 3–3.5) at room temperature. Both types of building blocks were generated in situ from $[XW_{11}O_{39}]^{8-}$ POM precursors, a copper(II) source and the pic ligand in its acidic form. For X = Si, different POM:Cu:pic ratios were tested and the solid products obtained upon evaporation were characterized preliminarily by IR spectroscopy. For a 1:3:2 ratio, a mixture of blue crystals of a copper(II)-picolinate complex and a white powder corresponding to a guanidinium salt of the plenary $[SiW_{12}O_{40}]^{4-}$ anion was obtained. Lowering the amount of Cu^{II} ions to a 1:2:2 ratio led to co-crystallization of the complex with compound **1** as the minor fraction. Crystallization of the former was avoided by using a 1:1:2 ratio. Formation of **1** was maximized in these conditions, but crystallization of a small amount of a second crystalline phase (compound **2**) was in turn observed. Crystals of **1** are formed before those of **2** and this fact could be explained on the basis of the different POM:Cu ratio in both compounds: 1:6 and 2:7, respectively. The initial POM:Cu ratio in the reaction mixture is 1:1, and hence the compound with the highest Cu^{II} content (**1**) tends to crystallize first. The amount of the copper(II)-picolinate complex in solution decreases with respect to that of the POM when **1** crystallizes and this in turn favors the formation of a small amount of **2** with the highest POM content to re-equilibrate the POM:Cu ratio. All attempts of improving the synthetic procedure to avoid formation of mixtures were unsuccessful. While **1** is obtained as the major phase in a mixture of crystals with the side-product **2**, the isostructural **3** is isolated as a single crystalline phase when $[GeW_{11}O_{39}]^{8-}$ is used under the same synthetic conditions and no traces of a hypothetical Ge-containing analogue of **2** are observed by powder X-ray diffraction analysis (Figure S1). It is also worth noting that we never obtained any spectroscopic indication of a compound containing copper(II)-monosubstituted $[XW_{11}O_{39}Cu(H_2O)]^{6-}$ species in spite of using monolacunary Keggin-type anions as precursors. These species are known to be metastable in weakly acidic conditions (typically in the pH range 4–6 for heteropolyoxotungstates), and hence slow conversion into the plenary clusters seems reasonable after considering the pH values of our reaction mixtures. Since **1–3** contain plenary Keggin-type anions as the inorganic building block, we also performed a set of reactions using $[XW_{12}O_{40}]^{4-}$ POMs as the precursors. In all cases, powders containing these clusters and copper(II)-picolinate complexes were obtained according to IR spectroscopy. These powders could not be recrystallized or unequivocally identified as compounds **1–3** on the basis of powder X-ray diffraction. Thus, the kinetically slow $[XW_{11}O_{39}]^{8-}$ to $[XW_{12}O_{40}]^{84-}$ conversion appears to be a key factor in isolating our compounds as single crystals suitable for further structural characterization.

The infrared spectra of **1–3** (Figure S2) show the characteristic features of the $[\alpha\text{-}XW_{12}O_{40}]^{4-}$ Keggin-type anion in the region below 1000 cm^{-1} with bands of strong intensity corresponding to the antisymmetric stretching of the W–O$_t$ and W–O$_b$–W bonds that appear at 970 and *ca.* 800 cm^{-1} for X = Si and 966 and 787 cm^{-1} for X = Ge, respectively. The grafting of the copper(II)-picolinate complexes onto the POM surfaces shift the above signals by 10 cm^{-1} compared to those of the clusters in the potassium salts and leads also to the appearance of additional peaks in the 760–660 cm^{-1} range related to the Cu–O and Cu–N stretching among other vibrations. The metalorganic region above 1000 cm^{-1} is dominated by signals of medium to strong intensity that are observed in the 1160–1684 cm^{-1} range and associate to C=C and C=N stretching vibrations in the pyridine rings.

2.2. Crystal Structures of Compounds 1–3

Compounds **1** and **3** are isostructural and crystallize in the triclinic space group *P*–1 with the following molecules in the asymmetric unit: one half of a centrosymmetric {XW$_{12}$O$_{40}$} Keggin cluster (X = Si, Ge), one half of a centrosymmetric {Cu$_2$(pic)$_4$} dinuclear complex supported on the cluster, two halfs of an isolated dimeric unit [Cu$_2$(pic)$_4$(H$_2$O)], two [C(NH$_2$)$_3$]$^+$ cations and three H$_2$O molecules. Compound **2** also crystallizes in the space group *P*–1 and its asymmetric unit contains one {SiW$_{12}$O$_{40}$} Keggin cluster, one supported {Cu$_2$(pic)$_4$(H$_2$O)} dinuclear complex, one half of a centrosymmetric {Cu(pic)$_2$} monomeric unit connected to the cluster, another {Cu(pic)$_2$} complex also connected to the cluster, four [C(NH$_2$)$_3$]$^+$ cations and four H$_2$O molecules (Figure 1). The inorganic $[XW_{12}O_{40}]^{4-}$ building block in all compounds shows the characteristic structure of the α-Keggin isomer consisting of a central XO$_4$ tetrahedron surrounded by four edge-shared W$_3$O$_{13}$ trimers, all of them linked via corner-sharing in ideal *T$_d$* symmetry. In the case of **1** and **3**, the Keggin anion lies on a center of inversion with the tetrahedral XO$_4$ group disordered over two crystallographic positions, which leads to its observation as a XO$_8$ cube with half-occupied O sites (Figure 1, left). Table S1 displays ranges of W–O and X–O bond lengths compared to those of the DFT-optimized Keggin anion [56].

2.2.1. Copper(II)-Picolinate Complexes

The title compounds contain different types of neutral copper(II) complexes with the ligand 2-picolinate. In all of these complexes, the Cu^{II} atom shows axial-type coordination geometry with two *trans*-related organic ligands forming the basal or equatorial plane (Figure 2). Selected bond lengths compared to those found in the anhydrous phases **1a** and **3a** are listed in Table 1.

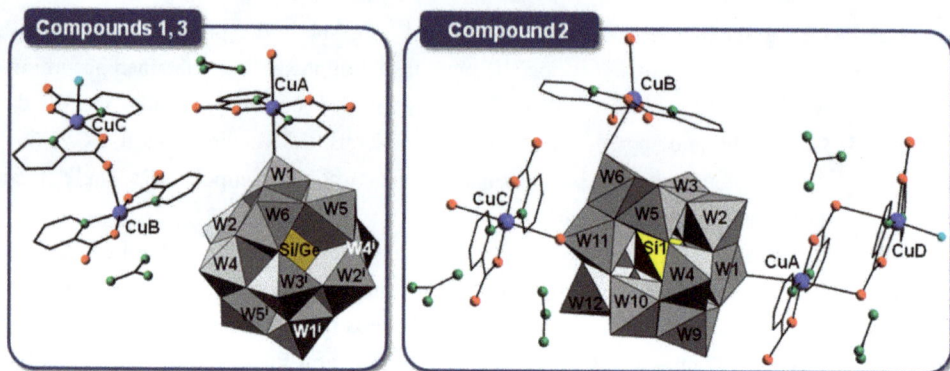

Figure 1. Connectivity between building blocks in the asymmetric units of **1–3**. Color code: W, gray polyhedra; Si/Ge, yellow polyhedra; Cu, blue spheres; N, green spheres; O, red spheres for O_{pic} or O_{POM} atoms and cyan spheres for terminal aqua ligands; C, black sticks. Symmetry code: (i) $-x, -y, -z$.

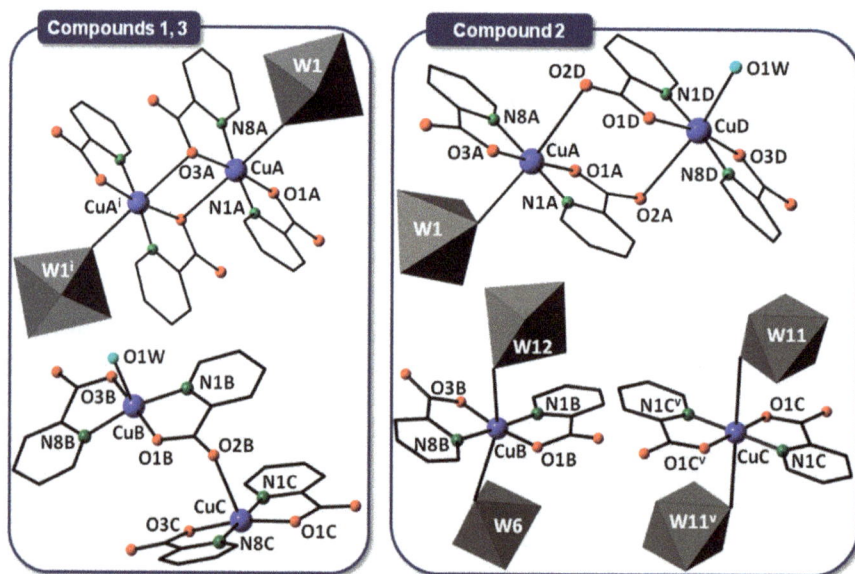

Figure 2. Copper(II)-picolinate complexes with atom labeling in **1–3** (for the symmetry codes i and v see Table 1).

Two different types of dinuclear complexes coexist in the structures of **1** and **3**. The metalorganic {$Cu_2(pic)_4$} subunit is composed of two centrosymmetrically related {$CuA(pic)_2$} fragments where the CuA atom is involved in a $CuN_2O_2O'_2$ chromophore with tetragonally elongated octahedral geometry. One of the axial positions in each CuA center is occupied by one of the O_{pic} atoms forming the equatorial CuN_2O_2 plane of the neighboring fragment, in such a way that a dimeric complex with *equatorial-axial* $Cu_2(\mu^2\text{-}O_{pic})_2$ rhomboid core is formed. A terminal O_{POM} atom (disordered over two sites, e.g., O1/O1Z in Table 1) occupies the second axial position, and hence the $\{Cu_2(pic)_4\}$ subunits link the Keggin clusters in hybrid $[\{XW_{12}O_{40}\}\{Cu_2(pic)_4\}]^{4-}$ chains with alternate inorganic and metalorganic building blocks. In contrast, the $[Cu_2(pic)_4(H_2O)]$ moiety is formed by one $\{CuC(pic)_2\}$ and one $\{CuB(pic)_2(H_2O)\}$ fragments where both Cu^{II} atoms show CuN_2O_2O' chromophores with distorted square-pyramidal geometry. A water molecule is located at the apical position of CuB, whereas that of CuC is occupied by one of the O_{pic} atoms that are not involved in the basal CuN_2O_2 plane of CuB. Therefore, the two Cu^{II} centers in this moiety are linked by a single pic bridging ligand acting in μ^2-κ^2N,O^1:κ^1O^2 coordination mode.

Table 1. Bond lengths and intradimeric Cu···Cu distances (Å) for the copper(II)-picolinate complexes in 1–3 compared to those in the anhydrous phases 1a and 3a.

Bond	1	1a	3	3a	Bond	2
CuA–N1A	1.964(8)	1.959(9)	1.965(8)	1.956(16)	CuA–N1A	1.98(2)
CuA–O1A	1.956(7)	1.962(7)	1.958(7)	1.972(14)	CuA–O1A	1.91(2)
CuA–N8A	1.965(9)	1.953(8)	1.954(8)	1.961(16)	CuA–N8A	1.96(2)
CuA–O3A	1.973(7)	1.970(7)	1.963(7)	1.975(14)	CuA–O3A	1.97(2)
CuA–O1/	2.459(18)/	2.527(17)/	2.468(17)/	2.48(3)/	CuA–O1	2.57(2)
–O1Z	2.349(18)	2.352(15)	2.346(17)	2.36(2)	CuA–O2D^{iii}	2.93(2)
CuA–O3A^i	2.778(8)	2.807(7)	2.774(7)	2.846(14)		
CuB–N1B	1.978(9)	1.956(8)	1.977(9)	1.972(16)	CuB–N1B	1.95(2)
CuB–O1B	1.972(8)	1.939(7)	1.970(7)	1.940(15)	CuB–O1B	1.96(2)
CuB–N8B	1.981(9)	1.956(9)	1.975(9)	1.944(17)	CuB–N8B	1.98(3)
CuB–O3B	1.969(8)	1.951(7)	1.968(7)	1.943(15)	CuB–O3B	1.89(2)
CuB–O1W	2.253(9)	–	2.258(8)	–	CuB–O12/	2.77(2)
					–O12Z	2.98(5)
					CuB–O6^{iv}	2.77(2)
CuC–N1C	1.973(9)	1.964(9)	1.972(9)	1.985(18)	CuC–N1C	1.958(18)
CuC–O1C	1.949(9)	1.953(8)	1.951(8)	1.931(16)	CuC–O1C	1.888(16)
CuC–N8C	1.971(9)	1.962(9)	1.966(9)	1.96(2)	CuC–N1C^v	1.958(18)
CuC–O3C	1.951(8)	1.928(8)	1.946(8)	1.945(16)	CuC–O1C^v	1.888(16)
CuC–O2B^{ii}	2.488(8)	3.310(8)	2.490(8)	3.314(18)	CuC–O11	2.909(19)
					CuC–O11^v	2.909(19)
					CuD–N1D	1.96(3)
					CuD–O1D	1.97(2)
					CuD–N8D	1.97(2)
					CuD–O3D	1.96(2)
					CuD–O1W	2.29(3)
					CuD–O2A^{iii}	2.97(3)
CuA···CuA^i	3.572(2)	3.554(2)	3.568(2)	3.617(3)	CuA···CuD^{iii}	5.305(6)
CuB···CuC^{ii}	5.538(3)	5.458(2)	5.540(2)	5.498(5)		

Note: Symmetry Codes: (i) 1–x, 1–y, –z; (ii) –x, 1–y, –z; (iii) 2–x, –y, –z; (iv) x, 1+y, z; (v) 1–x, –y, 1–z.

In the case of **2**, the structure contains one dinuclear and two crystallographically independent mononuclear complexes where the $CuN_2O_2O'_2$ coordination environment around all Cu^{II} centers is tetragonally elongated octahedral. In both monomeric subunits, the axial positions of the CuB and CuC atoms are occupied by terminal O_{POM} atoms. Thus, the {CuB(pic)$_2$} subunits link the Keggin clusters in a one-dimensional assembly of alternate inorganic and metalorganic building blocks, whereas the {CuC(pic)$_2$} subunits act as connectors between pairs of such hybrid chains to lead to the backbone of the $[\{SiW_{12}O_{40}\}_2\{Cu(pic)_2\}_3\{Cu_2(pic)_4(H_2O)\}_2]_n^{8n-}$ polymer. The dinuclear {Cu$_2$(pic)$_4$(H$_2$O)} subunit is made of one {CuA(pic)$_2$} and one {CuD(pic)$_2$(H$_2$O)} fragments linked in *equatorial-axial* fashion by two pic ligands in μ^2-κ^2N,O^1:κ^1O^2 bridging mode. Thus, each Cu^{II} center shows at axial positions one of the O_{pic} atoms that are not involved in the equatorial plane of the neighboring fragment. The coordination geometry of CuD is completed with one aqua ligand, whereas CuA axially anchors to a terminal O_{POM} atom, in such a way that the double-chained backbone of the hybrid polymer results decorated with antenna {Cu$_2$(pic)$_4$(H$_2$O)} subunits.

All of the dinuclear species mentioned above are new copper(II)-picolinate discrete complexes that have not been previously described in the literature. Nevertheless, the assembly modes between {Cu(pic)$_2$} fragments observed for the [Cu$_2$(pic)$_4$(H$_2$O)] moiety in **1** and **3** and the antenna {Cu$_2$(pic)$_4$(H$_2$O)} subunit in **2** are almost identical to those found in the polymeric derivatives {[Cu$_2$(pic)$_3$(H$_2$O)]X}$_n$ (X = ClO$_4^-$, BF$_4^-$) [52] and [Cu(pic)$_2$]$_n$ [57,58], respectively.

2.2.2. Crystal Packing of Compounds **1** and **3**

The crystal packing of **1** and **3** has a pronounced two-dimensional character with alternating hybrid and metalorganic layers stacked along the [001] direction (Figure 3). The hybrid layers consist of the $[\{XW_{12}O_{40}\}\{Cu_2(pic)_4\}]^{4-}$ chains running along the [110] direction and arranged in parallel fashion in the *xy* plane. The metalorganic sublattice contains the [Cu$_2$(pic)$_4$(H$_2$O)] dimers, all water molecules of hydration and all guanidinium cations. The interstitial water molecules do not appear to play a significant structural role as they only establish few hydrogen bonds that either connect Keggin clusters from different layers—the [Cu$_2$(pic)$_4$(H$_2$O)] dimers to the clusters or adjacent hybrid chains through the metalorganic subunits (Table S2). In contrast, the two guanidinium cations create an extended and massive network of N–H$\cdots$O$_{pic}$ hydrogen bonds with the carboxylate functionalities of the organic ligands. Each cation strongly interacts with the three crystallographically independent {Cu(pic)$_2$} fragments and arrange them in the (11−1) plane to lead to a corrugated double trihexagonal tiling of Cu^{II} atoms (Figure 4). The Keggin clusters are nested in the hexagonal motifs of this distorted Kagome-type double lattice, whereas the structure-directing guanidinium cations reside in the triangular cavities.

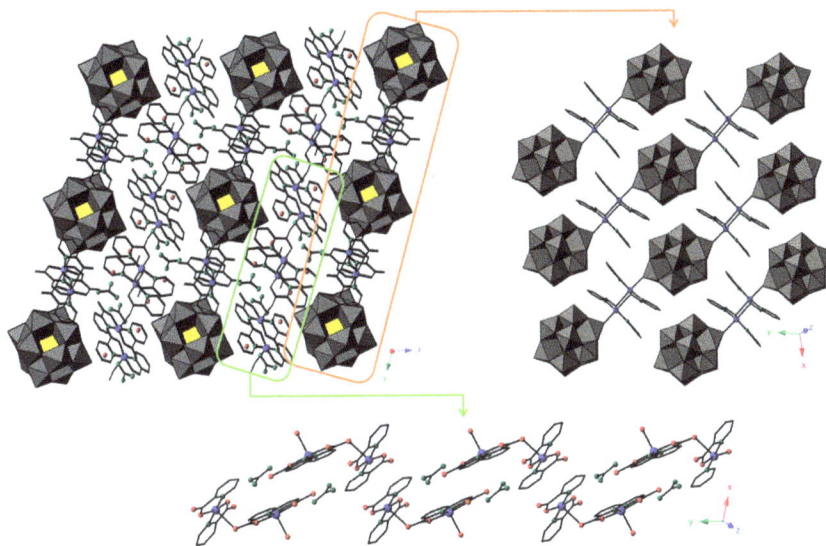

Figure 3. View of the crystal packing of **1** and **3** along the crystallographic *a* axis with details of the arrangement of the $[\{XW_{12}O_{40}\}\{Cu(pic)_2\}_2]^{4-}$ chains in the hybrid layers and the $[Cu_2(pic)_4(H_2O)]$ units and $[C(NH_2)_3]^+$ cations in the metalorganic regions.

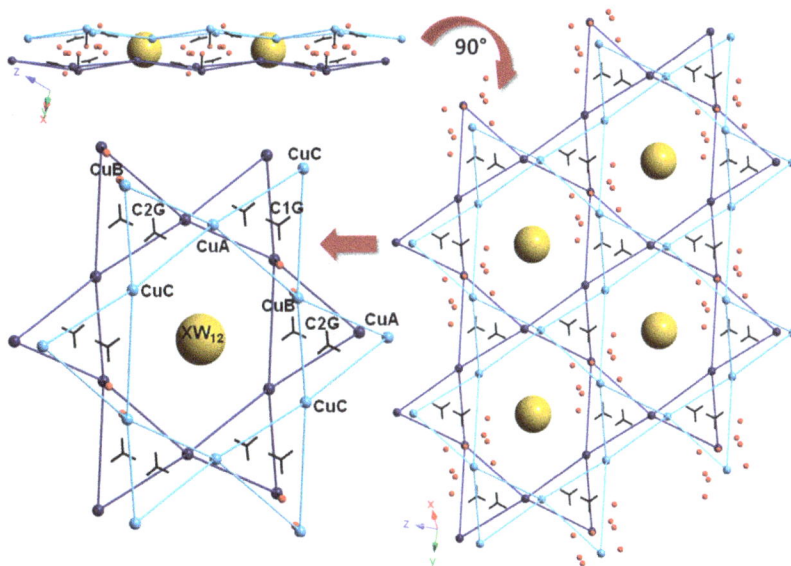

Figure 4. Schematic representation of the guanidinium-templated Kagome-type double lattice of copper(II)-picolinate complexes in **1** and **3**.

2.2.3. Crystal Packing of Compound **2**

The crystal packing of **2** shows also a two-dimensional character with hybrid layers parallel to *xy* plane (Figure 5). These layers are formed by a double sheet of Keggin clusters arranged in two

66

levels of z and held together by the monomeric {CuB(pic)$_2$} and {CuC(pic)$_2$} subunits. The former link clusters lying in the same z level to lead to a one-dimensional assembly of alternate inorganic and metalorganic building blocks parallel to the [010] direction. The connectivity between building blocks is such that the pyridinic ring of one of the picolinate ligands is sandwiched between the tetrameric {W$_4$O$_{18}$} faces of adjacent Keggin clusters with distances between the ring centroid and the average plane of the tetramers of 2.755 and 2.856 Å. These distances are comparable to those observed in related compounds with similar POM-aromatic interactions [30,31]. The {CuC(pic)$_2$} subunits connect in turn centrosymmetrically related clusters located at different z levels through long Cu–O$_{POM}$ bonds typical of semi-coordination (Table 1). In this case, the two aromatic rings interact with Keggin anions as they place almost parallel to tetrameric faces of the contiguous clusters with a centroid-tetramer plane distance of 2.781 Å. The linkage through the {CuC(pic)$_2$} subunits of pairs of one-dimensional hybrid assemblies running along the crystallographic b axis at different z levels results in double-chained [{SiW$_{12}$O$_{40}$}$_2${Cu(pic)$_2$}$_3$]$_n^{8n-}$ anions with rectangular cavities in the polymeric backbone and where each Keggin cluster is additionally decorated with a {Cu$_2$(pic)$_4$(H$_2$O)} dimer grafted as antenna subunit. Two of the guanidinium cations (C1G and C2G) are hosted in the rectangular cavities and they establish multiple N–H⋯O interactions with both the Keggin surfaces and the carboxylate groups of the picolinate ligands (Figure S3, N⋯O distances in the range 2.33(3)–3.28(3) Å). The layers pack with the antenna subunits directed to the interlamellar space to give rise to an alternate sequence of hybrid and metalorganic regions along the [001] direction. The antenna complexes in the metalorganic region are arranged in such a way that all picolinate ligands are almost parallel to the (10–2) plane and dimers grafted at contiguous POM sheets are hydrogen bonded by the water molecules of hydration and the guanidinium cations C3G and C4G.

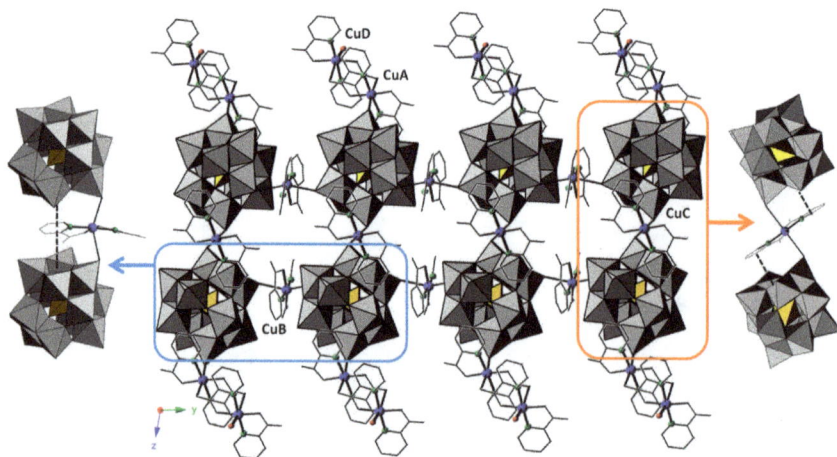

Figure 5. Projection of a [{SiW$_{12}$O$_{40}$}$_2${Cu(pic)$_2$}$_3$]$_n^{8n-}$ double chain decorated with {Cu$_2$(pic)$_4$(H$_2$O)} antenna complexes on the crystallographic bc plane and details of the POM–aromatic interactions involving the bridging {Cu(pic)$_2$} moieties.

2.3. Thermostructural Behavior

The thermostructural behavior of the title compounds was investigated by a combination of thermal analyses and variable temperature X-ray diffraction. Thermal analyses show that all compounds decompose via three mass loss stages (Figures 6 and S4). The first stage starts at room temperature and it is associated with two endothermic processes that originate from the release of the water molecules. For **1** and **3**, the dehydration stage extends up to *ca.* 130 °C and comprises the loss of a 2.24% of the total mass, which accounts for only 6 out of the 8 water molecules determined by single-crystal X-ray diffraction (calcd. for $6H_2O$: **1**, 2.13%; **3**, 2.11%). Analogously, dehydration of **2** is completed at *ca.* 95 °C with the release of eight out of the ten water molecules determined crystallographically [calcd. (found) for $8H_2O$: 1.69% (1.71)]. It is likely that these compounds lose some weakly bound interstitial water molecules when crystals are removed from their mother liquors and filtered at room temperature prior to be analyzed thermogravimetrically. The resulting anhydrous phases all show a wide range of thermal stability, up to *ca.* 300 °C for **1** and **3** and to 280 °C in the case of **2**. Above these temperatures, the anhydrous derivatives undergo further decomposition via two highly overlapping mass loss stages. The former originates from the combination of two endothermic and one exothermic consecutive processes that can be related to the release of $[C(NH_2)_3]^+$ cations as guanidine molecules and to the combustion of part of the picolinate ligands, respectively. For **1** and **3**, this stage extends up to *ca.* 355 °C and involves the loss of a 18.22 and 18.74% of the respective total mass, which roughly corresponds to 4 cations and 6 picolinate ligands (calcd. for $4(CH_6N_3) + 6(C_6H_4NO_2)$: **1**, $4.74 + 14.45 = 19.19\%$; **3**, $4.70 + 14.33 = 19.03\%$). In the case of **2**, the upper temperature limit and the mass loss are *ca.* 380 °C and 12.10%, which roughly accounts for 4 ligands besides 8 cations (calcd. for $8(CH_6N_3) + 4(C_6H_4NO_2)$: $5.64 + 5.73 = 11.37\%$). The final mass loss stage originates from a complex combination of exothermic processes that must associate with the combustion of the remaining organic matter and the crumbling of the Keggin framework. The final residues are obtained at temperatures in the 530–570 °C range and have been identified as mixtures of monoclinic WO_3 (PDF 88-269) [59] and triclinic $CuWO_4$ (PDF 43-1035) with Scheelite-type structure [60] according to powder X-ray diffraction (calcd. (found) for $aCuWO_4 + bWO_3 + cXO_2$: **1**, 65.5% (65.6), $a = b = 6$, $c = 1$; **2**, 73.2% (76.1), $a = 7$, $b = 17$, $c = 2$; **3**, 65.8% (65.9), $a = b = 6$, $c = 1$).

Variable temperature powder X-ray diffraction reveals that the title compounds retain crystallinity within the range of thermal stability upon dehydration (Figures 6 and S5). For **1** and **3**, well-defined diffraction patterns are obtained up to 310 °C, which is in full agreement with the upper temperature limit of the stability range in the TGA curves. The diffraction pattern is preserved with negligible variations in the positions and intensities of the diffraction maxima for the resulting anhydrous phases (**1a** and **3a**), and this fact indicates that dehydration does not result in drastic structural changes. Compound **2** also maintains crystallinity upon dehydration, but in contrast to **1** and **3**, a phase transformation is unequivocally observed between 50 °C and 70 °C. All compounds become amorphous solids in the temperature range corresponding to the release of the guanidinium cations and the combustion of the picolinate ligands. New crystalline phases corresponding to the final residue originated upon breakdown of the Keggin framework start appearing at 510 °C and they reach complete formation at temperatures slightly beyond the end of the third mass loss stage in the TGA curves (*ca.* 590 °C).

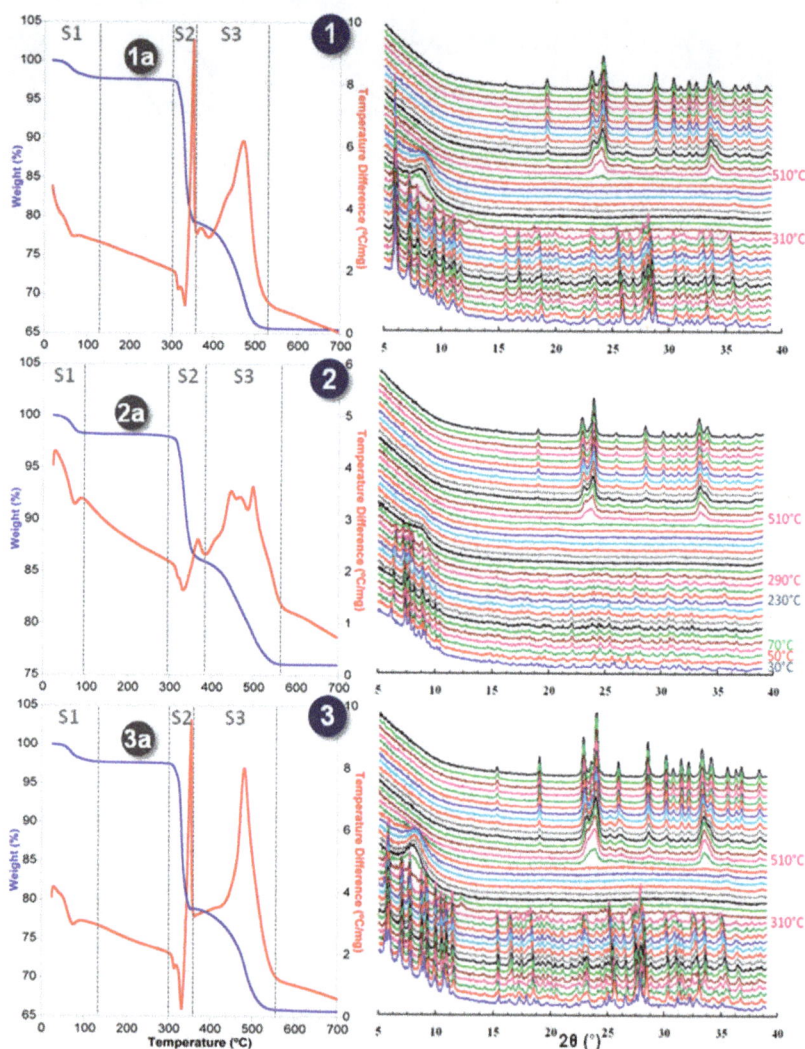

Figure 6. TGA/DTA curves and variable temperature X-ray diffraction patterns for **1–3**.

Analogous single-crystal X-ray diffraction studies were also carried out. Crystals of the title compounds were mounted at room temperature on a diffractometer and the temperature was raised at a rate of 1 °C min^{-1} to 140 °C for **1** and **3** and to 100 °C for **2**. The crystal of **1** preserved its integrity and crystallinity in the whole temperature range and darkening of its blue color was observed upon heating (Figure 7). This crystal stability allowed us to perform unit cell determinations at room temperature, 50, 80 and 140 °C (Table 2). In contrast, crystals of **3** cracked almost immediately after the temperature was ramped, but we could manually separate one of the resulting pieces to perform the experiment. Diffraction was of lower quality and much weaker than that observed for the isostructural **1**. Thus, the unit cell parameters determined for **3** are significantly less accurate than those of **1**, but nevertheless, they reproduce analogous trends acceptably. In the case of **2**, the laminar crystal also cracked when the temperature was ramped, but unfortunately, we could not apply the strategy followed for **3** because the extreme fragility of the resulting pieces prevented us from their manipulation. As shown in Table 2, a significant shortening of the parameter c and consequent contraction of the unit cell volume is observed for both **1** and **3** when going from 80 to 140 °C, which indicates formation of the corresponding anhydrous derivatives. At this point, we lowered the temperature to 100(2) K to carry out the full data collections for both compounds and the structures of **1a** and **3a** were determined.

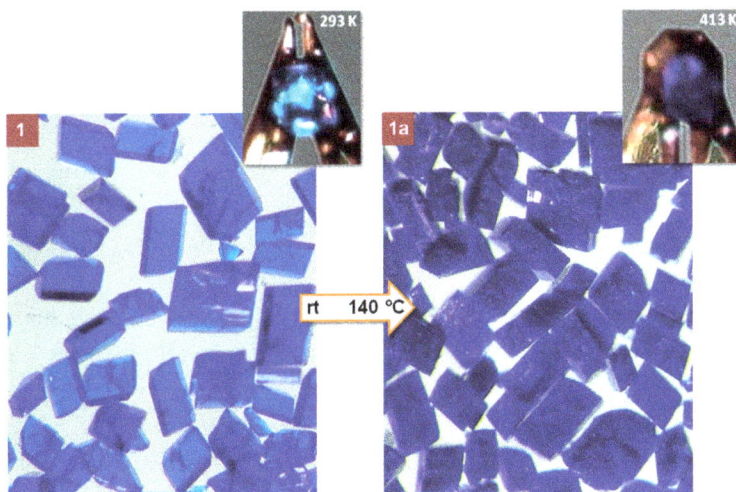

Figure 7. Photographs of single crystals of **1** taken at room temperature (left) and upon dehydration at 140 °C (right). Insets: images of the crystals used for performing the full single-crystal X-ray diffraction data collections of compouunds **1** and **1a**.

Table 2. Unit cell parameters of **1** and **3** at different temperatures.

Compounds	T (°C)	a (Å)	b (Å)	c (Å)	α (°)	β (°)	γ (°)	V (Å³)
1	r.t	11.805(6)	16.112(6)	16.443(5)	105.65(3)	101.97(3)	91.29(4)	2936(2)
	50	11.812(5)	16.098(5)	16.356(6)	105.21(3)	101.53(4)	91.60(3)	2958(2)
	80	11.861(6)	16.077(7)	16.292(7)	104.99(4)	101.29(4)	91.85(4)	2982(2)
	140	11.818(1)	16.123(5)	16.179(7)	104.62(3)	100.18(5)	92.07(5)	2926(2)
3	r.t	11.815(4)	16.050(6)	17.047(6)	106.90(3)	94.42(3)	100.20(3)	3016(2)
	50	11.81(1)	16.05(2)	17.06(2)	106.91(9)	94.40(7)	100.09(8)	3018(5)
	80	11.88(3)	16.35(6)	16.55(5)	106.0(3)	92.9(2)	101.3(3)	3012(15)
	140	11.75(3)	15.89(4)	16.21(4)	103.4(2)	91.8(2)	100.8(2)	2882(12)
% of indexed reflections for cells at $T > 50$ °C: above 97% for **1** and below 57% for **3**								

Simple TGA/DTA experiments were performed to determine the reversibility of the dehydration processes (Figure S6). Crystalline samples of **1** and **3** were heated at a rate of 2 °C min^{-1} up to 200 °C and the so-generated anhydrous samples were exposed to the room atmosphere for one day, and then, heated again at the same rate. The recorded TGA profiles are almost identical for both heating cycles, and this fact shows that the anhydrous phases **1a** and **3a** are fully rehydrated to the original compounds simply after being in contact with moisture for a few hours. These observations were confirmed by single-crystal X-ray diffraction. The crystals used for determining the structures of **1a** and **3a** were kept on the goniometer head in contact with the room environment and the intensity data were collected back at 100(2) K after a few days. The crystals still diffracted acceptably enough and the unit cells of the initial hydrated phases **1** and **3** were again obtained. The structural solutions were of poorer quality than those determined originally most likely due to disorder affecting the water molecules upon resorption (note the differences in the DTA profiles in Figure S6), but nevertheless, we could locate the inorganic and metalorganic building blocks in their original positions. These observations demonstrate that dehydration of both **1** and **3** proceeds via SCSC transformations, but furthermore, that this process is reversible and the anhydrous **1a** and **3a** phases also undergo SCSC transformations promoted by consequent rehydration.

2.4. SCSC Transformations of Compounds 1 and 3 into the Anhydrous Phases 1a and 3a

Dehydration of compounds **1** and **3** into the phases **1a** and **3a** does not equally affect the hybrid layers and the metalorganic regions. The hybrid layers remain virtually unaltered: for example, variations in the bond lengths within the {$CuA_2(pic)_4$} subunit are negligible (Table 1) and the relative arrangement of Keggin anions in the hybrid [{$XW_{12}O_{40}$}{$Cu(pic)_2$}$_2$]$^{4-}$ chains is only affected by a subtle lengthening of 0.1 Å in the X$\cdots$X distance between adjacent clusters. In contrast, significant changes take place in the metalorganic sublattice because all water molecules of coordination and hydration reside in this area (Figure 8). Their removal promotes a reorganization of the dimeric [$Cu_2(pic)_4(H_2O)$] moieties, each of which split into two independent [$Cu(pic)_2$] square-planar complexes as evidenced by the remarkable lengthening of the CuB–O4C distance from *ca.* 2.49 to 3.31 Å. The contraction of the parameter *c* by 1 Å is also consequence of this splitting. While the original metalorganic regions consist in a corrugated lattice of {$Cu(pic)_2$} fragments in the crystallographic *xy* plane, the release of the water molecules force the newly generated [$Cu(pic)_2$] monomers to spread on the plane in such a way that the corrugation degree decreases and the Cu^{II} atoms become nearly coplanar. In spite of this rearrangement of complexes, the structure-directing network of N–H$\cdots$O hydrogen bonds remains almost intact upon dehydration because of a slight reorientation of the guanidinium cations that preserves almost all contacts (Table S2). Thus, the double trihexagonal tiling described above is maintained without noticeable alterations (Figure 9).

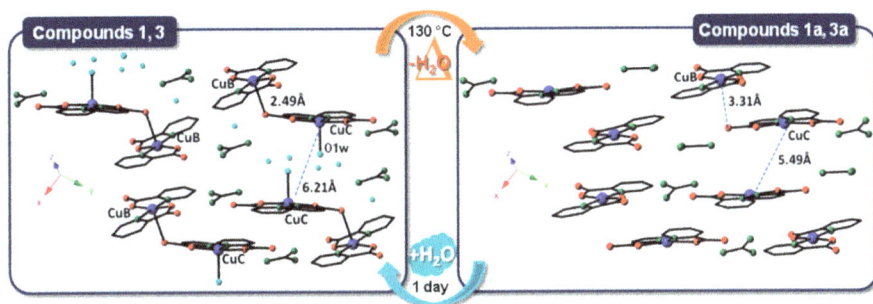

Figure 8. Arrangement of {Cu(pic)$_2$} fragments and [C(NH$_2$)$_3$]$^+$ cations in the metalorganic region of **1** and **3** compared to that found in the anhydrous phases **1a** and **3a**. Note the splitting of the dimer [Cu$_2$(pic)$_4$(H$_2$O)] into monomers upon dehydration.

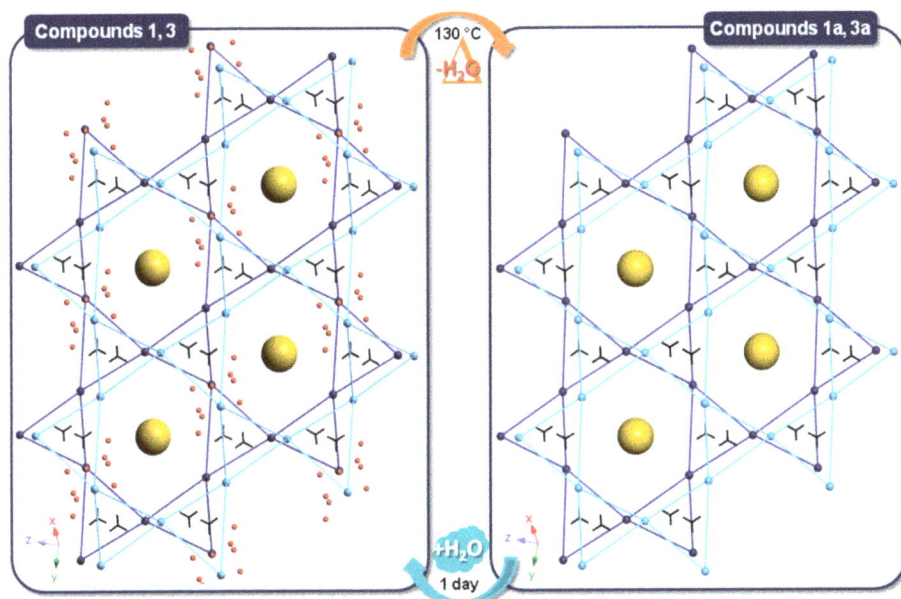

Figure 9. Guanidinium-templated Kagome-type double lattice of Cu^{II} atoms in **1** and **3** compared to that found in the anhydrous phases **1a** and **3a**.

2.5. Electron Paramagnetic Resonance Spectroscopy for Compounds 1 and 1a

The EPR spectra of **1** (Figure 10) and **3** (Figure S7) are virtually identical in good agreement with the isostructurality of the compounds. The only difference worth to be mentioned is the largest line width observed for **3**, which is likely due to the lower crystallinity of this compound when compared to **1**. The spectra are relatively complex as a result of the overlapping of the contributions from the different copper(II)-picolinate dimeric entities coexisting in the crystal packing. The X-band spectra show multiple resonances in the range 2300–3800 Gauss and a half-field signal corresponding to the $\Delta M_S = \pm 2$ forbidden transition centered at *ca.* 1600 Gauss, which indicates the presence of a magnetically isolated triplet state ($S = 1$). Moreover, a partially resolved hyperfine structure originating from the interaction of an electron spin with a limited number of non-zero nuclear spins is also observed in both spectra. The number of detectable lines in this hyperfine structure is above the 4 lines that would correspond to a spin doublet interacting with a single $I = 3/2$ nucleus and this confirms the presence of an isolated $S = 1$ state. Both the X- and Q-band spectra display at least one signal for which the apparent g value is substantially lower than that of the free electron (3600 and 12200 Gauss, respectively). As all Cu^{II} atoms in **1** and **3** are in octahedral or square-pyramidal coordination environments, the presence of such signals can only be attributed to a noticeable zero-field splitting (ZFS) within a multiplet state.

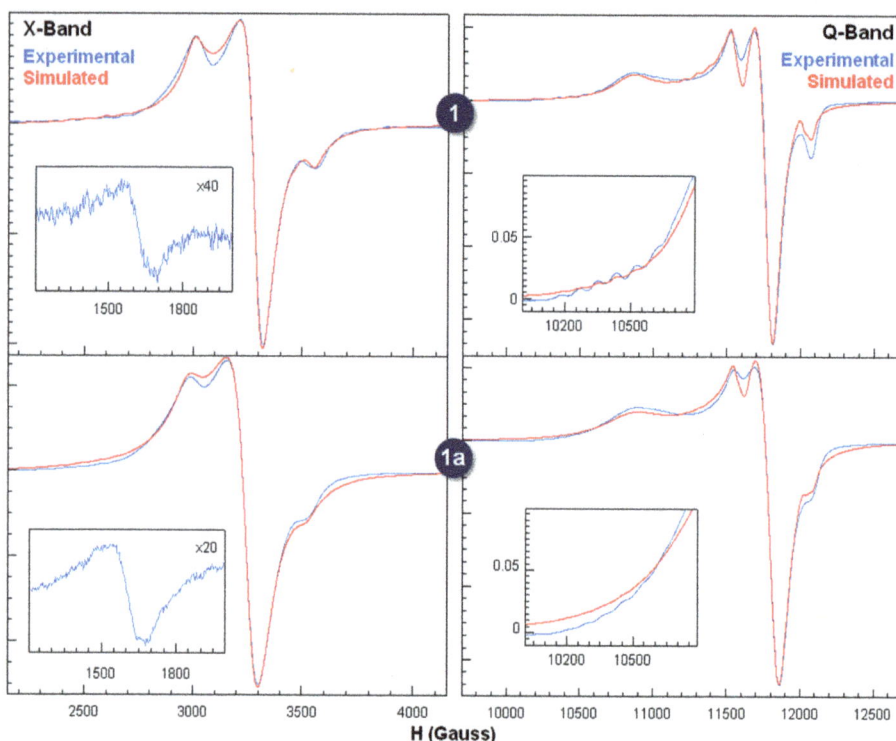

Figure 10. Experimental and simulated X-band ($\upsilon = 9.49$ GHz) and Q-band ($\upsilon = 34.05$ GHz) EPR spectra of **1** and **1a** at room temperature.

The observation of such multiplet states can be well correlated with the coexistence of the dinuclear entities {Cu_2(pic)$_4$} and [Cu_2(pic)$_4$(H_2O)]. In both complexes, intradimeric magnetic exchange takes place through axial-equatorial pathways: μ^2-O bridging atoms for {Cu_2(pic)$_4$} and O–C–O linkages in the case of [Cu_2(pic)$_4$(H_2O)]. The spectroscopic features suggest that the former are magnetically isolated by the bulky diamagnetic Keggin clusters and give rise to a common signal with the typical features of a triplet state with significant zero-field splitting, whereas the latter are coupled in an extended system of long-range, weak magnetic interactions that average their individual signals. Thus, the spectra were initially simulated as the sum of the following individual contributions: one signal of axial symmetry corresponding to a cooperative exchange **g** tensor (signal 1) and another signal originating from an $S = 1$ spin state with collinear **D** and **g** tensors of axial symmetry (Figure S8). The fit of signal 1 to the experimental profile was improved by breaking the axial symmetry with some equatorial anisotropy, whereas that of signal 2 required the introduction of certain ZFS also in the equatorial plane. The value of the parameter E used during the fitting should be considered only as a simple approximation or as a maximum value for the equatorial ZFS effect.

The calculated spin Hamiltonian parameters are shown in Table 3, together with those of the isostructural **3** and the anhydrous derivative **1a**. The g values are consistent with those expected for the topology shown by the copper(II) chromophores in the title compounds and confirms that the ground state is mainly of $d(x^2-y^2)$ type. Therefore, the intradimeric coupling interactions must be very small in both cases considering the axial-equatorial pathways. The calculated hyperfine coupling constant (87×10^{-4} cm^{-1}) is actually half of that expected for a Cu^{II} chromophore with $g = 2.25$ and N$_2$O$_2$-type equatorial plane, which implies that each unpaired electron must interact with the nuclear spins of two different Cu^{II} ions as corresponds to the presence of magnetically isolated dinuclear entities. The D parameter obtained is relatively small and this fact is in good agreement with the strong deviation of the axial component of the **g** tensors with respect to the intradimeric Cu–Cu axis, which reduces the anisotropic exchange contribution.

Table 3. Spin Hamiltonian parameters g, A, D and E ($\times10^{-4}$ cm^{-1}) for compounds **1**, **1a**, **2** and **3**.

Compounds	signal 1				signal 2				
	g_1	g_2	$g_\perp$	$g_3 = g_\parallel$	$g_\perp$	$g_\parallel$	A	D	E
1	2.061(1)	2.073(1)	2.067(1)	2.243(2)	2.061(1)	2.240(1)	87(1)	450(5)	20(5)
1a	2.056(1)	2.068(1)	2.062(1)	2.246(2)	2.061(1)	2.240(1)	87(1)	430(5)	20(5)
3	2.062(1)	2.071(1)	2.066(1)	2.243(2)	2.060(1)	2.240(1)	87(1)	450(5)	20(5)
2	-	-	2.063(1)	2.251(2)	2.060(2)	2.236(2)	88(3)	470(5)	-

Dehydration of **1** into **1a** does not significantly affect the EPR spectra in spite of the fact that the release of the water molecules results in the splitting of one of the dimeric entities into independent monomers with consequent modification of the coordination geometry around the Cu^{II} centers from square-pyramidal to square-planar. The two main contributions corresponding to the isolated dimer and the extended system can still be well appreciated in the spectra of **1a**. For the former, the lines of the fine structure approach each other due to a decrease of the D parameter that may associate with a reduction of the anisotropy around the CuA centers. All of the fine and hyperfine lines become wider, resulting in a loss of resolution that might a priori be related to an increase of dipolar interactions or to a reduction of the exchange coupling. However, we believe that this phenomenon is simply due to a loss of crystallinity in the solid sample upon heating the starting material in an oven. It is worth highlighting that EPR spectroscopy finely demonstrates that rehydration of **1a** into the initial **1** is fully achieved in a very short time under standard atmospheric conditions. Using a freshly prepared sample of **1a**, the original experimental profiles of **1** were recovered within a few minutes when an open sample holder was used, but fortunately, sealing the latter slowed the process down enough for allowing us to collect the spectra of the anhydrous derivative.

Figure 11 displays the X-band (9.40 GHz) and Q-band (34.10 GHz) EPR spectra of **2** recorded at room temperature on a grained polycrystalline sample. The spectra are closely related to those of **1** and **3** as contributions of two different magnetic systems are also observed: a magnetically isolated $S = 1$ state with significant zero-field splitting and a poorly resolved hyperfine structure in the parallel region and a more intense rhombic signal that must originate from a cooperative exchange g tensor after considering its lack of hyperfine lines. The calculated g, A and D values are similar to those determined for **1** and **3** (Table 3). Taking into account the structural features of **2**, it should be assumed that the signal of the isolated triplet state corresponds to the dinuclear $\{Cu_2(pic)_4(H_2O)\}$ antenna subunits, and hence the rhombic signal must then be ascribed to the presence of long-range, weak magnetic interactions involving the monomeric $\{Cu(pic)_2\}$ bridging subunits. Therefore, magnetic exchange pathways between the CuB and CuC ions must exist through the Keggin clusters, or most likely, through the strong N–H···O hydrogen bonds established with the guanidiniums cations.

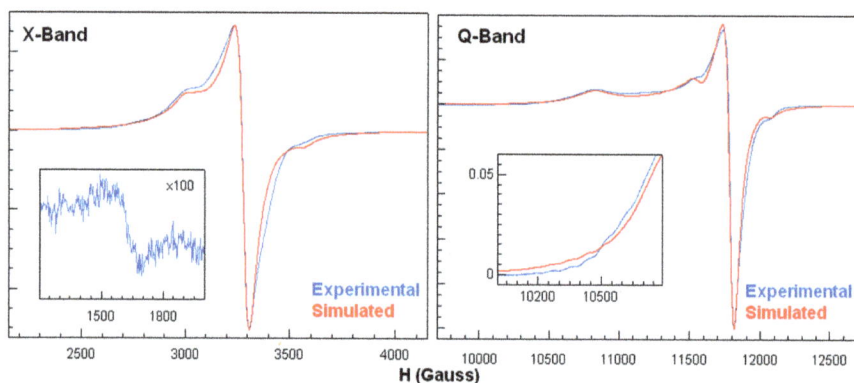

Figure 11. Experimental and simulated X-band ($\upsilon = 9.49$ GHz) and Q-band ($\upsilon = 34.06$ GHz) EPR spectra of **2** at room temperature.

3. Experimental Section

3.1. Materials and Methods

The precursors $K_8[\alpha\text{-}SiW_{11}O_{39}]\cdot 13H_2O$ and $K_8[\alpha\text{-}GeW_{11}O_{39}]\cdot 13H_2O$ were prepared according to the literature [61,62] and identified by IR spectroscopy. All other chemicals were obtained from commercial sources and used without further purification. Carbon, hydrogen, and nitrogen were determined on a Perkin-Elmer 2400 CHN analyzer (Perkin Elmer, Waltham, MA, USA). Infrared spectra for solid samples were obtained as KBr pellets on a SHIMADZU FTIR-8400S spectrometer (Shimadzu, Kyoto, Japan). Thermogravimetric and Differential Thermal Analyses were carried out from room temperature to 750 °C at a rate of 5 °C min^{-1} on a TA Instruments 2960 SDT thermobalance (TA Instruments, New Castle, DE, USA) under a 100 cm^3·min^{-1} flow of synthetic air. Electron Paramagnetic Resonance (EPR) spectra were recorded on Bruker ELEXSYS 500 (superhigh-Q resonator ER-4123-SHQ, (Bruker, Karlsruhe, Germany) and Bruker EMX (ER-510-QT resonator, Bruker, Karlsruhe, Germany) continuous wave spectrometers for Q- and X- bands, respectively.

3.2. Synthesis of [C(NH$_2$)$_3$]$_4$[{SiW$_{12}$O$_{40}$}{Cu$_2$(pic)$_4$}]·[Cu$_2$(pic)$_4$(H$_2$O)]$_2$·6H$_2$O (1) and [C(NH$_2$)$_3$]$_8$[{SiW$_{12}$O$_{40}$}$_2${Cu(pic)$_2$}$_3${Cu$_2$(pic)$_4$(H$_2$O)}$_2$]·8H$_2$O (2)

To a solution of K$_8$[α-SiW$_{11}$O$_{39}$]·13H$_2$O (322 mg, 0.10 mmol) in water (30 mL), CuCl$_2$·2H$_2$O (17 mg, 0.10 mmol) was added. After stirring the reaction mixture at room temperature for 30 min, picolinic acid (24 mg, 0.20 mmol) was added. The resulting solution was stirred for one additional hour and then aqueous 1M guanidinium chloride (1 mL) was added dropwise. A mixture of prismatic blue crystals of **1** as the major fraction and laminar blue crystals of **2** as a side product was obtained upon slow evaporation of the final solution for *ca.* five days. The two compounds were manually separated using an optical microscope for their full characterization and structural determination.

Compound **1**. Yield: 32% based on W. Elemental Analyses (%): Calcd. (found) for C$_{76}$H$_{88}$Cu$_6$N$_{24}$O$_{72}$SiW$_{12}$: C, 17.88 (18.02); H, 1.74 (1.70); N, 6.58 (6.57). IR (cm^{-1}): 3379 s, 2924 w, 2853 w, 1645 vs, 1603 s, 1572 s, 1478 m, 1447 w, 1362 s, 1350 s, 1292 m, 1265 w, 1167 w, 1096 w, 1051 m, 1015 w, 970 m, 924 vs, 883 m, 853 m, 804 vs, 758 s, 712 m, 694 m, 660 m, 525 m, 457 w.

Compound **2**. Yield: less than 5% based on W. Elemental Analyses (%): Calcd. (found) for C$_{92}$H$_{124}$Cu$_7$N$_{38}$O$_{118}$Si$_2$W$_{24}$: C, 12.9 (13.22); H, 1.45 (1.43); N, 6.21 (6.34). IR (cm^{-1}): 3366 s, 2926 w, 2853 w, 1642 vs, 1603 s, 1570 s, 1476 m, 1449 w, 1364 s, 1350 s, 1290 m, 1263 w, 1165 w, 1096 w, 1051 m, 1015 w, 970 m, 922 vs, 883 m, 854 m, 800 vs, 754 s, 712 m, 693 m, 660 m, 523 m, 455 w.

3.3. Synthesis of [C(NH$_2$)$_3$]$_4$[{GeW$_{12}$O$_{40}$}{Cu$_2$(pic)$_4$}]·[Cu$_2$(pic)$_4$(H$_2$O)]$_2$·6H$_2$O (3)

The synthetic procedure above was followed but for using a solution of K$_4$[α-GeW$_{11}$O$_{39}$]·13H$_2$O (329 mg, 0.10 mmol) in water (20 mL). Prismatic blue crystals of **3** suitable for X-ray diffraction were obtained as a single solid phase by slow evaporation of the final solution for *ca.* 5 days. Yield: 31% based on W. Elemental Analyses (%): Calcd. (found) for C$_{76}$H$_{88}$Cu$_6$GeN$_{24}$O$_{72}$W$_{12}$: C, 17.73 (18.05); H, 1.72 (1.60); N, 6.53 (6.94). IR (cm^{-1}): 3370 s, 2924 w, 2853 w, 1642 vs, 1603 s, 1570 s, 1478 m, 1445 w, 1367 s, 1350 s, 1290 m, 1263 w, 1165 w, 1094 w, 1051 m, 966 s, 883 vs, 853 m, 831 s, 787 vs, 754 s, 712 m, 692 m, 660 m, 557 w, 461 s.

3.4. X-ray Crystallography

Crystallographic data for **1–3** and the anhydrous phases **1a** and **3a** are given in Table 4. Intensity data were collected at 100(2) K on an Agilent Technologies SuperNova diffractometer (Santa Clara, CA, USA) equipped with an Oxford Cryostream 700 PLUS temperature device (Oxford, UK). Mirror-monochromated Mo Kα radiation ($λ = 0.71073$ Å) and an Eos CCD detector (Santa

Clara, CA, USA) was used in all cases with the exception of **1** and **3a**, for which data collection involved mirror-monochromated Cu Kα radiation (λ = 1.5418 Å) and an Atlas CCD detector (Santa Clara, CA, USA). In the case of the anhydrous phases **1a** and **3a**, a single crystal of the corresponding hydrated compound was mounted on the goniometer and a preliminary data collection was performed at room temperature to check that its diffraction was of sufficient quality. The temperature was then ramped at a rate of 2 K min^{-1} and unit cell measurements were carried out at 323, 353 and 413(2) K to ensure whether the sample maintained its integrity as a single crystal during the structural transformation associated to dehydration. Once the temperature reached 413(2) K, it was lowered to 100(2) K at a rate of 6 K min^{-1} for performing a full data collection of the so generated anhydrous phases. The crystals were kept on the goniometer head and exposed to room atmosphere for several days, after which routine full data collections corresponding to the initial hydrated forms were carried out at 100(2) K.

Table 4. Crystallographic data for **1–3** and for the anhydrous phases **1a** and **3a**.

Parameters	1	1a	2	3	3a
Formula	$C_{76}H_{88}Cu_6N_{24}$ $O_{72}SiW_{12}$	$C_{76}H_{72}Cu_6N_{24}$ $O_{64}SiW_{12}$	$C_{92}H_{124}Cu_7N_{38}$ $O_{118}Si_2W_{24}$	$C_{76}H_{88}Cu_6Ge$ $N_{24}O_{72}W_{12}$	$C_{76}H_{72}Cu_6Ge$ $N_{24}O_{64}W_{12}$
Fw (g mol^{-1})	5105.2	4961.1	8563.6	5149.7	5005.6
Crystal system	triclinic	triclinic	triclinic	triclinic	triclinic
Space group	P–1	P–1	P–1	P–1	P–1
a (Å)	11.7014(4)	11.6039(3)	11.9426(2)	11.7110(3)	11.6025(7)
b (Å)	15.9523(5)	15.9379(5)	12.8151(3)	15.9628(6)	15.9736(9)
c (Å)	17.0285(5)	15.9984(5)	30.0533(6)	17.0341(5)	15.9849(10)
α (°)	107.102(3)	104.292(3)	101.508(2)	107.057(3)	104.371(5)
β (°)	94.393(3)	91.168(2)	90.346(2)	94.372(2)	91.222(5)
γ (°)	101.008(3)	100.599(2)	105.544(2)	100.971(2)	100.534(5)
V (Å^3)	2952.2(2)	2811.6(1)	4333.6(2)	2958.9(2)	2814.5(3)
Z	1	1	1	1	1
ρ_{calcd} (g cm^{-3})	2.872	2.930	3.281	2.890	2.953
μ (mm^{-1})	23.180	13.447	16.822	13.025	24.412
Reflections:					
Collected	22568	19291	34946	19292	20323
Unique	11493	11071	17052	11018	10854
Observed [$I > 2\sigma(I)$]	10892	10022	15317	10485	7649
R_{int}	0.032	0.021	0.023	0.022	0.049
Parameters	514	494	788	518	494
$R(F)$ [a] [$I > 2\sigma(I)$]	0.050	0.046	0.086	0.044	0.079
$wR(F^2)$ [a] [all data]	0.117	0.085	0.176	0.093	0.231
GoF	1.278	1.363	1.273	1.280	1.039

[a] $R(F) = \Sigma||F_o - F_c||/\Sigma|F_o|$; $wR(F^2) = \{\Sigma[w(F_o^2 - F_c^2)_2]/\Sigma[w(F_o^2)^2]\}^{1/2}$.

Data frames were processed (unit cell determination, intensity data integration, correction for Lorentz and polarization effects, and analytical absorption correction with face indexing) using the CrysAlis Pro software package (Agilent Technologies UK Ltd., Oxford, UK) [63]. The structures were solved using OLEX (OlexSys Ltd in Durham University, Durham, UK) [64] and refined by full-matrix least-squares with SHELXL-97 (University of Goettingen, Goettingen, Germany) [65]. Final geometrical calculations were carried out with PLATON (Utrecht University, Utrecht, The Netherlands) [66] as integrated in WinGX (University of Glasgow, Glasgow, UK) [67]. Thermal vibrations were treated anisotropically for heavy atoms (W, Cu, Si). Hydrogen atoms of the organic ligands were placed in calculated positions and refined using a riding model with standard SHELXL parameters. In all cases, the Keggin clusters displayed disorder originated from slight tilting in the crystal packing. This tilting was modeled by disordering the O atoms of the Keggin clusters over two positions labeled as O/OZ. The population factor within the O/OZ pairs was initially refined as a single free variable, resulting in the following occupancies: 50/50 for **1**, 48/52 for **1a**, 65/35 for **2**, 51/49 for **3** and 52/48 for **3a**. CCDC-1058791 (**1**), -1058793 (**2**), -1058794 (**3**), -1058792 (**1a**), and -1058795 (**3a**) contain the supplementary crystallographic data for this paper. These data can be obtained free of charge from The Cambridge Crystallographic Data Centre via www.ccdc.cam.ac.uk/data_request/cif.

Powder X-ray diffraction patterns were collected on a Bruker D8 Advance diffractometer (Karlsruhe, Germany) operating at 30 kV and 20 mA and equipped with a Pt sample holder, Cu tube (λ = 1.5418 Å), Vantec-1 PSD detector (Karlsruhe, Germany), and Anton Parr HTK2000 high-temperature furnace (Graz, Austria). The patterns were recorded in 2θ steps of $0.033°$ in the $5 \leq 2\theta \leq 39°$ range using an exposure time of 0.3 s per step. Full data sets were recorded from 30 to 770 °C every 20 °C and a heating rate of 0.16 °C s^{-1} was applied between the temperatures.

4. Conclusions

The study presented herein represents a good indication of the fact that single-crystal to single-crystal transformations might be a common structural response to thermal dehydration in a wide scope of hybrid compounds composed of polyoxometalate anions and transition metal complexes bearing organic ligands. To date, such types of solid-state phase transition studies have only been developed for polyoxometalate-based hybrid compounds containing bis(aminopyridyl)-type ligands. In this work, we demonstrate that analogous behavior can also be found in related systems with completely different metalorganic subunits such as transition metal bis(picolinate) complexes. The aiding role of guanidinium cations as structure-directing agents appears to be a key factor in facilitating the crystal transformations because they are able to establish a massive network of intermolecular interactions that remains nearly unaltered upon dehydration.

Acknowledgments

This work was funded by Eusko Jaurlaritza/Gobierno Vasco (grant IT477-10 and predoctoral fellowship to A.P.), Ministerio de Economía y Competitividad (grant MAT2013-48366-C2-2P) and Universidad del País Vasco UPV/EHU (grant UFI11/53). Technical and human support provided by SGIker (UPV/EHU) is gratefully acknowledged.

Author Contributions

A.P. prepared the title compounds, performed their physicochemical characterization and analyzed the structures in close collaboration with A.I.; L.S.F. collected the single-crystal X-ray diffraction data and solved the structures; S.R. carried out the thermal analyses and prepared the manuscript; L.L. was in charge of collecting and interpreting the EPR spectra; and J.M.G.-Z. conceived the work and acted as the scientific coordinator together with L.L.

Conflicts of Interest

The authors declare no conflict of interest.

References

1. Reinoso, S.; Vitoria, P.; Gutiérrez-Zorrilla, J.M.; Lezama, L.; San Felices, L.; Beitia, J.I. Inorganic-Metalorganic Hybrids Based on Copper(II)-Monosubstituted Keggin Polyanions and Dinuclear Copper(II)-Oxalate Complexes. Synthesis, X-ray Structural Characterization, and Magnetic Properties. *Inorg. Chem.* **2005**, *44*, 9731–9742.
2. Reinoso, S.; Vitoria, P.; Gutiérrez-Zorrilla, J.M.; Lezama, L.; Madariaga, J.M.; San Felices, L.; Iturrospe, A. Coexistence of Five Different Copper(II)-Phenanthroline Species in the Crystal Packing of Inorganic-Metalorganic Hybrids Based on Keggin Polyoxometalates and Copper(II)-Phenanthroline-Oxalate Complexes. *Inorg. Chem.* **2007**, *46*, 4010–4021.
3. Zhang, Z.; Yang, J.; Liu, Y.-Y.; Ma, J.-F. Five Polyoxometalate-Based Inorganic-Organic Hybrid Compounds Constructed by a Multidentate N-Donor Ligand: Syntheses, Structures, Electrochemistry, and Photocatalysis Properties. *CrystEngComm* **2013**, *15*, 3843–3853.
4. Aoki, S.; Kurashina, T.; Kasahara, Y.; Nishijima, T.; Nomiya, K. Polyoxometalate (POM)-Based, Multi-Functional, Inorganic-Organic, Hybrid Compounds: Syntheses and Molecular Structures of Silanol- and/or Siloxane Bond-Containing Species Grafted on Mono- and Tri-Lacunary Keggin POMs. *Dalton Trans.* **2011**, *40*, 1243–1253.
5. Bai, Y.; Zhang, G.-Q.; Dang, D.-B.; Ma, P.-T.; Gao, H.; Niu, J.-Y. Assembly of Polyoxometalate-Based Inorganic-Organic Compounds from Silver-Schiff Base Building Blocks: Synthesis, Crystal Structures and Luminescent Properties. *CrystEngComm* **2011**, *13*, 4181–4187.
6. Dolbecq, A.; Dumas, E.; Mayer, C.R.; Mialane, P. Hybrid Organic-Inorganic Polyoxometalate Compounds: From Structural Diversity to Applications. *Chem. Rev.* **2010**, *110*, 6009–6048.
7. Liu, B.; Yu, Z.-T.; Yang, J.; Hua, W.; Liu, Y.-Y.; Ma, J.-F. First Three-Dimensional Inorganic-Organic Hybrid Material Constructed From an "Inverted Keggin" Polyoxometalate and a Copper(I)-Organic Complex. *Inorg. Chem.* **2011**, *50*, 8967–8972.
8. Zheng, L.M.; Wang, Y.S.; Wang, X.Q.; Korp, J.D.; Jacobson, A.J. Anion-Directed Crystallization of Coordination Polymers: Syntheses and Characterization of $Cu_4(2\text{-}pzc)_4(H_2O)_8(Mo_8O_{26})\cdot2H_2O$ and $Cu_3(2\text{-}pzc)_4(H_2O)_2(V_{10}O_{28}H_4)\cdot6.5H_2O$ (2-pzc = 2-Pyrazinecarboxylate). *Inorg. Chem.* **2001**, *40*, 1380–1385.

9. Hagrman, D.; Hagrman, P.J.; Zubieta, J. Solid-State Coordination Chemistry: The Self-Assembly of Microporous Organic-Inorganic Hybrid Frameworks Constructed from Tetrapyridylporphyrin and Bimetallic Oxide Chains or Oxide Clusters. *Angew. Chem.* **1999**, *38*, 3165–3168.

10. Lu, Y.; Xu, Y.; Wang, E.B.; Lü, J.; Hu, C.W.; Xu, L. Novel Two-Dimensional Network Constructed from Polyoxomolybdate Chains Linked through Copper-Organonitrogen Coordination Polymer Chains: Hydrothermal Synthesis and Structure of [H$_2$bpy][Cu(4,4'-bpy)]$_2$[HPCuMo$_{11}$O$_{39}$]. *Cryst. Growth Des.* **2005**, *5*, 257–260.

11. Shivaiah, V.; Nagaraju, M.; Das, S.K. Formation of a Spiral-Shaped Inorganic-Organic Hybrid Chain, [Cu^{II}(2,2'-bipy)(H$_2$O)$_2$Al(OH)$_6$Mo$_6$O$_{18}$]$_n^{n-}$: Influence of Intra- and Interchain Supramolecular Interactions. *Inorg. Chem.* **2003**, *42*, 6604–6606.

12. Yao, S.; Yan, J.-H.; Duan, H.; Zhang, Z.-M.; Li, Y.-G.; Han, X.-B.; Shen, J.-Q.; Fu, H.; Wang, E.-B. Integration of Ln-Sandwich POMs into Molecular Porous Systems Leading to Self-Assembly of Metal-POM Framework Materials. *Eur. J. Inorg. Chem.* **2013**, 4770–4774.

13. Kong, X.J.; Ren, Y.-P.; Zheng, P.-Q.; Long, Y.-X.; Long, L.-S.; Huang, R.-B.; Zheng, L.-S. Construction of Polyoxometalates-Based Coordination Polymers through Direct Incorporation between Polyoxometalates and the Voids in a 2D Network. *Inorg. Chem.* **2006**, *45*, 10702–10711.

14. Khan, M.I.; Yohannes, E.; Doedens, R.J. A Novel Series of Materials Composed of Arrays of Vanadium Oxide Container Molecules, [V$_{18}$O$_{42}$(X)] (X = H$_2$O, Cl$^-$, Br$^-$): Synthesis and Characterization of [M$_2$(H$_2$N(CH$_2$)$_2$NH$_2$)$_5$][(M(H$_2$N(CH$_2$)$_2$NH$_2$)$_2$)$_2$V$_{18}$O$_{42}$(X)]·9H$_2$O (M = Zn, Cd). *Inorg. Chem.* **2003**, *42*, 3125–3129.

15. An, H.Y.; Wang, E.B.; Xiao, D.R.; Li, Y.G.; Su, Z.M.; Xu, L. Chiral 3D Architectures with Helical Channels Constructed from Polyoxometalate Clusters and Copper-Amino Acid Complexes. *Angew. Chem.* **2006**, *45*, 904–908.

16. Dai, L.M.; You, W.S.; Wang, E.B.; Wu, S.X.; Su, Z.M.; Du, Q.H.; Zhao, Y.; Fang, Y. Two Novel One-Dimensional α-Keggin-Based Coordination Polymers with Argentophilic {Ag$_3$}$^{3+}$/{Ag$_4$}$^{4+}$ Clusters. *Cryst. Growth Des.* **2009**, *9*, 2110–2116.

17. Darling, K.; Smith, T.M.; Vargas, J.; O'Connor, C.J.; Zubieta, J. Polyoxometalate Clusters as Building Blocks for Oxide Materials: Synthesis and Structure of a Three-dimensional Copper-Pyrazinetetrazolate / Keggin Assembly. *Inorg. Chem. Commun.* **2013**, *32*, 1–4.

18. Hao, X.-L.; Ma, Y.-Y.; Wang, Y.-H.; Zhou, W.-Z.; Li, Y.-G. New Organic-Inorganic Hybrid Assemblies based on Metal-bis(betaine) Coordination Complexes and Keggin-type Polyoxometalates. *Inorg. Chem. Commun.* **2014**, *41*, 19–24.

19. Li, S.; Ma, H.; Pang, H.; Zhang, Z.; Yu, Y.; Liu, H.; Yu, T. Tuning the Dimension of POM-Based Inorganic-Organic Hybrids from 3D Self-Penetrating Framework to 1D Poly-Pendant Chain via Changing POM Clusters and Introducing Secondary Spacers. *CrystEngComm* **2014**, *16*, 2045–2055.

20. Henry, N.; Costenoble, S.; Lagrenee, M.; Loiseau, T.; Abraham, F. Lanthanide-Based 0D and 2D Molecular Assemblies with the Pyridazine-3,6-dicarboxylate Linker. *CrystEngComm* **2011**, *13*, 251–258.

21. Wang, X.; Qin, C.; Wang, E.; Li, Y.; Hao, N.; Hu, C.; Xu, L. Syntheses, Structures, and Photoluminescence of a Novel Class of d^{10} Metal Complexes Constructed from Pyridine-3,4-dicarboxylic Acid with Different Coordination Architectures. *Inorg. Chem.* **2004**, *43*, 1850–1856.

22. Cepeda, J.; Beobide, G.; Castillo, O.; Luque, A.; Pérez-Yánez, S.; Román, P. Structure-Directing Effect of Organic Cations in the Assembly of Anionic In(III)/Diazinedicarboxylate Architectures. *Cryst. Growth Des.* **2012**, *12*, 1501–1512.

23. Pinar, A.B.; Gómez-Hortiguela, L.; McCusker, L.B.; Pérez-Pariente, J. Controlling the Aluminum Distribution in the Zeolite Ferrierite via the Organic Structure Directing Agent. *Chem. Mater.* **2013**, *25*, 3654–3661.

24. Van Bommel, K.J.C.; Friggeri, A.; Shinkai, S. Organic Templates for the Generation of Inorganic Materials. *Angew. Chem.* **2003**, *42*, 980–999.

25. Decker, R.; Schlickum, U.; Klappenberger, F.; Zoppellaro, G.; Klyatskaya, S.; Ruben, M.; Barth, J.V.; Brune, H. Using Metal-Organic Templates to Steer the Growth of Fe and Co Nanocluster. *Appl. Phys. Lett.* **2008**, *93*, 243102 / 1–243102 / 3.

26. Abrahams, B.F.; Hawley, A.; Haywood, M.G.; Hudson, T.A.; Robson, R.; Slizys, D.A. Serendipity and Design in the Generation of New Coordination Polymers: An Extensive Series of Highly Symmetrical Guanidinium-Templated, Carbonate-Based Networks with the Sodalite Topology. *J. Am. Chem. Soc.* **2004**, *126*, 2894–2904.

27. Reinoso, S.; Dickman, M.H.; Kortz, U. Selective Crystallization of Dimeric *vs.* Monomeric Dimethyltin-Containing Tungstoarsenates(III) and -antimonates(III) with the Guanidinium Cation. *Eur. J. Inorg. Chem.* **2009**, 947–953.

28. Piedra-Garza, L.F.; Reinoso, S.; Dickman, M.H.; Sanguineti, M.M.; Kortz, U. The First 3-Dimensional Assemblies of Organotin-Functionalized Polyanions. *Dalton Trans.* **2009**, 6231–6234.

29. Reinoso, S.; Bassil, B.S.; Barsukova, M.; Kortz, U. pH-Controlled Assemblies of Dimethyltin-Functionalized 9-Tungstophosphates with Guanidinium as Structure-Directing Cation. *Eur. J. Inorg. Chem.* **2010**, 2537–2542.

30. Iturrospe, A.; Artetxe, B.; Reinoso, S.; San Felices, L.; Vitoria, P.; Lezama, L.; Gutiérrez-Zorrilla, J.M. Copper(II) Complexes of Tetradentate Pyridyl Ligands Supported on Keggin Polyoxometalates: Single-Crystal to Single-Crystal Transformations Promoted by Reversible Dehydration Processes. *Inorg. Chem.* **2013**, *52*, 3084–3093.

31. Iturrospe, A.; San Felices, L.; Reinoso, S.; Artetxe, B.; Lezama, L.; Gutiérrez-Zorrilla, J.M. Reversible Dehydration in Polyoxometalate-Based Hybrid Compounds: A Study of Single-Crystal to Single-Crystal Transformations in Keggin-Type Germanotungstates Decorated with Copper(II) Complexes of Tetradentate N-Donor Ligands. *Cryst. Growth Des.* **2014**, *14*, 2318–2328.

32. Wéry, A.S.J.; Gutiérrez-Zorrilla, J.M.; Luque, A.; Ugalde, M.; Román, P. Phase Transitions in Metavanadates. Polymerization of Tetrakis(tert-Butylammonium)-cyclo-Tetrametavanadate *Chem. Mater.* **1996**, *8*, 408–413.

33. Ritchie, C.; Streb, C.; Thiel, J.; Mitchell, S.G.; Miras, H.N.; Long, D.-L.; Boyd, T.; Peacock, R.D.; McGlone, T.; Cronin, L. Reversible Redox Reactions in an Extended Polyoxometalate Framework Solid. *Angew. Chem.* **2008**, *47*, 6881–6884.

34. Thiel, J.; Ritchie, C.; Streb, C.; Long, D.-L.; Cronin, L. Heteroatom-Controlled Kinetics of Switchable Polyoxometalate Frameworks. *J. Am. Chem. Soc.* **2009**, *131*, 4180–4181.

35. Uehara, K.; Mizuno, N. Heterolytic Dissociation of Water Demonstrated by Crystal-to-Crystal Core Interconversion from (µ-Oxo)divanadium to Bis(µ-hydroxo)divanadium Substituted Polyoxometalates. *J. Am. Chem. Soc.* **2011**, *133*, 1622–1625.

36. Shi, L.-X.; Zhao, W.-F.; Xu, X.; Tang, J.; Wu, C.D. From 1D to 3D Single-Crystal-to-Single-Crystal Structural Transformations Based on Linear Polyanion $[Mn_4(H_2O)_{18}WZnMn_2(H_2O)_2(ZnW_9O_{34})_2]^{4-}$. *Inorg. Chem.* **2011**, *50*, 12387–12389.

37. Uchida, S.; Takahashi, E.; Mizuno, N. Porous Ionic Crystals Modified by Post-Synthesis of $K_2[Cr_3O(OOCH)_6(etpy)_3]_2[\alpha\text{-}SiW_{12}O_{40}]\cdot8H_2O$ through Single-Crystal-to-Single-Crystal Transformation. *Inorg. Chem.* **2013**, *52*, 9320–9326.

38. Zhang, L.-Z.; Gu, W.; Liu, X.; Dong, Z.; Li, B. Solid-State Photopolymerization of a Photochromic Hybrid Based on Keggin Tungstophosphates. *CrystEngComm* **2008**, *10*, 652–654.

39. Zhang, L.-Z.; Gu, W.; Dong, Z.; Liu, X.; Li, B. Phase Transformation of a Rare-Earth Anderson Polyoxometalate at Low Temperature. *CrystEngComm* **2008**, *10*, 1318–1320.

40. Reinoso, S.; Dickman, M.H.; Praetorius, A.; Kortz, U. Low-Temperature Phase of Hexaguanidinium Heptamolybdate Monohydrate. *Acta Crystallogr.* **2008**, *E64*, m614–m615.

41. Barats-Damatov, D.; Shimon, L.J.W.; Feldman, Y.; Bendikov, T.; Neumann, R. Solid-State Crystal-to-Crystal Phase Transitions and Reversible Structure-Temperature Behavior of Phosphovanadomolybdic Acid, $H_5PV_2Mo_{10}O_{40}$. *Inorg. Chem.* **2015**, *4*, 628–634.

42. Chen, C.L.; Goforth, A.M.; Smith, M.D.; Su, C.Y.; zur Loye, H.-C. $[Co_2(ppca)_2(H_2O)(V_4O_{12})_{0.5}]$: A Framework Material Exhibiting Reversible Shrinkage and Expansion through a Single-Crystal-to-Single-Crystal Transformation Involving a Change in the Cobalt Coordination Environment. *Angew. Chem.* **2005**, *44*, 6673–6677.

43. Vittal, J.J. Supramolecular Structural Transformations Involving Coordination Polymers in the Solid State. *Coord. Chem. Rev.* **2007**, *251*, 1781–1795.

44. Abeysinghe, D.; Smith, M.D.; Yeon, J.; Morrison, G.; zur Loye, H.-C. Observation of Multiple Crystal-to-Crystal Transitions in a New Reduced Vanadium Oxalate Hybrid Material, $Ba_3[(VO)_2(C_2O_4)_5(H_2O)_6]\cdot(H_2O)_3$, Prepared via a Mild, Two-Step Hydrothermal Method. *Cryst. Growth Des.* **2014**, *14*, 4749–4758.

45. Tian, Y.; Allan, P.K.; Renouf, C.L.; He, X.; McCormick, L.J.; Morris, R.E. Synthesis and Structural Characterization of a Single-Crystal to Single-Crystal Transformable Coordination Polymer. *Dalton Trans.* **2014**, *43*, 1519–1523.

46. Hashemi, L.; Morsali, A.; Marandi, F.; Pantenburg, I.; Tehrani, A.A. Dynamic Crystal-to-Crystal Transformation of 1D to 2D Lead(II) Coordination Polymers by De- and Rehydration with No Change in the Morphology of Nano-Particles. *New J. Chem.* **2014**, *38*, 3375–3378.

47. Hanson, K.; Calin, N.; Bugaris, D.; Scancella, M.; Sevov, S.C. Reversible Repositioning of Zinc Atoms within Single Crystals of a Zinc Polycarboxylate with an Open-Framework Structure. *J. Am. Chem. Soc.* **2004**, *126*, 10502–10503.

48. Kitagawa, S.; Uemura, K. Dynamic Porous Properties of Coordination Polymers Inspired by Hydrogen Bonds. *Chem. Soc. Rev.* **2005**, *34*, 109–119.

49. Kitagawa, S.; Matsuda, R. Chemistry of Coordination Space of Porous Coordination Polymers. *Coord. Chem. Rev.* **2007**, *251*, 2490–2509.

50. Stamatatos, T.C.; Efthymiou, C.G.; Stoumpos, C.C.; Perlepes, S.P. Adventures in the Coordination Chemistry of Di-2-pyridyl Ketone and Related Ligands: From High-Spin Molecules and Single-Molecule Magnets to Coordination Polymers, and from Structural Aesthetics to an Exciting New Reactivity Chemistry of Coordinated Ligands. *Eur. J. Inorg. Chem.* **2009**, 3361–3368.

51. Huang, D.; Wang, W.; Zhang, X.; Chen, C.; Chen, F.; Liu, Q.; Liao, D.; Li, L.; Sun, L. Synthesis, Structural Characterizations and Magnetic Properties of a Series of Mono-, Di- and Polynuclear Manganese Pyridinecarboxylate Compounds. *Eur. J. Inorg. Chem.* **2004**, 1454–1464.

52. Biswas, C.; Mukherjee, P.; Drew, M.G.B.; Gómez-García, C.J.; Clemente-Juan, J.M.; Ghosh, A. Anion-Directed Synthesis of Metal-Organic Frameworks Based on 2-Picolinate Cu(II) Complexes: A Ferromagnetic Alternating Chain and Two Unprecedented Ferromagnetic Fish Backbone Chains. *Inorg. Chem.* **2007**, *46*, 10771–10780.

53. Pope, M.T.; Müller, A. Polyoxometalate Chemistry: An Old Field with New Dimensions in Several Disciplines. *Angew. Chem.* **1991**, *30*, 34–48.

54. Hervé, G.; Tézé, A.; Contant, R. General Principles of the Synthesis of Polyoxometalates in Aqueous Solution. In *Polyoxometalate Molecular Science*; Borrás-Almenar, J.J., Coronado, E., Müller, A., Pope, M.T., Eds.; Kluwer: Dordrecht, The Netherlands, 2003; NATO Science Series II. Volume 98, pp. 33–54.

55. Zhang, C.-J.; Pang, H.-J.; Tang, Q.; Chen, Y.-G. A Feasible Route to Approach 3D POM-Based Hybrids: Utilizing Substituted or Reduced Keggin Anions with High Charge Density. *Dalton Trans.* **2012**, *41*, 9365–9372.

56. San Felices, L.; Vitoria, P.; Gutiérrez-Zorrilla, J.M.; Lezama, L.; Reinoso, S. Hybrid Inorganic-Metalorganic Compounds Containing Copper(II)-Monosubstituted Keggin Polyanions and Polymeric Copper(I) Complexes. *Inorg. Chem.* **2006**, *45*, 7748–7757.

57. Żurowska, B.; Mroziński, J.; Ciunik, Z. One-Dimensional Copper(II) Compound with a Double Out-of-Plane Carboxylato-Bridge—Another Polymorphic Form of Cu(pyridine-2-carboxylate)$_2$. *Polyhedron* **2007**, *26*, 1251–1258.

58. Żurowska, B.; Mroziński, J.; Ślepokura, K. Structure and Magnetic Properties of a Double Out-of-Plane Carboxylato-Bridged Cu(II) Compound with Pyridine-2-carboxylate. *Polyhedron* **2007**, *26*, 3379–3387.

59. Woodward, P.M.; Sleight, A.W.; Vogt, T.J. Structure Refinement of Triclinic Tungsten Trioxide. *Phys. Chem. Solids* **1995**, *56*, 1305–1315.

60. Schofield, P.F.; Knight, K.S.; Redfern, S.A.T.; Cressey, G. Distortion Characteristics Across the Structural Phase Transition in $(Cu_{1-x}Zn_x)WO_4$. *Acta Crystallogr.* **1997**, *B53*, 102–112.

61. Tézé, A.; Hervé, G.; Finke, R.G.; Lyon, D.K. α-, β-, and γ-Dodecatungstosilicic Acids: Isomers and Related Lacunary Compounds. *Inorg. Synth.* **1990**, *27*, 85–96.

62. Hervé, G.; Tézé, A. Study of α and β-Enneatungstosilicates and Germanates. *Inorg. Chem.* **1977**, *16*, 2115–2117.

63. *CrysAlisPro Software System*, version 171.36.24; Agilent Technologies UK Ltd.: Oxford, UK, 2012.

64. Dolomanov, O.V.; Bourhis, L.J.; Gildea, R.J.; Howard, J.A.K.; Puschmann, H. A Complete Structure Solution, Refinement and Analysis Program. *J. Appl. Crystallogr.* **2009**, *42*, 339–341.

65. Sheldrick, G.M. A Short History of SHELX. *Acta Crystallogr.* **2008**, *A64*, 112–122.

66. Spek, A.L. Structure Validation in Chemical Crystallography. *Acta Crystallogr.* **2009**, *D65*, 148–155.

67. Farrugia, L.J. *WinGX* Suite for Small-Molecule Single-Crystal Crystallography. *J. Appl. Crystallogr.* **1999**, *32*, 837–836.

Understanding the Regioselective Hydrolysis of Human Serum Albumin by Zr(IV)-Substituted Polyoxotungstates Using Tryptophan Fluorescence Spectroscopy

Vincent Goovaerts, Karen Stroobants, Gregory Absillis and Tatjana N. Parac-Vogt

Abstract: The interaction between human serum albumin (HSA) and a series of Zr(IV)-substituted polyoxometalates (POMs) (Lindqvist type POM (("Bu$_4$N)$_6$[{W$_5$O$_{18}$Zr(μ-OH)}$_2$]·2H$_2$O, Zr2-L2), two Keggin type POMs ((Et$_2$NH$_2$)$_{10}$[Zr(PW$_{11}$O$_{39}$)$_2$]·7H$_2$O, Zr1-K2 and (Et$_2$NH$_2$)$_8$[{α-PW$_{11}$O$_{39}$Zr(μ-OH)(H$_2$O)}$_2$]·7H$_2$O, Zr2-K2), and two Wells-Dawson type POMs (K$_{15}$H[Zr(α_2-P$_2$W$_{17}$O$_{61}$)$_2$]·25H$_2$O, Zr1-WD2 and Na$_{14}$[Zr$_4$(P$_2$W$_{16}$O$_{59}$)$_2$(μ_3-O)$_2$(OH)$_2$(H$_2$O)$_4$]·10H$_2$O, Zr4-WD2) was investigated by tryptophan (Trp) fluorescence spectroscopy. The fluorescence data were analyzed using the Tachiya model, ideally suited for multiple binding site analysis. The obtained quenching constants have the same order of magnitude for all the measured POM:protein complexes, ranging from 1.9×10^5 M^{-1} to 5.1×10^5 M^{-1}. The number of bound POM molecules to HSA was in the range of 1.5 up to 3.5. The influence of the ionic strength was studied for the Zr1-WD2:HSA complex in the presence of NaClO$_4$. The calculated quenching constant decreases upon increasing the ionic strength of the solution from 0.0004 M to 0.5004 M, indicating the electrostatic nature of the interaction. The number of POM molecules bound to HSA increases from 1.0 to 4.8. ^{31}P NMR spectroscopy provided evidence for the stability of all investigated POM structures during the interaction with HSA.

Reprinted from *Inorganics*. Cite as: Goovaerts, V.; Stroobants, K.; Absillis, G.; Parac-Vogt, T.N. Understanding the Regioselective Hydrolysis of Human Serum Albumin by Zr(IV)-Substituted Polyoxotungstates Using Tryptophan Fluorescence Spectroscopy. *Inorganics* **2015**, *3*, 230-245.

1. Introduction

Amino acids, the individual building blocks of proteins, are linked via peptide bonds, which have an extremely high stability towards hydrolysis. Under physiological reaction conditions, the half-life for the uncatalyzed hydrolysis of the peptide bond is estimated to be between 350 and 600 years [1]. Nevertheless, the selective hydrolysis of proteins into smaller, workable fragments is necessary in many biochemical and biomedical procedures [2]. For instance, protein hydrolysis is of prime importance for a broad range of proteomics techniques ranging from protein sequencing to protein identification [3–5]. Moreover, efficient and selective protein hydrolysis is also highly used in the food industry [6], leather processing [7], the cleaning industry [8] and medicine [9]. These applications demand the use of reagents that are on the one hand capable of hydrolyzing the extremely inert peptide bond, and achieving this reactivity in a highly selective manner on the other hand. Because most of the organic molecules are not sufficiently strong nucleophiles at physiological pH, the hydrolytic activity of several metal salts and their corresponding complexes has been extensively investigated [2]. Although the improvements in this domain are considerable, most reported complexes need multistep reactions to be tethered to the protein, or can only be used in an extreme pH range, limiting their practical use as peptidases. Consequently, there is a need for new efficient peptidases.

Polyoxometalates (POMs) are a versatile and tunable class of negatively charged early transition metal-oxygen clusters [10–12]. These molecules are an assemblage of oxo-bridged early transition metals (mostly Mo, W, and V) in their highest oxidation state [13,14]. Their chemical and physical properties can be tuned, giving rise to a large number of possible applications in the fields of catalysis [15–17], material science [18–20] and medicine [21–24]. Moreover, the removal of one or more of the addenda atoms from the POM framework results in the formation of a lacunary species, which can coordinate different metal ions, resulting in the formation of metal-substituted POMs [11,25]. Metal-substituted POMs have been used as catalysts in several catalytic reactions, ranging from the epoxidation of olefins and alkanes to the splitting of water [26–29]. Additionally, POMs exhibit biological activity, which includes *in vitro* and *in vivo* anticancer, antiviral, antibiotic, antiprotozoal and antidiabetic activity [23,24,30]. Furthermore, inhibition of enzymes can be achieved as a result of the interaction of POMs with proteins [24,31], making POMs potential drugs for Alzheimer's disease [31]. Nonetheless, the exact molecular mechanism responsible for the medicinal activity of POMs has not yet been elucidated. Key components that influence the interaction between POMs and proteins include the size, shape and charge of the POM as well as the kind of incorporated metal ion [30,32–39].

It was previously reported by our research group that isopolyoxomolybdates [40–43], isopolyoxovandates [44,45] and metal-substituted POMs [46] are able to efficiently hydrolyze the phopho(di)ester bonds in DNA and RNA model compounds. Moreover, hydrolysis of the highly inert peptide bond in dipeptides [46–50], oligopeptides [51], and several model proteins [34,52,53] by different types of POMs has been demonstrated. The hydrolysis of hen egg white lysozyme (HEWL) by the Ce(IV)-substituted Keggin POM, $[Ce(PW_{11}O_{39})_2]^{10-}$, was the first reported example of protein hydrolysis promoted by a POM [34]. This study has shown that the negatively charged POM specifically interacts with positively charged patches on the protein surface, while at the same time it acts as a stabilizing ligand for the active metal ion, leading to regioselective hydrolysis. Interestingly, in preceding studies, it was found that the binding is not only directed by an electrostatic interaction between the negatively charged POM structure and positive patches on the surface of HEWL, but also by the coordination of the Ce(IV) ion to the carboxylate side chain of Glu or Asp amino acids [34].

Later we demonstrated that due to the high Lewis acidity of the zirconium(IV) ion, the Zr(IV)-substituted polytungstates were efficient catalysts for the hydrolysis of the peptide bond [49–51,54]. The hydrolytic activity of a series of Zr(IV)-substituted POMs toward human serum albumin (HSA) has been investigated in detail recently [52]. In this study, it was found that HSA was selectively hydrolyzed by Zr(IV)-substituted Lindqvist $(("Bu_4N)_6[\{W_5O_{18}Zr(\mu-OH)\}_2]\cdot 2H_2O$, Zr2-L2), Keggin $((Et_2NH_2)_{10}[Zr(PW_{11}O_{39})_2]\cdot 7H_2O$, Zr1-K2, and $(Et_2NH_2)_8[\{\alpha-PW_{11}O_{39}Zr(\mu-OH)(H_2O)\}_2]\cdot 7H_2O$, Zr2-K2), and Wells-Dawson $(K_{15}H[Zr(\alpha_2-P_2W_{17}O_{61})_2]\cdot 25H_2O$, Zr1-WD2, and $Na_{14}[Zr_4(P_2W_{16}O_{59})_2(\mu_3-O)_2(OH)_2(H_2O)_4]\cdot 10H_2O$, Zr4-WD2) type POMs. Although all POMs exhibited a similar hydrolysis pattern, the Wells-Dawson type POMs showed the highest reactivity among the investigated complexes towards the hydrolysis of HSA. Albumin proteins are frequently studied proteins in interaction studies because of their known primary sequence and the stability of their tertiary structure in solution [33,36–39,55]. HSA contains one tryptophan residue which is located at position 214, and the use of fluorescence quenching studies are an excellent tool to quantify the strength of the interaction and the number of binding sites for molecules that interact with HSA [56]. A detailed understanding of the parameters responsible for the interaction on a molecular level is necessary to tune the selectivity and rationally design metal-substituted POMs with explicit interaction properties. In addition, the establishment of the correlation between the rate of protein hydrolysis by metal-substituted POMs and the strength of their interaction that precedes the hydrolysis will result in a faster achievement of the application of metal-substituted POMs as artificial peptidases.

In this study, a series of Zr(IV)-substituted POMs, which were previously proven to be hydrolytically active towards HSA, were investigated with tryptophan (Trp) fluorescence spectroscopy in order to gain insight into the interaction between the POMs and HSA, which precedes the actual protein hydrolysis. The quenching of Trp214 fluorescence by Zr1-K2, Zr2-K2, Zr2-L2, Zr1-WD2, and Zr4-WD2 was investigated at physiological pH (pH 7.4) and room temperature to quantify the strength of the binding and identify the stoichiometry of this interaction.

2. Results and Discussion

2.1. Tryptophan Fluorescence Spectroscopy

Fluorescence quenching was used to study the interaction of a set of Zr(IV)-substituted POMs with HSA. The series consist of one Lindqvist type POM $(({}^{n}Bu_4N)_6[\{W_5O_{18}Zr(\mu\text{-}OH)\}_2]\cdot 2H_2O$, Zr2-L2), two Keggin type POMs $((Et_2NH_2)_{10}[Zr(PW_{11}O_{39})_2]\cdot 7H_2O$, Zr1-K2 and $(Et_2NH_2)_8[\{\alpha\text{-}PW_{11}O_{39}Zr(\mu\text{-}OH)(H_2O)\}_2]\cdot 7H_2O$, Zr2-K2), and two Wells-Dawson type POMs $(K_{15}H[Zr(\alpha_2\text{-}P_2W_{17}O_{61})_2]\cdot 25H_2O$, Zr1-WD2 and $Na_{14}[Zr_4(P_2W_{16}O_{59})_2(\mu_3\text{-}O)_2(OH)_2(H_2O)_4]\cdot 10H_2O$, Zr4-WD2), as depicted in Figure 1.

HSA is a 66.5 kDa protein and consists of 585 amino acid residues including 1 tryptophan (Trp) residue, 19 tyrosine (Tyr) residues and 17 disulfide bonds. Emission of HSA is dominated by the Trp residue, which absorbs at the longest wavelength and displays the largest extinction co-efficient. Moreover, energy absorbed by phenylalanine (Phe) and Tyr residues is often efficiently transferred to the Trp residues in the same protein [57]. In these steady state fluorescence experiments, the concentration of HSA was kept constant at 10 μM, while the concentration of the respective POMs was increased stepwise from 0 to 10 μM with increments of 1 μM. The quenching of the Trp fluorescence by Zr2-L2 is shown in Figure 2, and similar behavior was observed for the other POMs (Figures S1–S4). The Tachiya model, presented in Equation (1), was used for analysis of the strength and stoichiometry of the binding [56]:

$$\text{Log}(\frac{F_0 - F}{F}) = m \times \text{Log}(K_a) + m \times \text{Log}\left([Q] - [M]\frac{F_0 - F}{F}\right) \tag{1}$$

where F_0 is the unquenched fluorescence intensity, F the fluorescence in the presence of the quencher, $[Q]$ the concentration of the quencher, $[M]$ the concentration of protein and m the number of binding sites. This model is used since a stoichiometry higher than 1 is expected [52]. The key advantage of using this equation is that it offers the option to extract the quenching constant or association constant (K_a) as well as the number of bound molecules (m) directly. The plots of the equations for the different POM:HSA complexes can be seen in the insets of the respective figures. The calculated values of the quenching constants, their corresponding number of bound molecules, the HSA hydrolysis rate (%) [52], and the net charge of the POMs are given in Table 1.

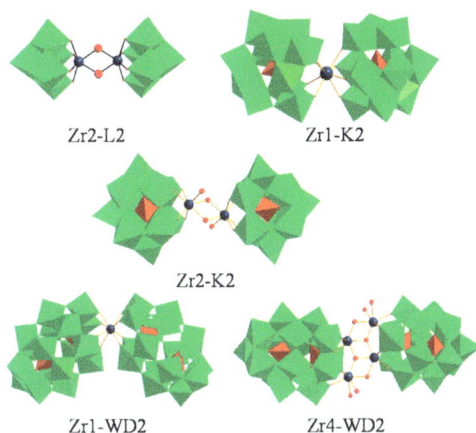

Figure 1. Combined polyhedral/ball-and-stick representations of Zr2-L2, Zr1-K2, Zr2-K2, Zr1-WD2, and Zr4-WD2. (Green octahedra: WO_6, red tetrahedra: PO_4, teal blue: Zr(IV) ion, red balls: oxygen atoms).

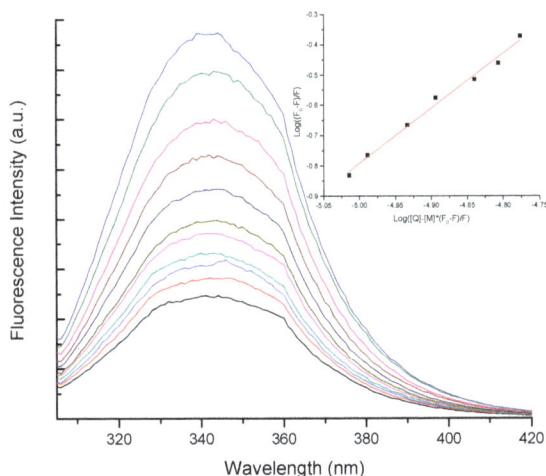

Figure 2. Emission fluorescence spectra of HSA in the absence and presence of different concentrations of Zr2-L2 ([HSA] = 10^{-5} M, pH = 7.4). From top to bottom, the concentration of Zr2-L2 increased stepwise from 0 to 10^{-5} M with increments of 10^{-6} M. In the inset, the plot of the Tachiya equation is given (with $R^2 = 0.99$). From the plot, K_a and m were calculated to be 2.0×10^5 M^{-1} and 3.02, respectively.

Table 1. Calculated values of the quenching constants, their corresponding number of bound molecules, the percentage of hydrolyzed human serum albumin (has) after 48 h incubation at pH 7.4 and 60 °C [52] and their polyoxometalates (POM) net charge for different POM:HSA complexes.

POM	K_a (M^{-1})	m	Hydrolysis (%) after 48 h [52]	POM Net Charge
Zr1-K2	1.9×10^5	3.44	<35	−10
Zr2-K2	2.5×10^5	2.23	<35	−6
Zr2-L2	2.0×10^5	3.02	<35	−4
Zr1-WD2	5.1×10^5	1.52	~75	−16
Zr4-WD2	2.8×10^5	2.05	~50	−4

From Table 1, it can be seen that the number of bound Zr(IV)-substituted POM complexes per HSA molecule varies between 1.52 and 3.44. This indication for the simultaneous binding of multiple POM molecules to HSA is not surprising since we have previously shown that up to four cleavage positions were observed for all five POM complexes [52]. Since HSA is a large protein, it is reasonable to assume that it can interact with more than one POM at the same time. In addition, the two largest POMs (Zr4-WD2 and Zr1-WD2) show the lowest value for the number of molecules bound to HSA, while a higher stoichiometry is obtained for the smaller POMs. This is reflected in the change of a 2:1 POM:HSA complex observed for (Zr4-WD2 and Zr1-WD2) to a 3:1 (for Zr2-L2) and nearly a 4:1 POM:HSA complex for Zr1-K2.

As is also shown in Table 1, the strength of the binding to HSA is in the same order of magnitude (10^5 M^{-1}) for all studied Zr(IV)-substituted POMs. The quenching of the fluorescence of HSA has been extensively studied with different types of POMs, including Keggin, Wells-Dawson, Lindqvist and wheel-shaped structured POMs [33,36,37,39]. The calculated values for the quenching constants of the Zr(IV)-substituted POMs are comparable to the reported values obtained for the lacunary Keggin (K$_7$PW$_{11}$O$_{39}$, 2.9×10^5 M^{-1}) [58], and the Ni(II)-substituted Wells-Dawson POM (α_2-[NiP$_2$W$_{17}$O$_{61}$]$^{8-}$, 1.15×10^5 M^{-1}) [37]. However, the values of the quenching constants for the Zr(IV)-substituted POMs are one order of magnitude larger than the corresponding values for the Eu(III)-substituted analogues of the Keggin (K$_4$EuPW$_{11}$O$_{39}$, 6.1×10^4 M^{-1}) [58], and Lindqvist (Na$_9$[Eu(W$_{10}$O$_{36}$)], 4.8×10^4 M^{-1}) [39] type POMs. In part, this might be a result of the larger POM net charge of the Zr(IV)-substituted POMs as compared to the Eu(III) analogues. Furthermore, the Zr(IV)-substituted POMs are larger than their Eu(III) counterparts, resulting in a larger negatively charged surface that is able to interact stronger with the positive patches on the HSA surface. For instance, the larger charge of the Zr(IV)-substituted Keggin type POMs evolves from the slight difference in structure; they all have two POM ligands, while the Eu(III) analogue only has one. The larger negative charge results in a stronger interaction and hence, a larger value for the quenching constant, since the POM:protein interactions are mainly electrostatic in nature [39].

A final important observation which is made from the data in Table 1 is that the reactivity of the studied POMs is correlated to the strength of their interaction with the HSA protein. Zr1-WD2 not only possesses the largest value of K_a, but also displayed the fastest rate of HSA hydrolysis (~75% after 48 h incubation) in our previous reaction study. The second strongest binding POM, Zr4-WD2, exhibited the second fastest hydrolysis of HSA (~50% after 48 h incubation). Conclusions for the other three POMs cannot be made, since it was not possible to distinguish between them in the hydrolysis experiments. However, the relationship between the reactivity and interaction strength can be expected to be valid for the other Zr(IV)-substituted POMs as well. In addition, it can be deduced that small changes in the interaction strength induce large changes in the hydrolysis reactivity toward the stable peptide bonds in HSA.

Although the net charge of the POM clearly plays an important role, it is not the only factor influencing the interaction with HSA. This can for example be seen for the value of the quenching constant for the Zr1-K2:HSA complex; this complex has a very similar value of K_a as compared to Zr2-L2, while the net charges of the POMs are -10 and -4, respectively. Consequently, it is proposed that the size and shape of the POM considerably affect the interaction as well. Moreover, different cations can cause changes in the protein aggregation and alter interaction processes. In addition, the substituted metal ion or the amount of substituted metal ions can play an important role in the interaction behavior. This is a result of the hydrolysis pattern of the Zr(IV)-substituted POMs. One cleavage site (Arg114-Leu115) is located in the central cleft of HSA. This cleft has a positive inner surface [59]. The three other cleavage sites (Ala257-Asp258, Lys313-Asp314, and Cys392-Glu393) are located near a positive surface patch with a negatively charged amino acid in the upstream position (Asp or Glu) [52], suggesting that the positively charged Zr(IV) ion is involved in the electronic interaction process as well.

To further investigate the electrostatic nature of the interactions between POMs and proteins, the Trp fluorescence quenching of HSA by Zr1-WD2 was studied at different ionic strength conditions. Since electrostatic interactions are known to diminish when the ionic strength (I) is increased, the effect of NaClO$_4$ addition on the interaction parameters was studied. The ionic strength can be calculated using Equation (2):

$$I = \frac{1}{2} \sum_{i=1}^{n} C_i Z_i^2 \tag{2}$$

where C_i is the molar concentration of the ion in M and Z_i is the charge of that ion. The ionic strength was varied in the range of 0.0004 M to 0.5004 M. The Trp fluorescence quenching of HSA was performed for five ionic strength values in this range and in the presence of the Zr1-WD2 POM, as this POM displayed the strongest interaction in the previous experiments. The analysis of the data was done using the Tachiya model [56] and the resulting parameters are shown in Table 2.

Table 2. Calculated values of the quenching constants and their corresponding number of bound molecules for the 1:2 Zr(IV)-substituted Wells-Dawson POM:HSA complex at different ionic strength conditions.

Ionic Strength (M)	K_a (M^{-1})	m
0.0004	5.1×10^5	1.52
0.0014	4.4×10^5	0.95
0.0504	2.4×10^5	2.44
0.1004	2.3×10^5	2.65
0.2504	1.9×10^5	2.95
0.5004	1.5×10^5	4.81

From Table 2, it can be concluded that the quenching constant decreases upon increasing the ionic strength of the solution. This observation can be explained by the thermodynamic parameters of the electrostatic interaction. The entropy becomes more unfavorable when the ionic strength increases [60], resulting in a weaker interaction. This is the result of the shielding of the high negative charge (-16) of Zr1-WD2 by the salt ions. Moreover, this lower quenching constant indicates that each binding is weaker because of the simultaneous disruption of the interaction with the positively charged residues of HSA. Although the interaction strength reduces with a factor larger than three, it remains in the same order of magnitude (10^5 M^{-1}) regardless of the change in ionic strength. The effect of the ionic strength is not as pronounced as it was the case for a Keggin shaped POM ($[H_2W_{12}O_{40}]^{6-}$) where a decrease in binding constant was observed of more than two orders of magnitude [61]. Therefore, it can be concluded that the interaction between Zr1-WD2 and HSA persists even in solutions with high ionic strengths.

A final important observation which is made from the data in Table 2 is that the number of bound molecules increases upon increasing the ionic strength. This result can be understood by considering the electrostatic nature of the POM:protein interaction. While the high negative charge of Zr1-WD2 results in the strongest interaction with HSA, it also leads to a lower number of possible binding positions as compared to the other POMs. This presumably is the result of repulsive forces with negatively charged residues on the HSA surface in close proximity to the positively charged regions to which the POMs tend to bind. When NaClO$_4$ is added, the high negative charge of Zr1-WD2 will partly be shielded, resulting in more possible binding positions. In previous studies performed in our research group, by using isothermal titration calorimetry, it was found that four Zr1-WD2 POMs were able to bind to HSA [53]. Therefore, it is expected that the number of bound molecules will plateau upon further increasing the ionic strength. As for the other Zr(IV)-substituted POMs, it is reasonable to assume the POMs will display similar behavior upon increasing the ionic strength; the strength of the binding will decrease as a result of the shielding of the charges by NaClO$_4$, while the number of bound molecules will increase for the same reason. The smaller Keggin and Lindqvist type POMs will probably be able to bind more POM molecules per HSA, as was seen in the fluorescence quenching experiments without added salt. The value of bound molecules could go up as high as eleven, as was reported for a Keggin shaped POM ($[H_2W_{12}O_{40}]^{6-}$) [61].

2.2. Polyoxometalate Stability in the Presence of HSA

^{31}P NMR spectra of the Zr1-K2, Zr2-K2, Zr1-WD2 and Zr4-WD2 POMs in phosphate buffer (10 mM, pD 7.4, 10% D$_2$O) were measured in the absence and presence of 0.1 equivalents of HSA. The Zr2-L2 POM could not be investigated in these experiments since this POM does not contain any phosphorus atoms. The Keggin structures are composed of one or two Zr(IV) metal ions, respectively, surrounded by two lacunary Keggin units with a PO$_4$ tetrahedron at the centre (Figure 1). The environment of the phosphorus atoms is symmetrical resulting in a single resonance for the phosphorus atoms. However, due to the formation of two isomers of the Zr1-K2 compound, two adjacent resonances were reported to characterize this POM [62]. The Wells-Dawson compounds contain a second PO$_4$ tetrahedron near the POM unit centre (Figure 1), they also are symmetrical and are characterized by two ^{31}P NMR signals. The resonance for the phosphorus atom close to the Zr(IV) metal falls within the ppm range where the characteristic peaks for metal-substituted Keggins are found due to the similar electrostatic environment. The second resonance for the phosphorus atom further away from the incorporated metal ion(s) is shifted downfield. As shown in Figure 3, there is no shift in the peak positions of the characteristic Zr1-WD2 resonances (at −9.33 and −13.94 ppm) in the presence of HSA, indicating the stability of the POM upon interaction with the protein. It should not be surprising that these polyoxotungstates are stable as this was already previously demonstrated by various techniques [63–65]. In addition, the slight decrease in the intensity of the POM peaks can be correlated with the formation of a large molecular complex with HSA which has a longer correlation time and thus a faster relaxation. The shifting of ^{31}P NMR resonances was also not observed for the Zr4-WD2 (resonances at −7.04 and −14.26 ppm), Zr1-K2 (resonances at −14.60 and −14.69 ppm) and Zr2-K2 (resonance at −13.60 ppm) POMs upon addition of 0.1 equivalents of HSA. Similarly to Zr1-WD2, an intensity decrease was observed for all POM resonances upon increasing the HSA concentration, giving an additional indication for the POM:HSA interactions. The absence of any new peaks in the ^{31}P NMR provides evidence for the stability of all investigated POM structures during the interaction with the albumin protein, indicating a fast exchange binding mechanism between the POM structures and HSA [66].

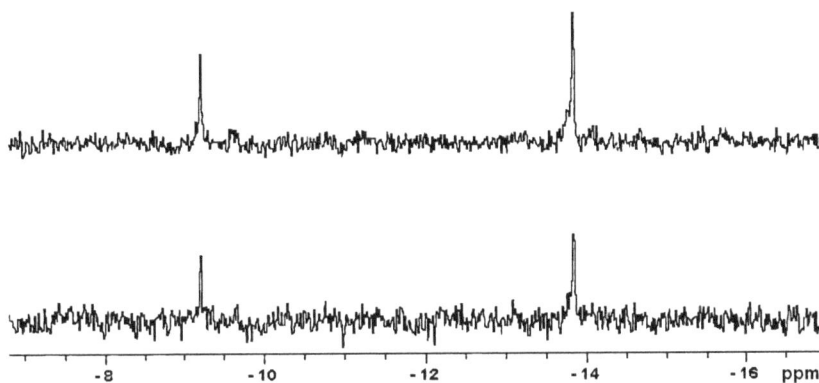

Figure 3. ^{31}P NMR spectra of Zr1-WD2 (2.0 mM) in the absence (**top**) and presence (**bottom**) of HSA (0.2 mM) in phosphate buffer (10 mM, pH 7.4, 10% D$_2$O). The peak positions at −9.33 and −13.94 ppm do not change upon addition of HSA.

Previous experiments in our research group on the mechanism of the hydrolysis of the peptide bond in di- and oligopeptides have demonstrated the need for the accessibility of the hydrolytically active metal ion. It was shown that the coordination of the peptide to Zr(IV) results in the polarization of the peptide carbonyl group, which speeds up the hydrolysis [49–51,54]. As a result of the saturated coordination sphere of Zr(IV) in the 1:2 POM structures, it is expected that these POMs display the slowest hydrolysis rate. However, equilibrium exists between the monomeric and dimeric species of Zr1-WD2, Zr1-K2, Zr2-K2 and Zr2-L2, depending on the pH and concentration of the solution where lower concentrations favored the formation of the monomeric species [49–51,67,68]. Moreover, the addition of protein can also shift the equilibrium towards the monomeric species [51]. These species have four free coordination sites around Zr(IV), resulting in faster hydrolysis as was demonstrated for Zr1-WD2 [52]. For this reason, the hydrolysis of the peptide bond in HSA is by a similar mechanism as for short peptides [51], although the interaction pattern was demonstrated to be very different. In HSA, it was shown that the POM structures are able to interact with positive surface patches with a supplementary interaction of Zr(IV) with an upstream negatively charged amino acid (Asp or Glu) and thereby influencing the selectivity [52].

3. Experimental Section

3.1. Materials

Human serum albumin (HSA) was purchased from Sigma-Aldrich (Schnelldorf, Germany) in the highest available purity and was used without further purification. $(Et_2NH_2)_{10}[Zr(PW_{11}O_{39})_2]\cdot 7H_2O$ (Zr1-K2) [62], $(Et_2NH_2)_8[\{\alpha-PW_{11}O_{39}Zr(\mu-OH)(H_2O)\}_2]\cdot 7H_2O$ (Zr2-K2) [69], $(^nBu_4N)_6[\{W_5O_{18}Zr(\mu-OH)\}_2]\cdot 2H_2O$ (Zr2-L2) [70], $K_{15}H[Zr(\alpha_2-P_2W_{17}O_{61})_2]\cdot 25H_2O$ (Zr1-WD2) [62] or $Na_{14}[Zr_4(P_2W_{16}O_{59})_2(\mu_3-O)_2(OH)_2(H_2O)_4]\cdot 10H_2O$ (Zr4-WD2) [71] were prepared by the published procedures.

3.2. Fluorescence Spectroscopy Studies

Steady state fluorescence experiments were recorded on a Photon Technology Quanta Master (North Edison, NJ, USA) QM-6/2005 spectrofluorimeter. Quartz cuvettes with 10.0 mm optical path length were used. Spectra were recorded in a buffered 10 μM protein concentration solution (phosphate buffer, pH = 7.4) at room temperature monitoring the emission from 305 nm to 420 nm, with a maximum at approximately 340 nm. Although several organic cations are used as counter ions for the polyoxoanions, no precipitation is observed as a result of the used phosphate buffer. To examine the influence of the ionic strength, $NaClO_4$ was added with concentrations in the range of 0.001 to 0.5 M. Excitation of the samples took place at 295 nm to avoid excitation of tyrosine residues. The emission and excitation slit widths were opened at 0.37 mm (resolution of 1.0 nm). The POM concentration was increased stepwise from 0 to 10 μM with increments of 1.0 μM. The fluorescence data of the POMs were analyzed using the Tachiya model, Equation (1) [56].

3.3. ^{31}P NMR Spectroscopy

Solutions (0.5 mL) containing 0.1 or 0.2 mM of HSA and 1.0 or 2.0 mM of POM were prepared in phosphate buffer (10.0 mM—pH 7.4—10% D_2O). ^{31}P NMR spectra were recorded directly after mixing on a Bruker (Karlsruhe, Germany) Avance 400 (161.98 MHz) spectrometer. As an external standard, 25% H_3PO_4 in D_2O in a sealed capillary was used.

4. Conclusions

The knowledge of the interaction between hydrolytically active POMs and proteins on a molecular level is necessary to further optimize the hydrolytic selectivity of metal-substituted POMs as artificial proteases. In this study, a direct correlation between the rate of protein hydrolysis by metal-substituted POMs and the strength of their interaction, preceding this reaction, was established for the first time.

Fluorescence spectroscopy was used to study the interaction between a series of Zr(IV)-substituted POMs and HSA. Analysis of the fluorescence quenching of HSA in the presence of these POMs has resulted in a binding stoichiometry ranging from 2:1 POM:HSA for the larger POMs up to 3:1 and even 4:1 for the smaller POMs and binding constants in the order of 10^5 M^{-1}. It was shown that the Wells-Dawson POM, Zr1-WD2, displays the strongest interaction with HSA, which correlates to its high hydrolysis reactivity, as was previously reported. This correlation between the hydrolysis reactivity and interaction strength was not established before and seems to be valid for the other Zr(IV)-substituted POMs as well.

The influence of the ionic strength was studied for the Zr1-WD2:HSA complex by adding $NaClO_4$. It was found that the value of the quenching constants decreases upon increasing the ionic strength of the solution from 0.0004 M to 0.5004 M. Nevertheless, it was found that the number of bound molecules increased from 0.95 to 4.81. While the increased number of bound molecules is the result of the shielding of the high negative charge of Zr1-WD2, resulting in more possible binding positions, the lower quenching constant indicates that each binding is weaker due to the simultaneous disruption of the interactions with positively charged residues. ^{31}P NMR spectroscopy was used to investigate the stability of all phosphorus containing POMs upon interaction with HSA. The lack of new peaks in the ^{31}P NMR spectra of all investigated POMs provides evidence for the stability during their interaction with HSA.

The presented interaction study complements the ongoing research in our group which mainly focuses on the hydrolysis of proteins by POMs, and greatly contributes to the establishment of a structure-activity relationship of metal-substituted polyoxotungstates as a new class of metalloproteases.

Acknowledgments

T.N.P.V. thanks KU Leuven and F.W.O. Flanders for the financial support. V.G. acknowledges KU Leuven (Belgium) for a doctoral fellowship. K.S. thanks F.W.O. Flanders for the doctoral fellowship and G.A. acknowledges F.W.O. Flanders for the post-doctoral fellowship.

Author Contributions

V.G.: Literature research, tryptophan fluorescence quenching experiments, analysis and discussion of the results; K.S. and G.A.: Synthesis of polyoxometalates and ^{31}P NMR spectroscopy experiments; T.N.P.-V.: careful follow-up and improvement of the manuscript.

Conflicts of Interest

The authors declare no conflict of interest.

94

References

1. Radzicka, A.; Wolfenden, R. Rates of uncatalyzed peptide bond hydrolysis in neutral solution and the transition state affinities of proteases. *J. Am. Chem. Soc.* **1996**, *118*, 6105–6109.
2. Grant, K.B.; Kassai, M. Major advances in the hydrolysis of peptides and proteins by metal ions and complexes. *Curr. Org. Chem.* **2006**, *10*, 1035–1049.
3. Delahunty, C.; Yates, J.R. Protein identification using 2D-LC-MS/MS. *Methods* **2005**, *35*, 248–255.
4. Meyer, B.; Papasotiriou, D.G.; Karas, M. 100% protein sequence coverage: A modern form of surrealism in proteomics. *Amino Acids* **2011**, *41*, 291–310.
5. Swaney, D.L.; Wenger, C.D.; Coon, J.J. Value of using multiple proteases for large-scale mass spectrometry-based proteomics. *J. Proteome Res.* **2010**, *9*, 1323–1329.
6. Bolumar, T.; Bindrich, U.; Toepfl, S.; Toldra, F.; Heinz, V. Effect of electrohydraulic shockwave treatment on tenderness, muscle cathepsin and peptidase activities and microstructure of beef loin steaks from holstein young bulls. *Meat Sci.* **2014**, *98*, 759–765.
7. Rai, S.K.; Mukherjee, A.K. Optimization of production of an oxidant and detergent-stable alkaline beta-keratinase from brevibacillus sp strain AS-S10-II: Application of enzyme in laundry detergent formulations and in leather industry. *Biochem. Eng. J.* **2011**, *54*, 47–56.
8. Polaina, J.; MacCabe, A.P. *Industrial Enzymes: Structure, Function and Applications*; Springer: Dordrecht, The Netherlands, 2007.
9. Katz, M.L.; Coates, J.R.; Sibigtroth, C.M.; Taylor, J.D.; Carpentier, M.; Young, W.M.; Wininger, F.A.; Kennedy, D.; Vuillemenot, B.R.; O'Neill, C.A.; *et al.* Enzyme replacement therapy attenuates disease progression in a canine model of late-infantile neuronal ceroid lipofuscinosis (CLN2 disease). *J. Neurosci. Res.* **2014**, *92*, 1591–1598.
10. Pope, M.T. *Heteropoly and Isopoly Oxometalates*; Springer-Verlag: New York, NY, USA, 1983.
11. Bassil, B.S.; Kortz, U. Recent advances in lanthanide-containing polyoxotungstates. *Z. Anorg. Allg. Chem.* **2010**, *636*, 2222–2231.
12. Cronin, L.; Mueller, A. From serendipity to design of polyoxometalates at the nanoscale, aesthetic beauty and applications. *Chem. Soc. Rev.* **2012**, *41*, 7333–7334.
13. Long, D.-L.; Tsunashima, R.; Cronin, L. Polyoxometalates: Building blocks for functional nanoscale systems. *Angew. Chem. Int. Ed.* **2010**, *49*, 1736–1758.
14. Proust, A.; Thouvenot, R.; Gouzerh, P. Functionalization of polyoxometalates: Towards advanced applications in catalysis and materials science. *Chem. Commun.* **2008**, 1837–1852.
15. Lv, H.; Geletii, Y.V.; Zhao, C.; Vickers, J.W.; Zhu, G.; Luo, Z.; Song, J.; Lian, T.; Musaev, D.G.; Hill, C.L.; *et al.* Polyoxometalate water oxidation catalysts and the production of green fuel. *Chem. Soc. Rev.* **2012**, *41*, 7572–7589.
16. Zakzeski, J.; Bruijnincx, P.C.A.; Jongerius, A.L.; Weckhuysen, B.M. The catalytic valorization of lignin for the production of renewable chemicals. *Chem. Rev.* **2010**, *110*, 3552–3599.
17. Chen, J.; Wang, S.; Huang, J.; Chen, L.; Ma, L.; Huang, X. Conversion of cellulose and cellobiose into sorbitol catalyzed by ruthenium supported on a polyoxometalate/metal-organic framework hybrid. *Chemsuschem* **2013**, *6*, 1545–1555.

18. Mueller, A.; Garai, S.; Schaeffer, C.; Merca, A.; Boegge, H.; Al-Karawi, A.J.M.; Prasad, T.K. Water repellency in hydrophobic nanocapsules-molecular view on dewetting. *Chem. Eur. J.* **2014**, *20*, 6659–6664.

19. Kourasi, M.; Wills, R.G.A.; Shah, A.A.; Walsh, F.C. Heteropolyacids for fuel cell applications. *Electrochim. Acta* **2014**, *127*, 454–466.

20. Zhang, W.-B.; Yu, X.; Wang, C.-L.; Sun, H.-J.; Hsieh, I.F.; Li, Y.; Dong, X.-H.; Yue, K.; Van Horn, R.; Cheng, S.Z.D.; *et al.* Molecular nanoparticles are unique elements for macromolecular science: From "nanoatoms" to giant molecules. *Macromolecules* **2014**, *47*, 1221–1239.

21. Yvon, C.; Surman, A.J.; Hutin, M.; Alex, J.; Smith, B.O.; Long, D.-L.; Cronin, L. Polyoxometalate clusters integrated into peptide chains and as inorganic amino acids: Solution-and solid-phase approaches. *Angew. Chem. Int. Ed.* **2014**, *53*, 3336–3341.

22. McGregor, D.; Burton-Pye, B.P.; Lukens, W.W.; Howell, R.C.; Francesconi, L.C. Insights into stabilization of the ^{99}Tc^{V}O core for synthesis of ^{99}Tc^{V}O compounds. *Eur. J. Inorg. Chem.* **2014**, *2014*, 1082–1089.

23. Aureliano, M.; Fraqueza, G.; Ohlin, C.A. Ion pumps as biological targets for decavanadate. *Dalton Trans.* **2013**, *42*, 11770–11777.

24. Iqbal, J.; Barsukova-Stuckart, M.; Ibrahim, M.; Ali, S.U.; Khan, A.A.; Kortz, U. Polyoxometalates as potent inhibitors for acetyl and butyrylcholinesterases and as potential drugs for the treatment of alzheimer's disease. *Med. Chem. Res.* **2013**, *22*, 1224–1228.

25. Briand, L.E.; Baronetti, G.T.; Thomas, H.J. The state of the art on wells-dawson heteropoly-compounds—A review of their properties and applications. *Appl. Catal. A* **2003**, *256*, 37–50.

26. Song, F.; Ding, Y.; Zhao, C. Progress in polyoxometalates-catalyzed water oxidation. *Acta Chim. Sin.* **2014**, *72*, 133–144.

27. Mizuno, N.; Nozaki, C.; Kiyoto, I.; Misono, M. Highly efficient utilization of hydrogen peroxide for selective oxygenation of alkanes catalyzed by diiron-substituted polyoxometalate precursor. *J. Am. Chem. Soc.* **1998**, *120*, 9267–9272.

28. Mizuno, N.; Nozaki, C.; Kiyoto, I.; Misono, M. Selective oxidation of alkenes catalyzed by di-iron-substituted silicotungstate with highly efficient utilization of hydrogen peroxide. *J. Catal.* **1999**, *182*, 285–288.

29. Orlandi, M.; Argazzi, R.; Sartorel, A.; Carraro, M.; Scorrano, G.; Bonchio, M.; Scandola, F. Ruthenium polyoxometalate water splitting catalyst: Very fast hole scavenging from photogenerated oxidants. *Chem. Commun.* **2010**, *46*, 3152–3154.

30. Stephan, H.; Kubeil, M.; Emmerling, F.; Mueller, C.E. Polyoxometalates as versatile enzyme inhibitors. *Eur. J. Inorg. Chem.* **2013**, 1585–1594.

31. Chen, Q.; Yang, L.; Zheng, C.; Zheng, W.; Zhang, J.; Zhou, Y.; Liu, J. Mo polyoxometalate nanoclusters capable of inhibiting the aggregation of a beta-peptide associated with alzheimer's disease. *Nanoscale* **2014**, *6*, 6886–6897.

32. Bashan, A.; Yonath, A. The linkage between ribosomal crystallography, metal ions, heteropolytungstates and functional flexibility. *J. Mol. Struct.* **2008**, *890*, 289–294.

96

33. Hungerford, G.; Hussain, F.; Patzke, G.R.; Green, M. The photophysics of europium and terbium polyoxometalates and their interaction with serum albumin: A time-resolved luminescence study. *Phys. Chem. Chem. Phys.* **2010**, *12*, 7266–7275.

34. Stroobants, K.; Moelants, E.; Ly, H.G.T.; Proost, P.; Bartik, K.; Parac-Vogt, T.N. Polyoxometalates as a novel class of artificial proteases: Selective hydrolysis of lysozyme under physiological ph and temperature promoted by a cerium(IV) keggin-type polyoxometalate. *Chem. Eur. J.* **2013**, *19*, 2848–2858.

35. Weinstein, S.; Jahn, W.; Glotz, C.; Schlunzen, F.; Levin, I.; Janell, D.; Harms, J.; Kolln, I.; Hansen, H.A.S.; Gluhmann, M.; *et al.* Metal compounds as tools for the construction and the interpretation of medium-resolution maps of ribosomal particles. *J. Struct. Biol.* **1999**, *127*, 141–151.

36. Zhang, G.; Keita, B.; Brochon, J.-C.; de Oliveira, P.; Nadjo, L.; Craescu, C.T.; Miron, S. Molecular interaction and energy transfer between human serum albumin and polyoxometalates. *J. Phys. Chem. B* **2007**, *111*, 1809–1814.

37. Zhang, G.; Keita, B.; Craescu, C.T.; Miron, S.; de Oliveira, P.; Nadjo, L. Molecular interactions between wells-dawson type polyoxometalates and human serum albumin. *Biomacromolecules* **2008**, *9*, 812–817.

38. Zheng, L.; Ma, Y.; Zhang, G.; Yao, J.; Bassil, B.S.; Kortz, U.; Keita, B.; de Oliveira, P.; Nadjo, L.; Craescu, C.T.; *et al.* Molecular interaction between a gadolinium-polyoxometalate and human serum albumin. *Eur. J. Inorg. Chem.* **2009**, 5189–5193.

39. Zheng, L.; Ma, Y.; Zhang, G.; Yao, J.; Keita, B.; Nadjo, L. A multitechnique study of europium decatungstate and human serum albumin molecular interaction. *Phys. Chem. Chem. Phys.* **2010**, *12*, 1299–1304.

40. Absillis, G.; Cartuyvels, E.; van Deun, R.; Parac-Vogt, T.N. Hydrolytic cleavage of an rna-model phosphodiester catalyzed by a highly negatively charged polyoxomolybdate $Mo_7O_{24}^{6-}$ cluster. *J. Am. Chem. Soc.* **2008**, *130*, 17400–17408.

41. Cartuyvels, E.; Absillis, G.; Parac-Vogt, T.N. Questioning the paradigm of metal complex promoted phosphodiester hydrolysis: $Mo_7O_{24}^{6-}$ polyoxometalate cluster as an unlikely catalyst for the hydrolysis of a DNA model substrate. *Chem. Commun.* **2008**, 85–87.

42. Cartuyvels, E.; van Hecke, K.; van Meervelt, L.; Gorller-Walrand, C.; Parac-Vogt, T.N. Structural characterization and reactivity of gamma-octamolybdate functionalized by proline. *J. Inorg. Biochem.* **2008**, *102*, 1589–1598.

43. Van Lokeren, L.; Cartuyvels, E.; Absillis, G.; Willem, R.; Parac-Vogt, T.N. Phosphoesterase activity of polyoxomolybdates: Diffusion ordered NMR spectroscopy as a tool for obtaining insights into the reactivity of polyoxometalate clusters. *Chem. Commun.* **2008**, 2774–2776.

44. Steens, N.; Ramadan, A.M.; Absillis, G.; Parac-Vogt, T.N. Hydrolytic cleavage of DNA-model substrates promoted by polyoxovanadates. *Dalton Trans.* **2010**, *39*, 585–592.

45. Steens, N.; Ramadan, A.M.; Parac-Vogt, T.N. When structural and electronic analogy leads to reactivity: The unprecedented phosphodiesterase activity of vanadates. *Chem. Commun.* **2009**, 965–967.

46. Vanhaecht, S.; Absillis, G.; Parac-Vogt, T.N. Hydrolysis of DNA model substrates catalyzed by metal-substituted wells-dawson polyoxometalates. *Dalton Trans.* **2012**, *41*, 10028–10034.

47. Ho, P.H.; Mihaylov, T.; Pierloot, K.; Parac-Vogt, T.N. Hydrolytic activity of vanadate toward serine-containing peptides studied by kinetic experiments and dft theory. *Inorg. Chem.* **2012**, *51*, 8848–8859.

48. Ho, P.H.; Stroobants, K.; Parac-Vogt, T.N. Hydrolysis of serine-containing peptides at neutral ph promoted by MoO_4^{2-} oxyanion. *Inorg. Chem.* **2011**, *50*, 12025–12033.

49. Hong Giang, T.L.; Absillis, G.; Bajpe, S.R.; Martens, J.A.; Parac-Vogt, T.N. Hydrolysis of dipeptides catalyzed by a zirconium(IV)-substituted lindqvist type polyoxometalate. *Eur. J. Inorg. Chem.* **2013**, 4601–4611.

50. Hong Giang, T.L.; Absillis, G.; Parac-Vogt, T.N. Amide bond hydrolysis in peptides and cyclic peptides catalyzed by a dimeric Zr(IV)-substituted keggin type polyoxometalate. *Dalton Trans.* **2013**, *42*, 10929–10938.

51. Absillis, G.; Parac-Vogt, T.N. Peptide bond hydrolysis catalyzed by the wells-dawson $Zr(\alpha_2-P_2W_{17}O_{61})_2$ polyoxometalate. *Inorg. Chem.* **2012**, *51*, 9902–9910.

52. Stroobants, K.; Absillis, G.; Moelants, E.; Proost, P.; Parac-Vogt, T.N. Regioselective hydrolysis of human serum albumin by Zr^{IV}-substituted polyoxotungstates at the interface of positively charged protein surface patches and negatively charged amino acid residues. *Chem. Eur. J.* **2014**, *20*, 3894–3897.

53. Stroobants, K.; Goovaerts, V.; Absillis, G.; Bruylants, G.; Moelants, E.; Proost, P.; Parac-Vogt, T.N. Molecular origin of the hydrolytic activity and fixed regioselectivity of a zriv-substituted polyoxotungstate as artificial protease. *Chem. Eur. J.* **2014**, *20*, 9567–9577.

54. Vanhaecht, S.; Absillis, G.; Parac-Vogt, T.N. Amino acid side chain induced selectivity in the hydrolysis of peptides catalyzed by a Zr(IV)-substituted wells-dawson type polyoxometalate. *Dalton Trans.* **2013**, *42*, 15437–15446.

55. Ajloo, D.; Behnam, H.; Saboury, A.A.; Mohamadi-Zonoz, F.; Ranjbar, B.; Moosavi-Movahedi, A.A.; Hasani, Z.; Alizadeh, K.; Gharanfoli, M.; Amani, M.; *et al.* Thermodynamic and structural studies on the human serum albumin in the presence of a polyoxometalate. *Bull. Korean Chem. Soc.* **2007**, *28*, 730–736.

56. Xiao, J.B.; Chen, X.Q.; Jiang, X.Y.; Hilczer, M.; Tachiya, M. Probing the interaction of trans-resveratrol with bovine serum albumin: A fluorescence quenching study with tachiya model. *J. Fluoresc.* **2008**, *18*, 671–678.

57. Lakowicz, J.R. *Principles of Fluorescence Spectroscopy*; Springer: New York, NY, USA, 2006.

58. Goovaerts, V.; Stroobants, K.; Absillis, G.; Parac-Vogt, T.N. Molecular interactions between serum albumin proteins and keggin type polyoxometalates studied using luminescence spectroscopy. *Phys. Chem. Chem. Phys.* **2013**, *15*, 18378–18387.

59. Sugio, S.; Kashima, A.; Mochizuki, S.; Noda, M.; Kobayashi, K. Crystal structure of human serum albumin at 2.5 angstrom resolution. *Protein Eng.* **1999**, *12*, 439–446.

60. Stroobants, K.; Saadallah, D.; Bruylants, G.; Parac-Vogt, T.N. Thermodynamic study of the interaction between hen egg white lysozyme and Ce(IV)-keggin polyoxotungstate as artificial protease. *Phys. Chem. Chem. Phys.* **2014**, *16*, 21778–21787.

61. Zhang, G.; Keita, B.; Craescu, C.T.; Miron, S.; de Oliveira, P.; Nadjo, L. Polyoxometalate binding to human serum albumin: A thermodynamic and spectroscopic approach. *J. Phys. Chem. B* **2007**, *111*, 11253–11259.

62. Kato, C.N.; Shinohara, A.; Hayashi, K.; Nomiya, K. Syntheses and X-ray crystal structures of zirconium(IV) and hafnium(IV) complexes containing monovacant wells-dawson and keggin polyoxotungstates. *Inorg. Chem.* **2006**, *45*, 8108–8119.

63. Lv, H.J.; Song, J.; Geletii, Y.V.; Vickers, J.W.; Sumliner, J.M.; Musaev, D.G.; Kogerler, P.; Zhuk, P.F.; Bacsa, J.; Zhu, G.B.; *et al.* An exceptionally fast homogeneous carbon-free cobalt-based water oxidation catalyst. *J. Am. Chem. Soc.* **2014**, *136*, 9268–9271.

64. Song, F.Y.; Ding, Y.; Ma, B.C.; Wang, C.M.; Wang, Q.; Du, X.Q.; Fua, S.; Song, J. $K_7[Co^{III}Co^{II}(H_2O)W_{11}O_{39}]$: A molecular mixed-valence keggin polyoxometalate catalyst of high stability and efficiency for visible light-driven water oxidation. *Energy Environ. Sci.* **2013**, *6*, 1170–1184.

65. Song, J.; Luo, Z.; Britt, D.K.; Furukawa, H.; Yaghi, O.M.; Hardcastle, K.I.; Hill, C.L. A multiunit catalyst with synergistic stability and reactivity: A polyoxometalate-metal organic framework for aerobic decontamination. *J. Am. Chem. Soc.* **2011**, *133*, 16839–16846.

66. Zerbe, O. *Bionmr in Drug Research*; Wiley-VCH: Weinheim, Germany, 2003; Volume 16.

67. Kholdeeva, O.A.; Maksimov, G.M.; Maksimovskaya, R.I.; Vanina, M.P.; Trubitsina, T.A.; Naumov, D.Y.; Kolesov, B.A.; Antonova, N.S.; Carbo, J.J.; Poblet, J.M.; *et al.* Zr(IV)-monosubstituted keggin-type dimeric polyoxometalates: Synthesis, characterization, catalysis of H_2O_2-based oxidations, and theoretical study. *Inorg. Chem.* **2006**, *45*, 7224–7234.

68. Kholdeeva, O.A.; Maksimovskaya, R.I. Titanium- and zirconium-monosubstituted polyoxometalates as molecular models for studying mechanisms of oxidation catalysis. *J. Mol. Catal. A* **2007**, *262*, 7–24.

69. Nomiya, K.; Saku, Y.; Yamada, S.; Takahashi, W.; Sekiya, H.; Shinohara, A.; Ishimaru, M.; Sakai, Y. Synthesis and structure of dinuclear hafnium(IV) and zirconium(IV) complexes sandwiched between 2 mono-lacunary alpha-keggin polyoxometalates. *Dalton Trans.* **2009**, 5504–5511; doi:10.1039/b902296a.

70. Carabineiro, H.; Villanneau, R.; Carrier, X.; Herson, P.; Lemos, F.; Ribeiro, F.R.; Proust, A.; Che, M. Zirconium-substituted isopolytungstates: Structural models for zirconia-supported tungsten catalysts. *Inorg. Chem.* **2006**, *45*, 1915–1923.

71. Gaunt, A.J.; May, I.; Collison, D.; Holman, K.T.; Pope, M.T. Polyoxometal cations within polyoxometalate anions. Seven-coordinate uranium and zirconium heteroatom groups in $[(UO_2)_{12}(\mu_3\text{-}O)_4(\mu_2\text{-}H_2O)_{12}(P_2W_{15}O_{56})_4]^{32-}$ and $[Zr_4(\mu_3\text{-}O)_2(\mu_2\text{-}OH)_2(H_2O)_4(P_2W_{16}O_{59})_2]^{14-}$. *J. Mol. Struct.* **2003**, *656*, 101–106.

Synthesis, Characterization and Study of Liquid Crystals Based on the Ionic Association of the Keplerate Anion [Mo$_{132}$O$_{372}$(CH$_3$COO)$_{30}$(H$_2$O)$_{72}$]$^{42-}$ and Imidazolium Cations

Nancy Watfa, Sébastien Floquet, Emmanuel Terazzi, William Salomon, Laure Guénée,

Kerry Lee Buchwalder, Akram Hijazi, Daoud Naoufal, Claude Piguet and Emmanuel Cadot

Abstract: A series of eight new materials based on the ionic association between 1-methyl-3-alkylimidazolium cations and the nanometric anionic Keplerate [Mo$_{132}$O$_{372}$(CH$_3$COO)$_{30}$(H$_2$O)$_{72}$]$^{42-}$ has been prepared and characterized in the solid state. The liquid crystal properties of these materials were investigated by the combination of Polarized Optical Microscopy, Differential Scanning Calorimetry and Small-angle X-Ray Diffraction showing a self-organization in lamellar (L) mesophases for the major part of them. From the interlamellar spacing h and the intercluster distance a_{hex}, we demonstrated that the cations are not randomly organized around the anionic cluster and that the alkyl chains of the cations are certainly folded, which limits the van der Waals interactions between the cations within the liquid crystal phase and therefore harms the quality of the mesophases.

Reprinted from *Inorganics*. Cite as: Watfa, N.; Floquet, S.; Terazzi, E.; Salomon, W.; Guénée, L.; Buchwalder, K.L.; Hijazi, A.; Naoufal, D.; Piguet, C.; Cadot, E. Synthesis, Characterization and Study of Liquid Crystals Based on the Ionic Association of the Keplerate Anion [Mo$_{132}$O$_{372}$(CH$_3$COO)$_{30}$(H$_2$O)$_{72}$]$^{42-}$ and Imidazolium Cations. *Inorganics* **2015**, *3*, 246-266.

1. Introduction

Liquid crystals constitute a fascinating example of functional self-assembled materials. Incorporation of some inorganic components into liquid crystalline phases appears particularly relevant to the elaboration of synergistic multifunctional materials according to a "bottom-up" approach [1]. Polyoxometalates (POMs), often described as polyanionic molecular oxides, exhibit various properties in all domains of chemistry [2–8]. Dietz and Wu demonstrated that the association of various POMs with phosphonium or ammonium derivatives lead to ionic liquids or ionic mesomorphic self-assemblies [9–17]. In this context, the "Keplerates" anions with the general formula [(Mo$_6$O$_{21}$)$_{12}$(linker)$_{30}$(Ligand)$_{30}$(H$_2$O)$_{72}$]$^{n-}$ (Figure 1) constitute an unique class of nanoscopic hollow spherical clusters which exhibit singular structural features which easily exchange the internal ligands/functionalities [18,19]. They therefore constitute a fascinating class of molecular materials which display a large variety of properties such as magnetism [20,21], catalysis [22,23], electric conductivity [24], non-linear optics [25,26] and are able to serve as nano-reactor for chemical reactions, such as stoichiometric transformations of substrates with unusual selectivity [27]. Several research groups have been interested to exploit this cluster in the context of materials science where many efforts have been done to organize and stabilize such materials. For example, the incorporation of Keplerates into suitable matrices such as polymers [28] polyelectrolyte [29] silica [30] or organic cations [31–34] can enhance stability and induce synergistic functionalities between POMs and matrices. For instance, the counter-cations of the Keplerate anion $[Mo_{132}O_{372}(CH_3COO)_{30}(H_2O)_{72}]^{42-}$, noted hereafter Mo132, can be replaced with dioctadecyl-dimethylammonium cations (DODA$^+$) resulting in functional nanodevices [35]. Motivated by this challenge, we reported that the wrapping of Keplerate-type capsules with DODA$^+$ cations lead to the formation of the materials (DODA)$_{36}$(NH$_4$)$_6$[Mo$_{132}$O$_{372}$(CH$_3$COO)$_{30}$(H$_2$O)$_{72}$]·75H$_2$O, (DODA)$_{44}$(NH$_4$)$_{14}$[Mo$_{132}$O$_{312}$S$_{60}$(SO$_4$)$_{23}$(H$_2$O)$_{86}$] and (DODA)$_{56}$(NH$_4$)$_{16}$[Mo$_{132}$O$_{312}$S$_{60}$(SO$_4$)$_{30}$(H$_2$O)$_{72}$] which exhibit liquid crystalline properties [36,37]. Interestingly, the DODA$^+$ cations are not randomly distributed on the surface of the Keplerate capsule, but organized within the sheets of Keplerates. In the following of these two previous works, we studied the association of another highly charged POM, namely $[K_2Na_xLi_yH_z\{Mo_4O_4S_4(OH)_2(H_2O)_3\}_2(P_8W_{48}O_{184})]^{(34-x-y-z)-}$ with a series of 1-methyl-3-alkylimidazolium cations [38], which present the particularities to be easily modified and to be easily prepared on a gram scale. This work evidenced that the chains of the cations are organized perpendicularly to the sheets of the POMs and that the structural parameters of the liquid crystal phases can be finely tuned as a function of the alkyl chain length of the imidazolium cations [38]. More recently, we evidenced that the Keplerate Mo132 interacts with 1-methyl-3-ethylimidazolium cations in aqueous phase with an association constant of 5800, a value significantly higher than that measured with tetramethylammonium cation (1550) taken as a model for the DODA$^+$ cation [39].

We report here on the synthesis and the characterizations of materials built from the ionic association of Keplerate Mo132 with 1-methyl-3-alkylimidazolium cations, noted mimC$_n^+$ (n = 12, 14, 16, 18, 20, see Figure 1). After having characterized the materials, the liquid crystal properties are studied by polarized optical microscopy (TD-POM), differential scanning calorimetry (DSC) and small angle variable temperature X-ray diffraction (SA-XRD) in order to determine if it is possible (i) to build liquid crystals materials resulting from the ionic associations Mo132 / 1-methyl-3-alkylimidazolium and (ii) to change the nature of the liquid crystal phases upon replacement of DODA$^+$ cations with 1-methyl-3-alkylimidazolium cations, for which the interaction with Mo132 is stronger in solution.

2. Results and Discussion

2.1. Synthesis and Characterization of the Mo132-Based Materials

2.1.1. Synthesis and Characterizations

The Keplerate-based materials were prepared by mixing an aqueous solution of the precursor compound abbreviated NH$_4$–Mo132 with a large excess (ca. 170 equivalents per Mo132 capsule) of 1-methyl-3-alkylimidazolium cations, noted mimC$_n^+$ (n = 12, 14, 16, 18, 20), in chloroform in order to replace the maximum of NH$_4^+$ countercations. After stirring the mixture for about 1 h, the Keplerate capsule Mo132 was totally transferred into the chloroform phase, indicating a rapid and quantitative transfer. The target materials were precipitated with a good yield by addition of ethanol into the chloroform phase and finally filtered, washed with ethanol and dried in air (see experimental section for more details). We were also interested to investigate the preparation of mixed salts of mimC$_{12}^+$ and mimC$_{20}^+$ by using different ratios during the synthesis. The materials were obtained as brown-black soft solids, no longer soluble in water, but soluble in organic solvents such as chloroform, acetonitrile, toluene and ether. These compounds were characterized by FT-IR, TGA, EDX and elemental analyses.

$$[Mo_{132}O_{372}(CH_3COO)_{30}(H_2O)_{72}]^{42-}$$

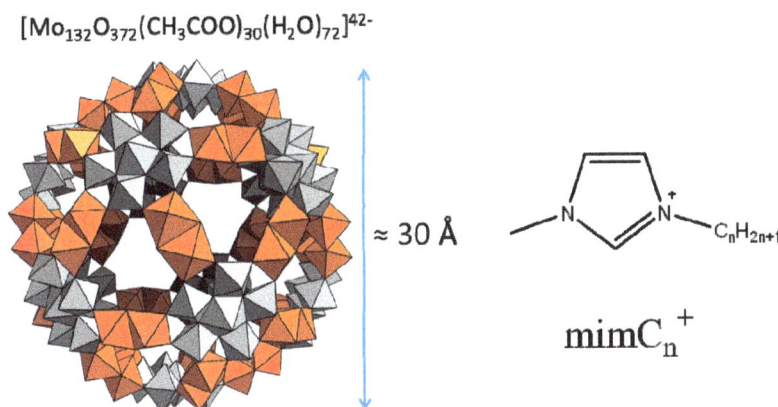

$$\approx 30\ \text{Å}$$

$$\text{mimC}_n^+$$

Figure 1. Representation of the Keplerate nano capsule Mo$_{132}$ and the imidazolium cations used in this study. The grey polyhedra correspond to the fragments {Mo$_6$O$_{21}$}, whereas the polyhedra in orange correspond to the linkers {Mo$_2$O$_4$}. Acetates ligands are found inside the cavity of the capsule and are not shown in the figure.

FT-IR spectra were performed using Diamond ATR technique and an ATR correction was applied. As shown in Figure 2 for the compound (mimC$_{18}$)$_{37}$–Mo$_{132}$, the comparison of its IR spectrum with those of the precursors confirm the conservation of the Mo$_{132}$ cluster and its association with the organic cation since the IR-spectrum exhibits the characteristic bands of both components. In addition the band at 1403 cm^{-1}, attributed to NH$_4^+$ counter-cations, is still present in the final material indicating that not all NH$_4^+$ have been replaced by the organic cations in agreement with previous studies [36,37].

Based on the data of the elemental analyses (C, H, N, Mo), EDX and TGA, the structural formula of the obtained materials is established. The results are summarized in Table 1 and more details are available in the experimental section. Firstly, EDX measurements evidence only the presence of the Mo and reveal the absence of Bromide ions, the fingerprint of the presence of the starting imidazolium salt. The number of acetate ligands was 30 for all synthesized compound. It should be noted that, despite the large excess of the organic cations, the replacement of the ammonium cations was not complete, in agreement with FT-IR spectra. The number of interacting organic cations with the Keplerate ion was found between 36 and 41, which is consistent with the previously published result with DODA salts [36,37]. This result suggests that some NH$_4^+$ are probably trapped within the **Mo$_{132}$** nanocapsule. It is worth noting that by mixing imidazolium salts bearing C$_{12}$ and C$_{20}$ alkyl chains, a purely statistical distribution is observed in the final compound where the number of each organic cations is controlled with the initial ratio. This demonstrates the possibility to tune the final composition of the mixed surfactant encapsulated clusters, a key parameter for the design of functional materials. Finally, the thermogravimetric analyses estimate the total amount of water molecules.

Table 1. Molecular formula of the Mo_{132}-based materials established from elemental analysis and TGA.

Samples	Estimated Molecular Formula
$(mimC_{12})_{36}-Mo_{132}$	$(mimC_{12}H_{25})_{36}(NH_4)_6[Mo_{132}O_{372}(CH_3COO)_{30}(H_2O)_{72}]\cdot38H_2O$
$(mimC_{14})_{38}-Mo_{132}$	$(mimC_{14}H_{29})_{38}(NH_4)_4[Mo_{132}O_{372}(CH_3COO)_{30}(H_2O)_{72}]\cdot30H_2O$
$(mimC_{16})_{41}-Mo_{132}$	$(mimC_{16}H_{33})_{41}(NH_4)_1[Mo_{132}O_{372}(CH_3COO)_{30}(H_2O)_{72}]\cdot38H_2O$
$(mimC_{18})_{37}-Mo_{132}$	$(mimC_{18}H_{37})_{37}(NH_4)_5[Mo_{132}O_{372}(CH_3COO)_{30}(H_2O)_{72}]\cdot48H_2O$
$(mimC_{20})_{37}-Mo_{132}$	$(mimC_{20}H_{41})_{37}(NH_4)_5[Mo_{132}O_{372}(CH_3COO)_{30}(H_2O)_{72}]\cdot43H_2O$
$(mimC_{12})_{20}(mimC_{20})_{20}-Mo_{132}$	$(mimC_{12}H_{25})_{20}(mimC_{20}H_{33})_{20}(NH_4)_2[Mo_{132}O_{372}(CH_3COO)_{30}(H_2O)_{72}]\cdot38H_2O$
$(mimC_{12})_{33}(mimC_{20})_7-Mo_{132}$	$(mimC_{12}H_{25})_{33}(mimC_{20}H_{33})_7(NH_4)_2[Mo_{132}O_{372}(CH_3COO)_{30}(H_2O)_{72}]\cdot48H_2O$
$(mimC_{12})_8(mimC_{20})_{33}-Mo_{132}$	$(mimC_{12}H_{25})_8(mimC_{20}H_{33})_{33}(NH_4)_1$ $[Mo_{132}O_{372}(CH_3COO)_{30}(H_2O)_{72}]\cdot38H_2O$

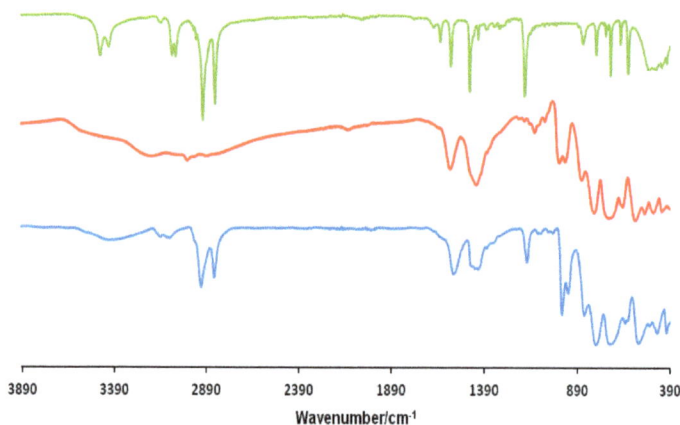

Figure 2. FT-IR spectra of $(mimC_{18})Br$ (**green**), NH_4-Mo_{132} (**red**) and $(mimC_{18})_{37}-Mo_{132}$ (**blue**).

2.1.2. Thermal Stability of the Mo_{132}-Based Materials

Chemical stability of the **Mo_{132}**-based materials was assessed by TGA under N_2 and O_2 and FT-IR measurements at variable temperature under N_2 and under O_2. TGA traces recorded under oxygen display a first stage between 20 °C and 180 °C corresponding to the removal of all the water molecules (hydration and coordination) followed by a large weight loss in the 250–800 °C temperature range, which is assigned to the decomposition of the cations and of the acetate ligands of the **Mo_{132}** cluster to give molybdenum oxides. This second step translates into a thermal stability of about 250 °C for our **Mo_{132}**-based materials.

Variable temperature FT-IR studies were performed on the compound **$(mimC_{18})_{37}-Mo_{132}$** arbitrarily chosen as a reference compound. The FT-IR spectra were recorded in diffuse reflectance mode on a sample of **$(mimC_{18})_{37}Mo_{132}$** dispersed into a KBr matrix. The experiments were done either under air or under a flow of nitrogen between 20 °C and 400 °C with a heating/cooling rate of 2 °C/min. The results obtained under air atmosphere are shown in Figure 3 for different temperatures. In these conditions, the disappearance of the bands characteristic of the water molecules around 3400 cm^{-1} and 1640 cm^{-1} below 180 °C agrees with TGA. If we focus on the vibration bands assigned to the Keplerate capsules in the 1000–400 cm^{-1} range, no significant modification are observed up to ≈125 °C. In contrast, dramatic changes are observed between 125 °C and 200 °C, especially for the strong bands located at 855, 727 and 574 cm^{-1}, which almost totally disappear, while the two bands assigned to the Mo=O stretching vibration located at 975 and 939 cm^{-1} give only one strong band at 953 cm^{-1} at 190 °C. Concomitantly, the color of the compound turns to blue and this phenomenon is not reversible upon returning room temperature, which indicates that the Keplerate capsule in the compound **(mimC$_{18}$)$_{37}$Mo$_{132}$** is not stable in air above 125 °C.

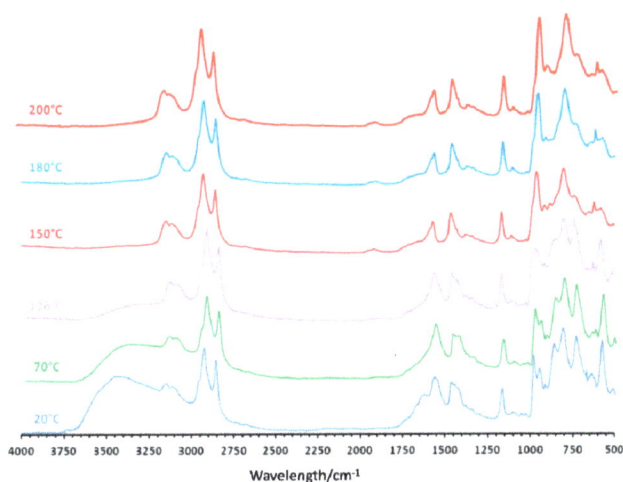

Figure 3. Selected FT-IR spectra recorded at various temperatures for compound **(mimC$_{18}$)$_{37}${Mo$_{132}$}** under air.

Similar experiments were performed under a flow of nitrogen and the results obtained under nitrogen are shown in Figure 4. In these conditions, the behavior is totally different. The removal of water molecules is observed between 20 °C and 180 °C, but only small changes in the Keplerate skeleton are observed on the FT-IR spectra, in contrast with experiment performed in air. From 20 °C to 250 °C, the main evolution observed concerns the vibration band at 855 cm^{-1}, for which the intensity decreases with increasing temperature, but this modification is reversible when returning room temperature after re-hydration of the material. Finally, heating the compound at temperature higher than 250 °C provokes the degradation of the organic cation and of the acetate ligands in agreement with TGA. This is associated with a strong modification of the Keplerate.

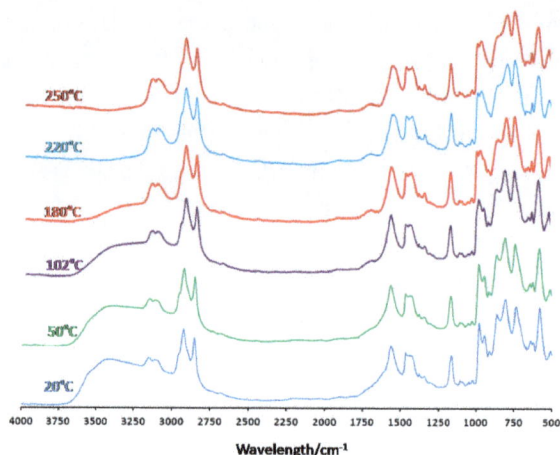

Figure 4. Selected FT-IR spectra recorded at various temperatures for compound **(mimC$_{18}$)$_{37}${Mo$_{132}$}** under nitrogen.

2.2. Liquid Crystal Properties

2.2.1. Polarized Optical Microscopy

The liquid crystal properties of all the compounds were examined by temperature dependent polarized optical microscopy (TD-POM) between two glass plates. Under heating our materials from room temperature to their decomposition temperatures, the TD-POM revealed the formation of relatively fluid birefringent and homogenous textures characteristic of the liquid crystalline nature for the samples **(mimC$_{12}$)$_{36}$–Mo$_{132}$, (mimC$_{14}$)$_{38}$–Mo$_{132}$, (mimC$_{16}$)$_{41}$–Mo$_{132}$, (mimC$_{18}$)$_{37}$–Mo$_{132}$,** and **(mimC$_{20}$)$_{37}$–Mo$_{132}$** in the 100–240 °C temperature range as shown in Figure 5. Furthermore, the samples did not show characteristic textures due to their decomposition before reaching the isotropic state and consequently it is difficult to identify the type of the mesophase from the TD-POM pictures. However, the three mixed compounds, **(mimC$_{12}$)$_{20}$(mimC$_{20}$)$_{20}$–Mo$_{132}$, (mimC$_{12}$)$_{33}$(mimC$_{20}$)$_7$–Mo$_{132}$** and **(mimC$_{12}$)$_8$(mimC$_{20}$)$_{33}$–Mo$_{132}$** do not show birefringent textures, even at high temperature, indicating the lack of liquid crystal properties for these compounds.

2.2.2. Differential Scanning Calorimetry

The DSC curves of the Keplerate-based materials were recorded under nitrogen at a scan rate of 5 °C/min between −40 °C and 220–260 °C. At least three thermal cycles were achieved and the obtained results for the second thermal cycles are shown in Figure 6, while the thermodynamic parameters are listed in Table 2.

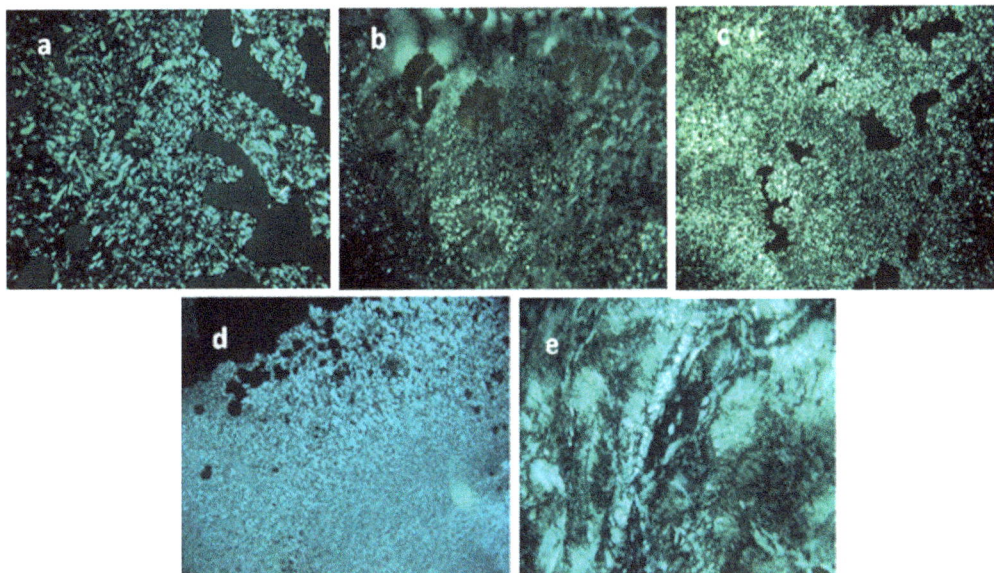

Figure 5. Polarized Optical Microphotographs of **(mimC$_{12}$)$_{36}$–Mo$_{132}$** at 113 °C (**a**), **(mimC$_{14}$)$_{38}$–Mo$_{132}$** at 220 °C (**b**), **(mimC$_{16}$)$_{41}$–Mo$_{132}$** at 160 °C (**c**), **(mimC$_{18}$)$_{37}$–Mo$_{132}$** at 100 °C (**d**), and **(mimC$_{20}$)$_{37}$–Mo$_{132}$** at 200 °C (**e**).

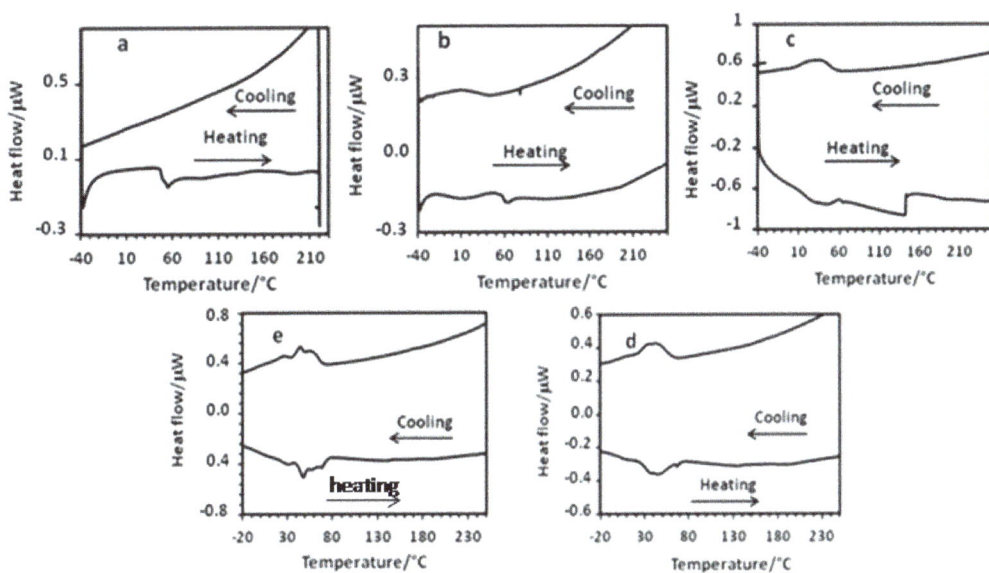

Figure 6. DSC traces of the second thermal cycle recorded under N$_2$ (5 °C min^{-1}) for **(mimC$_{12}$)$_{36}$–Mo$_{132}$** (**a**), **(mimC$_{14}$)$_{38}$–Mo$_{132}$** (**b**), **(mimC$_{16}$)$_{41}$–Mo$_{132}$** (**c**), **(mimC$_{18}$)$_{37}$–Mo$_{132}$** (**d**), **(mimC$_{20}$)$_{37}$–Mo$_{132}$** (**e**).

Table 2. Temperatures, melting enthalpy and entropy changes of the phase transitions between the glassy state and the liquid crystal phases observed for the materials exhibiting liquid crystalline properties. Temperatures are given for the peaks observed by DSC measurements during the second heating process.

Compounds	$T/^\circ C$ (Heating Mode)	MM g/mol	$n(CH_2)$	ΔH_m /kJ·mol^{-1}	ΔS_m /J·mol^{-1}·K^{-1}	$\Delta H'_m$ [a] /J·mol^{-1}	$\Delta S'_m$ [a] /J·mol^{-1}·K^{-1}
$(mimC_{12})_{36}$–Mo_{132}	+55	30982	396	−4.3	−13	−10.95	−0.03
$(mimC_{14})_{38}$–Mo_{132}	+13	32387	494	−46.3	−162	−94	−0.33
$(mimC_{16})_{41}$–Mo_{132}	+31	34889	615	−204.1	−671	−331.8	−1.1
$(mimC_{18})_{37}$–Mo_{132}	+43	35059	629	−158.5	−501	−252	−0.8
$(mimC_{20})_{37}$–Mo_{132}	+46	36006	703	−163.8	−514	−233	0.73
$DODA_{36}$–Mo_{132} [37]	+9	42982	1224	−849.0	−3009	−639	−2.45
$DODA_{44}$–$Mo_{132}S_{60}$ [36]	+9	47838	1548	−1333.0	−4724	−918	−3.25
$DODA_{56}$–$Mo_{132}S_{60}$ [36]	+9	55357	1904	−1649.0	−5848	−866	−3.07
DODACl	+18	586.5	34	−83.1	−278.5	−2445	−8.19

[a] $\Delta Y'_m = \Delta Y_m/n(CH_2)$ (Y = H, S) where $n(CH_2)$ corresponds to the total number of methylene groups of the alkyl chains borne by the cations.

Globally, the DSC curves exhibit very broad and often composite first-order phase transitions for the melting of the solid to give the mesophase. This contrasts with the relatively sharp first-order phase transitions reported for DODA$^+$ salts of Keplerate Mo_{132} and $Mo_{132}S_{60}$ [36,37]. These results suggest that the local molecular or supramolecular mobility is insufficient to allow the development of a network of intermolecular interactions sufficiently regular to obtain a well-developed mesophase . In addition, excepted for $(mimC_{12})_{36}$–Mo_{132}, the melting transitions appear reversible for the other compounds even if it can be broader for some compounds such as $(mimC_{14})_{38}$–Mo_{132} for example.

The average melting transition temperatures, which are determined in the heating mode, are summarized in Table 2. In short, disregarding the compound $(mimC_{12})_{36}$–Mo_{132}, the melting transition temperatures increase roughly linearly in the series $(mimC_{14})_{38}$–Mo_{132}, $(mimC_{16})_{41}$–Mo_{132}, $(mimC_{18})_{37}$–Mo_{132}, and $(mimC_{20})_{37}$–Mo_{132} from +13 °C to +46 °C, when the alkyl chain length of the imidazolium cation increases, and are significantly different from those obtained with the three previous DODA$^+$ salts (+9 °C) [36,37].

The corresponding melting enthalpy and entropy changes, ΔH_m and ΔS_m, are also gathered in Table 2. A comparison of these values remains difficult since the number of cations can vary in the series of Mo_{132}-based materials. The weighted $\Delta H'_m$ and $\Delta S'_m$ values correspond to the ratios $\Delta H_m/n(CH_2)$ and $\Delta S_m/n(CH_2)$ respectively, where $n(CH_2)$ is the total number of methylene groups in the compounds, and is more convenient. The plot of $-\Delta H'_m$ versus $-\Delta S'_m$ including data previously reported for DODA salts and DODACl alone is given in Figure 7. It displays a linear relationship typical for H/S compensation [40] and evidences two sets of points : one corresponding to the imidazolium salts of Mo_{132} of this study, $(mimC_{14})_{38}$–Mo_{132}, $(mimC_{16})_{41}$–Mo_{132}, $(mimC_{18})_{37}$–Mo_{132}, and $(mimC_{20})_{37}$–Mo_{132} and one corresponding to the DODA$^+$ salts $(DODA)_{36}$–Mo_{132}, $(DODA)_{44}$–$Mo_{132}S_{60}$, $(DODA)_{56}$–$Mo_{132}S_{60}$ bearing liquid crystalline properties [36,37], whereas DODACl appears clearly different. For DODACl, the alkyl chains are weakly constrained by the presence of the counter chloride anion and the melting of the chains follows the trend expected for simple alkanes [40]. Figure 7 suggests that the alkyl chains of the cations associated to Mo_{132} are strongly perturbed by the presence of the Keplerate capsule. From a macroscopic point of view, it results in a drastic lowering of the fluidity of the samples. This phenomenon is more pronounced for the series of imidazolium salts than for the three $DODA^+$ salts. Furthermore, as seen in the insert of the Figure 7, the ratio $\Delta H'_m/\Delta S'_m$ appears globally correlated for the series of imidazolium salts of Mo_{132}. The magnitude of the $\Delta H_m/nCH_2$ and $\Delta S_m/nCH_2$ ratios seems to be strongly and linearly correlated with the average closeness of the $-CH_2-$ with the surface of the anions. In other words, the $-CH_2-$ located closer to the surface are more constrained than those which are more distant. It is as if the $-CH_2-$ located in a crown close to the surface were "frozen" and consequently no more available for melting. This may explain why the fluidity of the samples bearing long aliphatic chains is higher.

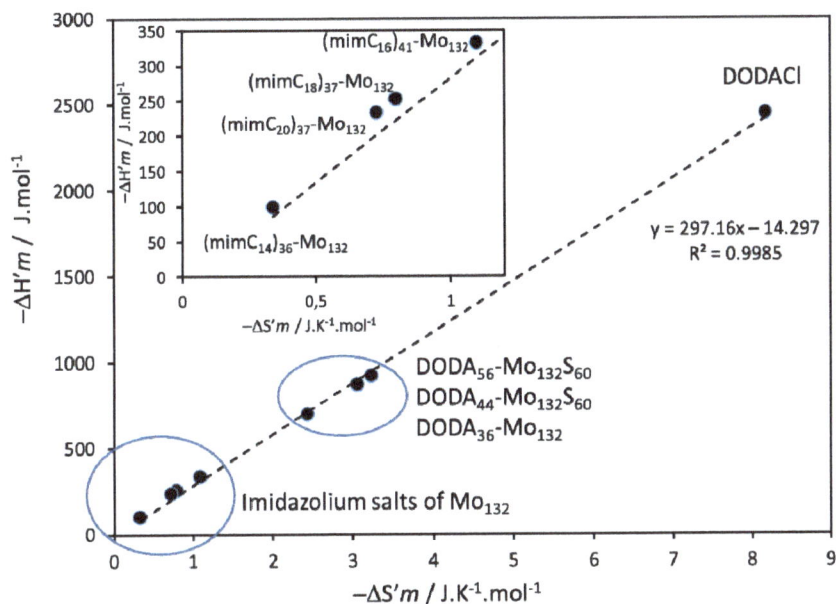

Figure 7. $\Delta H'_m$ *versus* $\Delta S'_m$ plot.

2.2.3. Small-Angle X-ray Diffraction Studies

Small-angle X-ray Diffraction (SA-XRD) experiments (from −40 to 200 °C temperature range) were carried out to elucidate the nature of the liquid crystalline phases. Arbitrarily, we have chosen to present the SA-XRD patterns obtained at 200 °C, when possible, for two main reasons: (i) to compare with the previous data published at the same temperature [36,37], (ii) at this temperature the fluidity is relatively high and results in sharper peaks. The resulting SA-XRD patterns are depicted in Figure 8, whereas the corresponding data are summarized in Table 3.

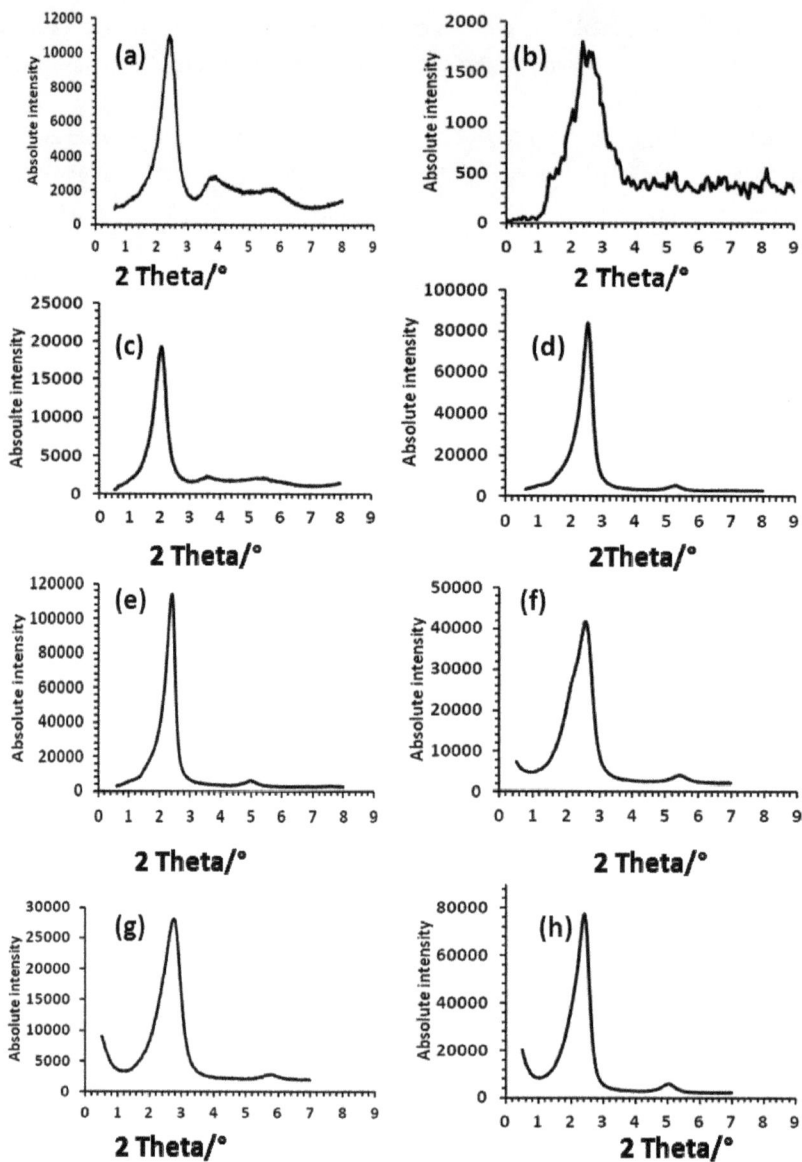

Figure 8. SA-XRD patterns recorded at 200 °C for **(mimC$_{12}$)$_{36}$–Mo$_{132}$ (a)**, **(mimC$_{14}$)$_{38}$–Mo$_{132}$ (b)**, **(mimC$_{16}$)$_{41}$–Mo$_{132}$ (c)**, **(mimC$_{18}$)$_{37}$–Mo$_{132}$ (d)**, **(mimC$_{20}$)$_{37}$–Mo$_{132}$ (e)**, **(mimC$_{12}$)$_{20}$(mimC$_{20}$)$_{20}$–Mo$_{132}$ (f)**, **(mimC$_{12}$)$_{33}$(mimC$_{20}$)$_{7}$–Mo$_{132}$** at 140 °C **(g)**, **(mimC$_{12}$)$_{8}$(mimC$_{20}$)$_{33}$–Mo$_{132}$ (h)**.

Table 3. Indexation at 200 °C during the cooling mode for the reflections detected in the liquid-crystalline phase by SA-XRD.

Compounds	d_{hkl} Measured/Å	I/a.u.	Indexation	Cell Parameters/Å
(mimC$_{12}$)$_{36}$–Mo$_{132}$	36.90	VS	c	
	23.19	MW (broad)	c	
	15.45	MW (broad)	c	
(mimC$_{14}$)$_{38}$–Mo$_{132}$ [a]	35.57	S	001	$h = 35.57$ $a_{hex} = 41.79$
(mimC$_{16}$)$_{41}$–Mo$_{132}$	40.71	VS	c	
	24.55	MW (broad)	c	
	16.52	MW (broad)	c	
(mimC$_{18}$)$_{37}$–Mo$_{132}$	32.78	VS	001	$h = 32.78$
	16.34	W	002	$a_{hex} = 45.29$
(mimC$_{20}$)$_{37}$–Mo$_{132}$	35.34	VS	001	$h = 35.34$
	17.67	W	002	$a_{hex} = 44.21$
(mimC$_{12}$)$_{20}$(mimC$_{20}$)$_{20}$–Mo$_{132}$	32.91	VS	001	$h = 32.91$
	16.26	W	002	$a_{hex} = 44.98$
(mimC$_{12}$)$_{33}$(mimC$_{20}$)$_7$–Mo$_{132}$ [b]	31.34	VS	001	$h = 31.34$
	15.22	W	002	$a_{hex} = 45.23$
(mimC$_{12}$)$_8$(mimC$_{20}$)$_{33}$–Mo$_{132}$	35.30	VS	001	$h = 35.30$
	17.60	W	002	$a_{hex} = 43.42$
DODA$_{36}${Mo$_{132}$} [37]	26.9	VS	001	$h = 26.9$
	13.3	M	002	$a_{hex} = 54.2$
	9.1	W	003	
DODA$_{44}$–{Mo$_{132}$S$_{60}$} [36]	34.20	VS	001	$h = 34.12$
	16.98	M	002	$a_{hex} = 51.64$
	11.40	W	003	
DODA$_{56}$–{Mo$_{132}$S$_{60}$} [36]	34.91	VS	001	$h = 34.55$
	17.09	M	002	$a_{hex} = 55.21$

[a] measured at 190 °C; [b] measured at 140 °C; [c] not determined. I corresponds to the intensity of the reflections (VS: very strong, S: strong, M: medium, W: weak, VW: very weak); h is the lattice parameter of the lamellar phase; a_{Hex} is the local hexagonal organization within the layers calculated with Equation 1.

Globally, in contrast with the previous DODA$^+$ salts, which exhibited up to three sharp equidistant reflections indexed as the 001, 002 and 003 Miller indices in addition to a broad and diffuse signal at approximately 4.5 Å associated with the liquid-like molten chains of the DODA$^+$, the imidazolium salts display broader reflections with no signal which can be attributed to the molten chains.

The three compounds (mimC$_{14}$)$_{38}$–Mo$_{132}$, (mimC$_{18}$)$_{37}$–Mo$_{132}$ and (mimC$_{20}$)$_{37}$–Mo$_{132}$ unambiguously display two lines indexed as 001 and 002 reflections which are characteristic of a 1D lamellar ordering, whereas the two compounds (mimC$_{12}$)$_{36}$–Mo$_{132}$ and (mimC$_{16}$)$_{41}$–Mo$_{132}$ exhibits three broad lines for which we were not able to give a realistic indexation. Interestingly, the three mixed compounds (mimC$_{12}$)$_{20}$(mimC$_{20}$)$_{20}$–Mo$_{132}$, (mimC$_{12}$)$_{33}$(mimC$_{20}$)$_7$–Mo$_{132}$ and

$(mimC_{12})_8(mimC_{20})_{33}–Mo_{132}$ also display the 001 and 002 lines, whereas no liquid crystals behaviors and no first-order phase transition were detected by TD-POM and DSC, respectively. For the two latter mixed salts, we also noticed that increasing temperature above 160–180 °C produces unidentified additional reflections, which suggests either a degradation of the product or the existence of a mixture of different compounds in the solid. The results obtained for the mixed salts demonstrate the propensity for this family of salts to self-organize in a lamellar fashion even in the solid state.

Focusing on the major part of our materials, a 1D lamellar ordering is evidenced and the interlayer spacing h can be deduced since it corresponds to the d_{001} distance. These values are listed in Table 3 and are found in the range 31.34 to 35.57 Å, in agreement with the values found for the DODA$^+$ salts $(DODA)_{36}–Mo_{132}$, $(DODA)_{44}–Mo_{132}S_{60}$ and $(DODA)_{56}–Mo_{132}S_{60}$ for which h was found between 26.90 Å and 34.55 Å [36,37]. As mentioned previously, such a lamellar organization could appear surprising if we consider the spherical shape of the cluster. Nevertheless, previous studies clearly confirm the capability of Mo_{132} to assemble into lamellar aggregates [34,36,37].

Regarding the ionic character of these materials, the positive charges the imidazolium cations are necessarily located close to the spherical surface of the keplerate Mo_{132}. Logically, taking into account the isotropic shape of the Keplerate, imidazolium or DODA$^+$ cations should be randomly distributed onto the surface. Interestingly, the situation can be very different in the solid or the liquid crystalline states. In particular, if a deformation is required to minimize the electric multipolar interactions, the system can be distorted. When the starting object is a sphere, an elongation or a compression may occur, thus leading to the formation of ellipsoids, which are compatible with lamellar organizations. As the central Mo_{132} core is rigid, the deformation can only come from the non-uniform distribution of the cations around the spherical Keplerate [37].

In all cases, the diameter of the anionic inorganic cores (≈ 30 Å) is of the same order of the lamellar periodicity h (31.34 to 35.57 Å range), which implies that the layers of the lamellar phases are composed of oblate clusters where cations are globally distributed in a plan around the Mo_{132} sphere corresponding to the plans of the sheets. Taking into account a realistic density of $d = 1.0$ g cm^{-3} in the mesophase [36,37,41] and assuming that clusters are locally ordered in a compact hexagonal lattice with one cluster per unit cell ($Z = 1$) [36,37], we can thus calculate the hexagonal lattice parameter a_{hex} by using Equation (1), where MM_c is the molar mass of the cluster, h the interlayer spacing, d the density and N_{AV}, the Avogadro number.

$$a_{Hex} = \left(\frac{2 \cdot Z \cdot MM_C}{h \cdot d \cdot N_{AV} \cdot 10^{-24} \cdot 3^{1/2}} \right)^{1/2} \tag{1}$$

We calculate $41.79 \leq a_{hex} \leq 45.29$ Å, which are smaller than the parameters found for DODA salts [36,37], but in good agreement with $a_{hex} = 45$ Å reported by Volkmer and Müller [34] for a hexagonal monolayer of $(DODA)_{36}–Mo_{132}$.

Considering the fixed size of the inorganic cluster (≈ 30 Å) and the size of cations in a linear configuration ranging from 20 Å for $mimC_{12}^+$ to 30 Å for $mimC_{20}^+$, the a_{hex} distances which correspond to the distance between two clusters within the layers implies some folding of the alkyl chains of the cations. Indeed, the hypothesis of linear arrangements of the alkyl chains of the imidazolium cations accompanied by full interdigitation of the alkyl chains with the neighboring clusters would lead to a_{hex} distances of at least 50–55 Å as found for the DODA$^+$ salts ($51.54 \leq a_{hex} \leq 55.21$ Å). In contrast to the DODA$^+$ cations, in which the two neighboring octadecyl chains favor the organization of the cations through van der Waals interactions, the alkyl chains of the imidazolium cations are probably folded, prohibiting the interdigitation with the neighboring cations, a situation not favorable for the developpement of well-ordered mesophases.

3. Experimental Section

3.1. Fourier Transformed Infrared (FT-IR) Spectra

Fourier Transformed Infrared (FT-IR) spectra were recorded on a 6700 FT-IR Nicolet spectrophotometer (Les Ulis, France), using diamond ATR technique. The spectra were recorded on undiluted samples and ATR correction was applied. The variable temperature FT-IR spectra were recorded on an IRTF Nicolet iS10 spectrometer (Les Ulis, France) in diffuse reflectance mode by using high temperature diffuse reflectance environmental chamber. The background was recorded using dry KBr at 150 °C and the samples were diluted into a KBr matrix (about 10% of compound) before heating. The FT-IR spectra were recorded in the 20–500 °C temperature range under air or under nitrogen with a heating rate of 2 °C min^{-1}.

3.2. Elemental Analyses

Elemental analyses were performed by the service central d'analyses du CNRS, Vernaison, France and by the service d'analyses du CNRS, ICSN, Gif sur Yvette, France.

3.3. Water Content

Water content was determined by thermal gravimetric (TGA) analysis with a Seiko TG/DTA 320 thermogravimetric balance (5 °C min^{-1}, under air) (Chiba, Japan).

3.4. Nuclear Magnetic Resonance (NMR)

Solution 1H NMR measurements were performed on a Bruker Avance 300 instrument (Wissembourg, France) operating at 300 MHz in 5 mm o.d. tubes. Chemical shifts were referenced to TMS.

3.5. Differential Scanning Calorimetry (DSC)

DSC traces were obtained with a Mettler Toledo DSC1 Star Systems differential scanning calorimeter (Greifensee, Swizerland) from 3 to 5 mg samples (5 °C min^{-1}, under N$_2$). Several thermal cycles were performed between 40 °C and 220 °C, the first one allowing the removal of water and the organization of the solid, the following cycles explored the reproducibility and the thermal stability of the materials in this temperature range.

3.6. Temperature Dependent Polarized Optical Microscopy (TD-POM)

Temperature dependent polarized optical microscopy (TD-POM) characterizations of the optical textures of the mesophases were performed with a Leitz Orthoplan Pol polarizing microscope (Brugg, Switzerland) with a Leitz LL 20°/0.40 polarizing objective and equipped with a Linkam THMS 600 variable temperature stage (Brugg, Switzerland).

3.7. Small Angle X-ray Diffraction (SA-XRD)

The crude powder was filled in Lindemann capillaries of 0.8 mm diameter. For almost all our materials, the diffraction patterns were performed on an Empyrean (PANalytical) diffractometer (Zurich, Switzerland) in capillary mode, with a focusing X-ray mirror for Cu radiation and a PIXcel3D area detector. For the compound $(mimC_{14})_{36}$–Mo_{132} experiments were performed with a STOE transmission powder diffractometer system STADI P (Darmstadt, Germany) using a focused monochromatic Cu-$K_{\alpha1}$ beam obtained from a curved Germanium monochromator (Johann-type, STOE, Darmstadt, Germany) and collected on a curved image plate position-sensitive detector (IP-PSD). A calibration with silicon and copper laurate standards, for high and low angle domains, respectively, was preliminarily performed. Sample capillaries were placed in the high-temperature furnace for measurements in the range of desired temperatures (from −40 up to 240 °C) within 0.05 °C. Periodicities up to 50 Å could be measured. The exposure times were of 15 min.

3.8. Synthesis of Mo_{132}-Based Materials

The precursor compound $(NH_4)_{42}[Mo_{132}O_{372}(CH_3COO)_{30}(H_2O)_{72}]\cdot300H_2O.ca.10CH_3COONH_4$ noted NH_4–Mo_{132} was prepared as described by Müller et al. [42]. 1-methyl-3-alkylimidazolium bromides salts were prepared as described previously and characterized by routine ^1H NMR in $CDCl_3$ [38].

3.9. General Preparation of Mo_{132}-Based Materials

The Keplerate-based materials were prepared as follows: NH_4–Mo_{132} (300 mg, 0.01 mmol) was dissolved in 30 mL of water and then a chloroform solution containing the organic cations ($\approx$168 equivalents / NH_4–Mo_{132}, 1.72 mmol). The mixture was stirred for 1 h, where the Keplerate was totally transferred into the organic phase as indicated by the colorless aqueous phase. The organic phase was then separated and the target materials were precipitated from the organic phase by addition of an excess of ethanol, isolated by filtration, washed with ethanol, dried in air and characterized by FT-IR, EDX, Elemental Analyses, TGA and ^1H NMR.

3.9.1. $(mimC_{12}H_{25})_{36}(NH_4)_4[Mo_{132}O_{372}(CH_3COO)_{30}(H_2O)_{72}]\cdot38H_2O$, $(mimC_{12})_{36}$–Mo_{132}

It was prepared using $mimC_{12}Br$ (582 mg, 1.76 mmol). Yield 200 mg, 63%. IR/cm^{-1}: 2922 (vs), 2852 (s), 1553 (s), 1442 (m), 1163 (m), 975 (vs), 940 (s), 855(s) 801 (vs), 727 (s), 635 (m), 573 (s). Elemental analysis calcd. (%) for $(mimC_{12}H_{25})_{36}(NH_4)_6[Mo_{132}O_{372}(CH_3COO)_{30}(H_2O)_{72}]\cdot38H_2O$ (M = 31529 g·mol^{-1}) C 24.23; H, 4.64; N, 3.47; Mo, 40.17. Found C, 24.30; H, 4.45; N, 3.39;

Mo, 39.28. EDX only evidenced Mo and shows no traces of Br which could be due to an excess of the starting salt. Thermogravimetric analysis (TGA) suggests a mass loss of 6% from room temperature to 188 °C corresponding to crystallization and coordinated water molecules (calcd.: 6.3%).

3.9.2. $(mimC_{14}H_{29})_{38}(NH_4)_4[Mo_{132}O_{372}(CH_3COO)_{30}(H_2O)_{72}]\cdot30H_2O$, $(mimC_{14})_{38}-Mo_{132}$

It was prepared using $mimC_{14}Br$ (634 mg, 1.76 mmol). Yield 269 mg, 80%. IR/cm^{-1}: 2920 (m), 2850 (m), 1547 (m), 1440 (m), 1161 (m), 974 (vs), 939 (s), 789 (vs), 711 (vs), 563 (vs), 410 (vs). Elemental analysis calcd. (%) for $(mimC_{14}H_{29})_{38}(NH_4)_4[Mo_{132}O_{372}(CH_3COO)_{30}(H_2O)_{72}]\cdot30H_2O$ (M = 32461 g·mol^{-1}) C 26.12; H, 5.09; N, 3.45; Mo, 39.01. Found C, 26.22; H, 4.90; N, 3.30; Mo, 39.28. EDX only evidenced Mo and shows no traces of Br which could be due to an excess of the starting salt. Thermogravimetric analysis (TGA) suggests a mass loss of 5.4% from room temperature to 135 °C corresponding to crystallization and coordinated water molecules (calcd.: 5.6%).

3.9.3. $(mimC_{16}H_{33})_{41}(NH_4)_1[Mo_{132}O_{372}(CH_3COO)_{30}(H_2O)_{72}].38H_2O$, $(mimC_{16})_{41}-Mo_{132}$

It was prepared using $mimC_{16}Br$ (683 mg, 1.76 mmol). Yield 350 mg, quantitative yield. IR/cm^{-1}: 3451 (m), 2923 (s), 2853 (m), 1560 (m), 1466 (m), 1164 (m), 976 (vs), 941 (m), 807 (vs), 729 (vs), 577 (s), 415 (vs). Elemental analysis calcd. (%) for $(mimC_{16}H_{29})_{40}(NH_4)_3[Mo_{132}O_{372}(CH_3COO)_{30}(H_2O)_{72}]\cdot48H_2O$ (M = 34889.2 g·mol^{-1}) C 29.61; H, 5.62; N, 3.25; Mo, 36.33. Found C, 29.60; H, 5.48; N, 3.39; Mo, 36.3. EDX only evidenced Mo and shows no traces of Br which could be due to an excess of the starting salt. Thermogravimetric analysis (TGA) suggests a mass loss of 6.4% from room temperature to 150 °C corresponding to crystallization and coordinated water molecules (calcd.: 6.2%).

3.9.4. $(mimC_{18}H_{37})_{37}(NH_4)_5[Mo_{132}O_{372}(CH_3COO)_{30}(H_2O)_{72}]\cdot48H_2O$, $(mimC_{18})_{37}-Mo_{132}$

It was prepared using $mimC_{18}Br$ (732 mg, 1.76 mmol). Yield 355 mg, 96%. IR/cm^{-1}: 2919 (s), 2849 (m), 1555 (m), 1424 (m), 1161 (m), 975 (s), 941 (s), 854 (s), 791 (vs), 713 (vs), 565 (vs), 411 (vs). Elemental analysis calcd. (%) for $(mimC_{18}H_{29})_{40}(NH_4)_3[Mo_{132}O_{372}(CH_3COO)_{30}(H_2O)_{72}]\cdot48H_2O$ (M = 35058.68 g·mol^{-1}) C 29.88; H, 5.57; N, 3.11; Mo, 36.28 Found C, 29.94; H, 5.58; N, 3.15; Mo, 36.13. EDX only evidenced Mo and shows no traces of Br which could be due to an excess of the starting salt. Thermogravimetric analysis (TGA) suggests a mass loss of 6.25% from room temperature to 188 °C corresponding to crystallization and coordinated water molecules (calcd.: 6.16%).

3.9.5. $(mimC_{20}H_{41})_{37}(NH_4)_5[Mo_{132}O_{372}(CH_3COO)_{30}(H_2O)_{72}]\cdot43H_2O$, $(mimC_{20})_{37}-Mo_{132}$

It was prepared using $mimC_{20}Br$ (780 mg, 1.76 mmol). Yield 0.35 mg, 92%. IR/cm^{-1}: 2919 (s), 2850 (s), 1541 (m), 1443 (m), 1162 (m), 973 (s), 938 (s), 854 (s), 792 (vs), 632 (vs), 565 (vs), 411 (vs). Elemental analysis calcd. (%) for $(mimC_{20}H_{41})_{37}(NH_4)_5[Mo_{132}O_{372}(CH_3COO)_{30}(H_2O)_{72}]\cdot43H_2O$ (M = 36006.57 g·mol^{-1}) C 31.62; H, 5,81; N, 3.07; Mo, 35.17. Found C, 31.36; H, 5.51; N, 3.07; Mo, 35.12. EDX only evidenced Mo and shows no traces of Br which could be due to an excess of the starting salt. Thermogravimetric analysis (TGA) suggests a mass loss of 5.5%

from room temperature to 160 °C corresponding to crystallization and coordinated water molecules (calcd.: 5.7%).

3.9.6. $(mimC_{12}H_{25})_{20}(mimC_{20}H_{33})_{20}(NH_4)_2[Mo_{132}O_{372}(CH_3COO)_{30}(H_2O)_{72}]\cdot38H_2O$, $(mimC_{12})_{20}(mimC_{20})_{20}-Mo_{132}$

It was prepared using $mimC_{20}Br$ (390 mg, 0.882 mmol) and $mimC_{12}Br$ (290 mg, 0.882 mmol). Yield 100 mg, 28%. IR/cm^{-1}: 2920 (vs), 2849 (s), 1557 (m), 1442 (m), 1162 (m), 972 (vs), 940 (s), 854 (s), 798 (vs), 570 (s), 412 (s). Elemental analysis calcd. (%) for $(mimC_{12}H_{25})_{20}(mimC_{20}H_{33})_{20}$ $(NH_4)_2[Mo_{132}O_{372}(CH_3COO)_{30}(H_2O)_{72}]\cdot38H_2O$ (M = 34709.05 g·mol^{-1}) C 29.75; H, 5.45; N, 3.30. Found C, 29.70; H, 5.63; N, 3.25. EDX atomic ratios calculated for $(mimC_{12})_{20}(mimC_{20})_{20}-$ $\{Mo_{132}\}$ shows no Br or traces. Thermogravimetric analysis (TGA) suggests a mass loss of 5.7% from room temperature to 170 °C corresponding to crystallization and coordinated water molecules (calcd.: 5.7%) and % org: 42.5% (cald. 40.6%).

3.9.7. $(mimC_{12}H_{25})_{33}(mimC_{20}H_{33})_7(NH_4)_2[Mo_{132}O_{372}(CH_3COO)_{30}(H_2O)_{72}]\cdot48H_2O$, $(mimC_{12})_{33}(mimC_{20})_7-Mo_{132}$

It was prepared using $mimC_{12}Br$ (462 mg, 1.41 mmol) and $mimC_{20}Br$ (156 mg, 0.34 mmol). Yield 120 mg, 34.3%. IR/cm^{-1}. 2922 (vs), 2851 (s), 1557 (s), 1444 (m), 1163 (m), 973 (vs), 941 (s), 855 (s), 799 (s), 724 (vs), 570 (vs), 412 (vs). Elemental analysis calcd. (%) for $(mimC_{12}H_{25})_{33}(mimC_{20}H_{33})_7(NH_4)_2[Mo_{132}O_{372}(CH_3COO)_{30}(H_2O)_{72}]\cdot38H_2O$ (M = 33430.45 g·mol^{-1}) C 27.16; H, 5.09; N, 3.43. Found C, 27.30; H, 5.01; N, 3.14. EDX only evidenced Mo and shows no traces of Br which could be due to an excess of the starting salt. Thermogravimetric analysis (TGA) suggests a mass loss of 6.6% from room temperature to 170 °C corresponding to crystallization and coordinated water molecules (calcd.: 6.4%) and % org: 37.4 (calcd. 37.4%).

3.9.8. $(mimC_{12}H_{25})_8(mimC_{20}H_{33})_{33}(NH_4)_1[Mo_{132}O_{372}(CH_3COO)_{30}(H_2O)_{72}]\cdot38H_2O$, $(mimC_{12})_8(mimC_{20})_{33}-Mo_{132}$

It was prepared using $mimC_{12}Br$ (120 mg, 0.35 mmol) and $mimC_{20}Br$ (620 mg, 1.4 mmol). Yield 110 mg, 30.5%. IR/cm^{-1}: 2923 (vs), 2852 (s), 1559 (s), 1465 (m), 1163 (s), 976 (vs), 941 (vs), 859 (vs), 732 (vs), 576 (s), 412 (vs). Elemental analysis calcd. (%) for $(mimC_{12}H_{25})_8(mimC_{20}H_{33})_{33}(NH_4)_2[Mo_{132}O_{372}(CH_3COO)_{30}(H_2O)_{72}]\cdot38H_2O$ (M = 34709.05 g·mol^{-1}) C 32.17; H, 5.87; N, 3.17. Found C, 32.11; H, 5.71; N, 3.24. EDX atomic ratios calculated for $(mimC_{12})(mimC_{20})-\{Mo_{132}\}$ shows no Br or traces. Thermogravimetric analysis (TGA) suggests a mass loss of 5.7% from room temperature to 170 °C corresponding to crystallization and coordinated water molecules (calcd.: 5.7%) and org: 46.8% (calcd. 49%).

4. Conclusions

In the following of our previous works describing the liquid crystal properties of Keplerates associated to DODA$^+$ cations or of the nanoscopic inorganic cluster

$[K_2Na_xLi_yH_z\{Mo_4O_4S_4(OH)_2\cdot(H_2O)_3\}_2(P_8W_{48}O_{184})]^{(34-x-y-z)-}$ combined with alkylmethylimidazolium cations, we prepared in this study a series of eight new compounds resulting from the ionic association of the nanoscopic inorganic cluster $[Mo_{132}O_{372}(CH_3COO)_{30}(H_2O)_{72}]^{42-}$ with various alkylmethylimidazolium cations, denoted $mimC_n^+$ ($n = 12–20$). The results obtained in this study confirm that the strategy which consists in associating highly charged polyoxometalate clusters with very simple organic cations bearing only alkyl chains of variable sizes can provide liquid crystalline phases. The latter were investigated by TD-POM, DSC and SA-XRD. The major part of the synthesized materials exhibit lamellar organization in the solid state or in the liquid crystalline phase. However, DSC shows very broad phase transitions between the solid and the mesophase which could be explained by insufficient supramolecular mobility to develop network of intermolecular interactions. The small angle-X ray diffraction studies provide the interlayer spacing h and the hexagonal lattice parameter a_{hex}, which evidence a non-uniform distribution of the imidazolium cations around the spherical clusters Mo_{132} and a folding of the alkyl chains. This lead to less interdigitation between the alkyl chains of the cations and therefore to poorly organized mesophases. This observation contrasts with our previous results collected for the $DODA^+$ salts of Keplerates [36,37] and of similar salts obtained with the anisotropic POM $[K_2Na_xLi_yH_z\{Mo_4O_4S_4(OH)_2(H_2O)_3\}_2(P_8W_{48}O_{184})]^{(34-x-y-z)-}$ [38]. In the former case, the larger number and the vicinity of alkyl chains of the $DODA^+$ cations within the solid favor the interdigitation of the alkyl chains driven by van der waals interactions. In the second case, the anisotropic shape of the POM associated with the cationic character of its cavity promotes the localization of the cations on the anionic ring of the P_8W_{48}–POM. This favors the closeness of the cations and thus the van der Waals interactions between alkyl chains and therefore the organization within the solid and the liquid crystal phase. This strategy to combine POMs and very simple cations like alkylimidazolium or $DODA^+$ cation then appears efficient if we can induce interaction between alkyl chains especially by forcing the cations to be close by using a limited surface for the POM or by using cations bearing several alkyl chains like $DODA^+$ cations.

Acknowledgments

We acknowledge the Centre National de la Recherche Scientifique (CNRS), the Ministère de l'Education Nationale de l'Enseignement Supérieur et de la Recherche (MENESR) and the University of Versailles Saint Quentin for their financial support. SF gratefully acknowledges the "Institut Universitaire de France, IUF" for financial support. CP and LG gratefully acknowledges the financial support from the Swiss National Science Foundation.

Author Contributions

The syntheses and characterizations were performed by NW, WS and AH. The study of the liquid crystals phases by TD-POM, DSC and SA-XRD were ensured by ET, LG and KLB. CP contributed for the interpretation of the liquid crystal data. EC and DN contributed to this work as supervisors of NW and WS. SF ensured the coordination of this work, the co-supervision of NW, WS and AH and the FT-IR experiments with temperature.

Conflicts of Interest

The authors declare no conflict of interest.

References

1. Binnemans, K.; Gorller-Walrand, C. Lanthanide-Containing Liquid Crystals and Surfactants. *Chem. Rev.* **2002**, *102*, 2303–2346.
2. Vila-Nadal, L.; Mitchell, S.G.; Markov, S.; Busche, C.; Georgiev, V.; Asenov, A.; Cronin, L. Towards Polyoxometalate-Cluster-Based Nano-Electronics. *Chem.-Eur. J.* **2013**, *19*, 16502–16511.
3. Evangelisti, F.; Guttinger, R.; More, R.; Luber, S.; Patzke, G.R. Closer to Photosystem II: A Co_4O_4 Cubane Catalyst with Flexible Ligand Architecture. *J. Am. Chem. Soc.* **2013**, *135*, 18734–18737.
4. Carraro, M.; Modugno, G.; Zamolo, V.; Bonchio, M.; Fabbretti, E. Polyoxometalate-Based Conjugates for Biological Targeting. *J. Biol. Inorg. Chem.* **2014**, *19*, S406–S406.
5. Ibrahim, M.; Xiang, Y.X.; Bassil, B.S.; Lan, Y.H.; Powell, A.K.; de Oiveira, P.; Keita, B.; Kortz, U. Synthesis, Magnetism, and Electrochemistry of the Ni_{14} and Ni_5-Containing Heteropolytungstates $[Ni_{14}(OH)_6(H_2O)_{10}(HPO_4)_4(P_2W_{15}O_{56})_4]^{34-}$ and $[Ni_5(OH)_4(H_2O)_4(\beta\text{-}GeW_9O_{34})(\beta\text{-}GeW_8O_{30}(OH))]^{13-}$. *Inorg. Chem.* **2013**, *52*, 8399–8408.
6. Absillis, G.; Parac-Vogt, T.N. Peptide Bond Hydrolysis Catalyzed by the Wells–Dawson $[Zr(\alpha_2\text{-}P_2W_{17}O_{61})_2]$ Polyoxometalate. *Inorg. Chem.* **2012**, *51*, 9902–9910.
7. Riflade, B.; Oble, J.; Chenneberg, L.; Derat, E.; Hasenknopf, B.; Lacote, E.; Thorimbert, S. Hybrid Polyoxometalate Palladacycles: DFT Study and Application to the Heck Reaction. *Tetrahedron* **2013**, *69*, 5772–5779.
8. Wang, Y.F.; Weinstock, I.A. Polyoxometalate-Decorated Nanoparticles. *Chem. Soc. Rev.* **2012**, *41*, 7479–7496.
9. Rickert, P.G.; Antonio, M.R.; Firestone, M.A.; Kubatko, K.-A.; Szreder, T.; Wishart, J.F.; Dietz, M.L. Tetraalkylphosphonium Polyoxometalate Ionic Liquids: Novel, Organic-Inorganic Hybrid Materials. *J. Phys. Chem. B* **2007**, *111*, 4685–4692.
10. Lin, X.K.; Li, W.; Zhang, J.; Sun, H.; Yan, Y.; Wu, L.X. Thermotropic Liquid Crystals of a Non-Mesogenic Group Bearing Surfactant-Encapsulated Polyoxometalate Complexes. *Langmuir* **2010**, *26*, 13201–13209.
11. Yin, S.Y.; Sun, H.; Yan, Y.; Li, W.; Wu, L.X. Hydrogen-Bonding-Induced Supramolecular Liquid Crystals and Luminescent Properties of Europium-Substituted Polyoxometalate Hybrids. *J. Phys. Chem. B* **2009**, *113*, 2355–2364.
12. Yin, S.Y.; Li, W.; Wang, J.F.; Wu, L.X. Mesomorphic Structures of Protonated Surfactant-Encapsulated Polyoxometalate Complexes. *J. Phys. Chem. B* **2008**, *112*, 3983–3988.
13. Li, W.; Yin, S.Y.; Wang, J.F.; Wu, L.X. Tuning Mesophase of Ammonium Amphiphile-Encapsulated Polyoxometalate Complexes through Changing Component Structure. *Chem. Mater.* **2008**, *20*, 514–522.

14. Li, W.; Bu, W.F.; Li, H.L.; Wu, L.X.; Li, M. A Surfactant-Encapsulated Polyoxometalate Complex towards a Thermotropic Liquid Crystal. *Chem. Commun.* **2005**, 3785–3787.

15. Li, B.; Zhang, J.; Wang, S.; Li, W.; Wu, L.X. Nematic Ion-Clustomesogens from Surfactant-Encapsulated Polyoxometalate Assemblies. *Eur. J. Inorg. Chem.* **2013**, 1869–1875.

16. Jiang, Y.X.; Liu, S.X.; Zhang, J.; Wu, L.X. Phase Modulation of Thermotropic Liquid Crystals of Tetra-*n*-Alkylammonium Polyoxometalate Ionic Complexes. *Dalton Trans.* **2013**, *42*, 7643–7650.

17. Yin, S.Y.; Sun, H.; Yan, Y.; Zhang, H.; Li, W.; Wu, L.X. Self-Assembly and Supramolecular Liquid Crystals based on Organic Cation Encapsulated Polyoxometalate Hybrid Reverse Micelles and Pyridine Derivatives. *J. Colloid Interface Sci.* **2011**, *361*, 548–555.

18. Müller, A.; Gouzerh, P. Capsules with Highly Active Pores and Interiors: Versatile Platforms at the Nanoscale. *Chem.-Eur. J.* **2014**, *20*, 4862–4873.

19. Müller, A.; Gouzerh, P. From Linking of Metal-Oxide Building Blocks in a Dynamic Lbrary to Giant Clusters with Unique Properties and Towards Adaptive Chemistry. *Chem. Soc. Rev.* **2012**, *41*, 7431–7463.

20. Kögerler, P.; Tsukerblat, B.; Müller, A. Structure-Related Frustrated Magnetism of Nanosized Polyoxometalates: Aesthetics and Properties in Harmony. *Dalton Trans.* **2010**, *39*, 21–36.

21. Botar, B.; Kögerler, P.; Müller, A.; Garcia-Serres, R.; Hill, C.L. Ferrimagnetically Ordered Nanosized Polyoxomolybdate-Based Cluster Spheres. *Chem. Commun.* **2005**, 5621–5623.

22. Rezaeifard, A.; Haddad, R.; Jafarpour, M.; Hakimi, M. Catalytic Epoxidation Activity of Keplerate Polyoxomolybdate Nanoball toward Aqueous Suspension of Olefins under Mild Aerobic Conditions. *J. Am. Chem. Soc.* **2013**, *135*, 10036–10039.

23. Kopilevich, S.; Gil, A.; Garcia-Rates, M.; Bonet-Avalos, J.; Bo, C.; Müller, A.; Weinstock, I.A. Catalysis in a Porous Molecular Capsule: Activation by Regulated Access to Sixty Metal Centers Spanning a Truncated Icosahedron. *J. Am. Chem. Soc.* **2012**, *134*, 13082–13088.

24. Ostroushko, A.A.; Grzhegorzhevskii, K.V. Electric Conductivity of Nanocluster Polyoxomolybdates in the Solid State and Solutions. *Russ. J. Phys. Chem. A* **2014**, *88*, 1008–1011.

25. Zhou, Y.S.; Shi, Z.H.; Zhang, L.J.; ul Hassan, S.; Qu, N.N. Notable Third-Order Optical Nonlinearities of a Keplerate-Type Polyoxometalate in Solution and in Thin Films of PMMA. *Appl. Phys. A* **2013**, *113*, 563–568.

26. Zhang, L.J.; Shi, Z.H.; Zhang, L.H.; Zhou, Y.S.; ul Hassan, S. Fabrication and Optical Nonlinearities of Ultrathin Composite Films Incorporating a Keplerate Type Polyoxometalate. *Mater. Lett.* **2012**, *86*, 62–64.

27. Besson, C.; Schmitz, S.; Capella, K.M.; Kopilevich, S.; Weinstock, I.A.; Kögerler, P. A Regioselective Huisgen Reaction inside a Keplerate Polyoxomolybdate Nanoreactor. *Dalton Trans.* **2012**, *41*, 9852–9854.

28. Zhang, Q.; He, L.P.; Wang, H.; Zhang, C.; Liu, W.S.; Bu, W.F. Star-Like Supramolecular Polymers Fabricated by a Keplerate Cluster with Cationic Terminated Polymers and their Self-Assembly into Vesicles. *Chem. Commun.* **2012**, *48*, 7067–7069.

29. Caruso, F.; Kurth, D.G.; Volkmer, D.; Koop, M.J.; Müller, A. Ultrathin Molybdenum Polyoxometalate-Polyelectrolyte Multilayer Films. *Langmuir* **1998**, *14*, 3462–3465.

30. Cazacu, A.; Mihai, S.; Nasr, G.; Mahon, E.; van der Lee, A.; Meffre, A.; Barboiu, M. Lipophilic Polyoxomolybdate Nanocapsules in Constitutional Dynamic Hybrid Materials. *Inorg. Chim. Acta* **2010**, *363*, 4214–4219.

31. Kurth, D.G.; Volkmer, D.; Ruttorf, M.; Richter, B.; Müller, A. Ultrathin Composite Films Incorporating the Nanoporous Isopolyoxomolybdate "Keplerate" $(NH_4)_{42}[Mo_{132}O_{372}(CH_3COO)_{30}(H_2O)_{72}]$. *Chem. Mater.* **2000**, *12*, 2829–2831.

32. Kurth, D.G.; Lehmann, P.; Volkmer, D.; Müller, A.; Schwahn, D. Biologically Inspired Polyoxometalate-Surfactant Composite Materials. Investigations on the Structures of Discrete, Surfactant-Encapsulated Clusters, Monolayers, and Langmuir-Blodgett Films of $(DODA)_{40}(NH_4)_2(H_2O)_n[Mo_{132}O_{372}(CH_3CO_2)_{30}(H_2O)_{72}]$. *Dalton Trans.* **2000**, 3989–3998.

33. Kurth, D.G.; Lehmann, P.; Volkmer, D.; Colfen, H.; Koop, M.J.; Müller, A.; Du Chesne, A. Surfactant-Encapsulated Clusters (SECs): $(DODA)_{20}(NH_4)[H_3Mo_{57}V_6(NO)_6O_{183}(H_2O)_{18}]$, a Case Study. *Chem.-Eur. J.* **2000**, *6*, 385–393.

34. Volkmer, D.; Du Chesne, A.; Kurth, D.G.; Schnablegger, H.; Lehmann, P.; Koop, M.J.; Müller, A. Toward Nanodevices: Synthesis and Characterization of the Nanoporous Surfactant-Encapsulated Keplerate $(DODA)_{40}(NH_4)_2(H_2O)_n[Mo_{132}O_{372}(CH_3CO_2)_{30}(H_2O)_{72}]$. *J. Am. Chem. Soc.* **2000**, *122*, 1995–1998.

35. Clemente-Leon, M.; Ito, T.; Yashiro, H.; Yamase, T. Two-Dimensional Array of Polyoxomolybdate Nanoball Constructed by Langmuir-Blodgett Semiamphiphilic Method. *Chem. Mater.* **2007**, *19*, 2589–2594.

36. Floquet, S.; Terazzi, E.; Korenev, V.S.; Hijazi, A.; Guénée, L.; Cadot, E. Layered Ionic Liquid-Crystalline Organisations Built from Nano-Capsules $[Mo_{132}O_{312}S_{60}(SO_4)_x (H_2O)_{(132-2x)}]^{(12+2x)-}$ and $DODA^+$ cations. *Liq. Cryst.* **2014**, *41*, 1000–1007.

37. Floquet, S.; Terazzi, E.; Hijazi, A.; Guénée, L.; Piguet, C.; Cadot, E. Evidence of Ionic Liquid Crystal Properties for a $DODA^+$ Salt of the Keplerate $[Mo_{132}O_{372}(CH_3COO)_{30}(H_2O)_{72}]^{42-}$. *New J. Chem.* **2012**, *36*, 865–868.

38. Watfa, N.; Floquet, S.; Terazzi, E.; Haouas, M.; Salomon, W.; Korenev, V.S.; Taulelle, F.; Guénée, L.; Hijazi, A.; Naoufal, D.; et.al. Synthesis, Characterization, and Tuning of the Liquid Crystal Properties of Ionic Materials based on the Cyclic Polyoxothiometalate $[\{Mo_4O_4S_4(H_2O)_3(OH)_2\}_2(P_8W_{48}O_{184})]^{36-}$. *Soft Matter* **2015**, *11*, 1087–1099.

39. Watfa, N.; Melgar, D.; Haouas, M.; Taulelle, F.; Hijazi, A.; Naoufal, D.; Bonet Avalos, J.; Floquet, S.; Bo, C.; Cadot, E. Hydrophobic Effect as Driving Force for Host-Guest Chemistry of a Multireceptor Keplerate-Type Capsule. *J. Am. Chem. Soc.* **2015**, *137*, 5845–5851.

40. Dutronc, T.; Terazzi, E.; Guénée, L.; Buchwalder, K.L.; Spoerri, A.; Emery, D.; Mareda, J.; Floquet, S.; Piguet, C. Enthalpy-Entropy Compensation Combined with Cohesive Free-Energy Densities for Tuning the Melting Temperatures of Cyanobiphenyl Derivatives. *Chem.-Eur. J.* **2013**, *19*, 8447–8456.

41. Deschenaux, R.; Donnio, B.; Guillon, D. Liquid-Crystalline Fullerodendrimers. *New J. Chem.* **2007**, *3s1*, 1064–1073.

42. Müller, A.; Krickemeyer, E.; Bögge, H.; Schmidtmann, M.; Peters, F. Organizational Forms of Matter: An Inorganic Super Fullerene and Keplerate based on Molybdenum Oxide. *Angew. Chem.* **1998**, *37*, 3359–3363.

Synthesis and Characterization of 8-Yttrium(III)-Containing 81-Tungsto-8-Arsenate(III), [Y$_8$(CH$_3$COO)(H$_2$O)$_{18}$(As$_2$W$_{19}$O$_{68}$)$_4$(W$_2$O$_6$)$_2$(WO$_4$)]$^{43-}$

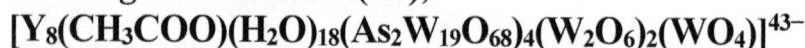

Masooma Ibrahim, Bassem S. Bassil and Ulrich Kortz

Abstract: The 8-yttrium(III)-containing 81-tungsto-8-arsenate(III) [Y$_8$(CH$_3$COO)(H$_2$O)$_{18}$(As$_2$W$_{19}$O$_{68}$)$_4$(W$_2$O$_6$)$_2$(WO$_4$)]$^{43-}$ (**1**) has been synthesized in a one-pot reaction of yttrium(III) ions with [B-α-AsW$_9$O$_{33}$]$^{9-}$ in 1 M NaOAc/HOAc buffer at pH 4.8. Polyanion **1** is composed of four {As$_2$W$_{19}$O$_{68}$} units, two {W$_2$O$_{10}$} fragments, one {WO$_6$} group, and eight Y^{III} ions. The hydrated cesium-sodium salt of **1** (**CsNa-1**) was characterized in the solid-state by single-crystal XRD, FT-IR spectroscopy, thermogravimetric and elemental analyses.

Reprinted from *Inorganics*. Cite as: Ibrahim, M.; Bassil, B.S.; Kortz, U. Synthesis and Characterization of 8-Yttrium(III)-Containing 81-Tungsto-8-Arsenate(III), [Y$_8$(CH$_3$COO)(H$_2$O)$_{18}$ (As$_2$W$_{19}$O$_{68}$)$_4$(W$_2$O$_6$)$_2$(WO$_4$)]$^{43-}$. *Inorganics* **2015**, *3*, 267-278.

1. Introduction

Polyoxometalate (POM) chemistry has gained much attention in recent years, due to potential applications in catalysis, magnetism, material science, and even biomedical systems [1–10]. Most of the work in this area has been focused on *3d* transition metal-containing derivatives [11–17]. However, also the area of rare earths-containing POMs has undergone a significant expansion in recent years [18–23]. Lanthanide ions usually have larger coordination numbers (8, 9 or more) than *d*-block metal ions, and exhibit various coordination geometries [24]. In addition, the larger size of these lanthanide ions usually restricts their full incorporation into the lacunary POM sites. Such coordination behavior renders lanthanide ions frequently as linkers of two or more lacunary POM units, forming large structures [18–23,25–27]. Yttrium is formally not a lanthanide, but can be considered as a pseudo-lanthanide, due to its comparable size and coordination number. Lanthanide-containing POMs possess interesting physicochemical properties such as luminescence and magnetism, as well as Lewis acidity relevant for catalysis [28–30]. For example, the terminal, labile coordination sites of the lanthanide (Ln) ions in [(Ln)(H$_2$O)$_4$P$_2$W$_{17}$O$_{61}$]$^{6-}$ and [(Ln)(H$_2$O)$_4$PW$_{11}$O$_{39}$]$^{4-}$ (Ln = Y^{III}, La^{III}, Eu,^{III} Sm^{III}, or Yb^{III}) are known to catalyze aldol and imino Diels–Alder reactions [29,30]. Mizuno's group has recently reported the high catalytic activity of the yttrium-containing tungstosilicate dimer [{Y(H$_2$O)$_2$}$_2$(γ-SiW$_{10}$O$_{36}$)$_2$]$^{10-}$ for the cyanosilylation of ketones and aldehydes with trimethylsilyl cyanide (TMS)CN [31,32].

The number of reports on yttrium-containing POMs is significantly lower than for classical lanthanides. In 1971, Peacock and Weakley reported [Y(W$_5$O$_{18}$)$_2$]$^{9-}$ [33], which was later used as a catalyst for alcohol oxidations and alkene epoxidations by H$_2$O$_2$ [34]. Francesconi's group reported the Y^{III}-containing polyanion [(PY$_2$W$_{10}$O$_{38}$)$_4$(W$_3$O$_{14}$)]$^{30-}$, which consists of four [PW$_9$O$_{34}$]$^{9-}$ units linked by a central [Y$_8$W$_7$O$_{30}$]$^{6+}$ group [35]. Hill and co-workers reported two sandwich-type polyanions, [(YOH$_2$)$_3$(CO$_3$)(A-α-PW$_9$O$_{34}$)$_2$]$^{11-}$ [36], and [{Y$_4$(μ_3-OH)$_4$(H$_2$O)$_8$}(α-P$_2$W$_{15}$O$_{56}$)$_2$]$^{16-}$ [37]. Very recently, Boskovic's group reported on two Y^{III}-containing polyanions, $[As^{III}_4(Y^{III}W^{VI}_3)W^{VI}_{44}Y^{III}_4O_{159}(Gly)_8(H_2O)_{14}]^{9-}$ and $[As^{III}_4(Mo^V_2Mo^{VI}_2)W^{VI}_{44}Y^{III}_4O_{160}(Nle)_9(H_2O)_{11}]^{18-}$, which are functionalized by incorporation of glycine or norleucine [38]. A mixed Ni^{II}-Y^{III}-containing polyanion, $[Y(H_2O)_5\{Ni(H_2O)\}_2As_4W_{40}O_{140}]^{21-}$, based on the large, cyclic POM ligand $[As_4W_{40}O_{140}]^{28-}$ has also been reported [39]. Niu's group reported the solid-state structure of the one-dimensional chain $\{Y(GeW_{11}O_{39})(H_2O)_2\}_n$ [40], composed of Y^{III} and $[GeW_{11}O_{39}]^{8-}$ units. The same group also reported an oxalate-bridged $[\{(\alpha\text{-}PW_{11}O_{39})Y(H_2O)\}_2(C_2O_4)]^{10-}$ [41] and very recently a tartrate-bridged yttrium-containing tungstoarsenate(III) $[\{Y_2(C_4H_4O_6)(C_4H_2O_6)(AsW_9O_{33})\}_2]^{18-}$ [42].

Our group has also been active in the synthesis of novel yttrium-containing POMs. We reported the solid state and solution structure of the Weakley-Peacock type dimer $[YW_{10}O_{36}]^{9-}$, which consists of two monolacunary Lindqvist-based $[W_5O_{18}]^{6-}$ fragments encapsulating a central Y^{III} ion with a square-antiprismatic coordination geometry [43]. Later, we reported the Y^{III}-containing isopolytungstate $[Y_2(H_2O)_{10}W_{22}O_{72}(OH)_2]^{8-}$ as a member of a large family of lanthanide-containing compounds. These isostructural polyanions $[Ln_2(H_2O)_{10}W_{22}O_{72}(OH)_2]^{8-}$ consist of a $\{W_{22}\}$ isopolyanion unit and two lanthanide ions with five terminal water ligands each. [44] We also synthesized the acetate-bridged, dimeric polyanion $[\{Y(\alpha\text{-}GeW_{11}O_{39})(H_2O)\}_2(\mu\text{-}CH_3COO)_2]^{12-}$ [45], as well as the trimeric polyanion $[\{Y(SbW_9O_{31}(OH)_2)(CH_3COO)(H_2O)\}_3(WO_4)]^{17-}$, which is composed of three $\{SbW_9O_{33}\}$ units connected by three Y^{3+} ions and a capping, tetrahedral tungstate group $\{WO_4\}$ [46]. Interestingly, we were also able to encapsulate yttrium(III), and many lanthanide ions, in polyoxopalladates(II), $[Ln^{III}Pd^{II}_{12}(AsPh)_8O_{32}]^{5-}$ (Ln = Y, Pr, Nd, Sm, Eu, Gd, Tb, Dy, Ho, Er, Tm, Yb, Lu). These cuboid-shaped polypalladates(II) are composed of a Pd_{12}-oxo cage, capped by eight phenylarsonate heterogroups, and encapsulating a central guest ion Ln [47]. As part of our continuous effort to incorporate yttrium(III) ions in POMs, we describe herein the synthesis and structural characterization of an 8-yttrium(III)-containing 81-tungsto-8-arsenate(III).

2. Results and Discussion

Synthesis and Structure. The polyanion $[Y_8(CH_3COO)(H_2O)_{18}(As_2W_{19}O_{68})_4(W_2O_6)_2(WO_4)]^{43-}$ (**1**) has been synthesized using a simple, one-pot procedure by reacting the trilacunary POM precursor $[B\text{-}\alpha\text{-}AsW_9O_{33}]^{9-}$ with Y^{III} ions in 1 M NaOAc/AcOH buffer at pH 4.8, and was characterized in the solid state by IR spectroscopy, as well as thermogravimetric and elemental analysis. The title polyanion crystallizes as a hydrated mixed sodium-cesium salt (**CsNa-1**) in the triclinic space group $P\bar{1}$. Single-crystal XRD on **CsNa-1** revealed that polyanion **1** is composed of four $\{As_2W_{19}O_{68}\}$ units, two $\{W_2O_{10}\}$ fragments, one $\{WO_6\}$ group, and eight yttrium(III) ions (see Figures 1 and 2). The mono- and di-tungsten fragments are most likely formed *in situ* by fragmentation of some of the lacunary $[AsW_9O_{33}]^{9-}$ precursor during the course of the reaction, also seen previously for other large POM architectures [18,21]. The $\{As_2W_{19}\}$ units, also formed *in situ*, are composed of two $\{AsW_9\}$ units connected by an octahedral $\{WO_6\}$ group via corner-shared oxo-bridges. The structure of the $\{As_2W_{19}\}$ units in **1** is different from the known $[As_2W_{19}O_{67}(H_2O)]^{14-}$ POM precursor, where the linking tungsten center has terminal *trans*-oxo-aqua ligands [48]. The linking tungsten(VI) in **1** bridges the two $\{AsW_9\}$ units in a way that the terminal oxo groups are *cis* to each other, non-protonated, and linking the tungsten(VI) to two Y^{III} ions. A total of eight such oxo-bridges are present in **1**, corresponding to the number of yttrium(III) ions. These yttrium(III) ions are all eight-coordinate with distorted square-antiprismatic coordination geometries and Y–O bond distances in the range of 2.3–2.5 Å. The coordination sphere of each Y^{III} center is composed of bridging oxo and terminal aqua ligands (*vide infra*). Two $\{Y_2As_2W_{19}\}$ units are linked to each other by an edge-shared $\{W_2O_6\}$ group to form two dimeric subunits (see Figures 1 and 2). Finally, a tungsten center (W81, corresponding to the orange octahedron in Figures 1 and 2) links the two subunits to each other via two W-O-W (namely to W79 and W80) and four W-O-Y bridges (one to Y6 and Y7, and two to Y8), leading to a 'dimer of dimers'-type assembly. The two subunits are also linked by two Y-O-W bridges, namely Y7-O-W30 and Y4-O-W25. The yttrium centers are all octa-coordinated as stated above, with the coordination sphere filled by Y-O-W bridges or terminal oxo ligands. The two bridging yttriums Y7 and Y8, which are connected to W81, have two terminal aqua ligands each. The rest of the yttrium ions, apart from Y1, have three terminal aqua ligands each. For Y1, an acetate group (from the buffer medium) and two aqua ligands complete the coordination sphere. The existence of such kind of acetate groups has been previously observed, for example in $[Gd_6As_6W_{65}O_{229}(OH)_4(H_2O)_{12}(OAc)_2]^{38-}$ and $[Yb_{10}As_{10}W_{88}O_{308}(OH)_8(H_2O)_{28}(OAc)_4]^{40-}$ [18]. Bond valence sum (BVS) calculations indicate that no bridging oxygen atoms in **1** are protonated [49]. The total charge of **1** is therefore 43-, and is balanced in the solid state by 34.5 sodium and 8.5 cesium counter cations, hence resulting in the formula unit $Cs_{8.5}Na_{34.5}[Y_8(CH_3COO)(H_2O)_{18}(As_2W_{19}O_{68})_4(W_2O_6)_2(WO_4)]\cdot230H_2O$. This formula is supported by elemental and thermogravimetric analyses (see Figure 3).

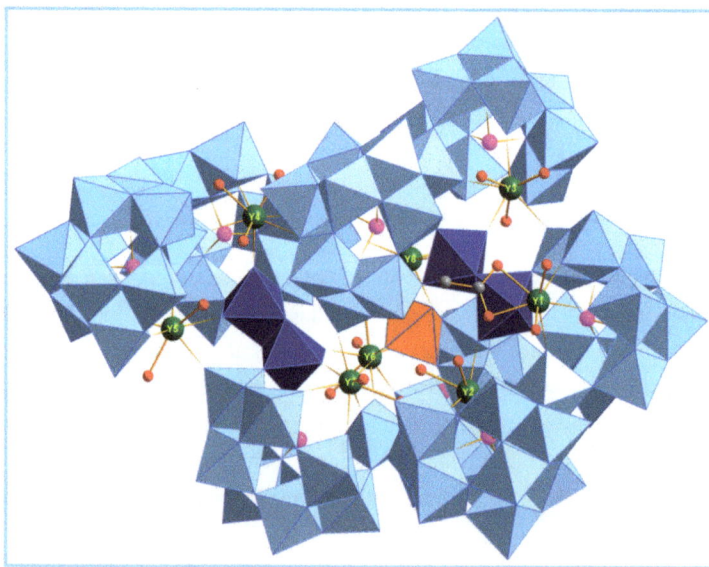

Figure 1. Combined polyhedral/ball-and-stick representation of **1**. Color code: WO_6 octahedra pale blue/dark blue/orange, As pink balls, Y green balls, O red balls, C grey balls.

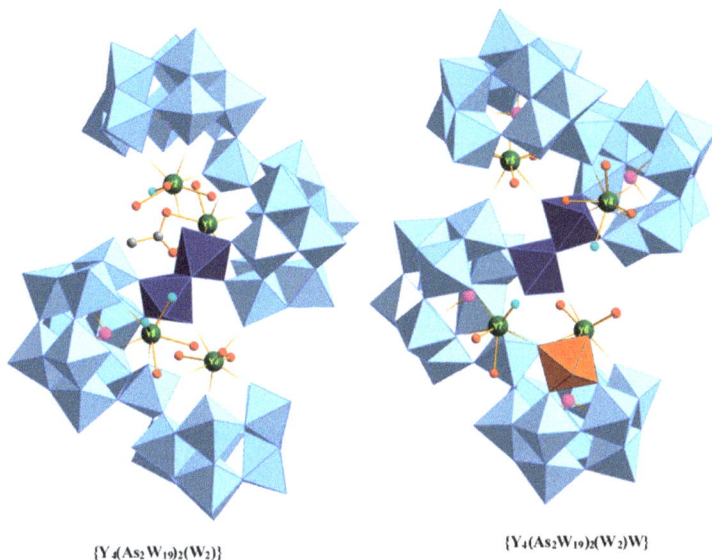

{Y₄(As₂W₁₉)₂(W₂)}   {Y₄(As₂W₁₉)₂(W₂)W}

Figure 2. Combined polyhedral/ball-and-stick representation of the two half-units in **1**. The turquoise balls represent bridging oxygens (O$_{bridge}$). The color code is the same as in Figure 1.

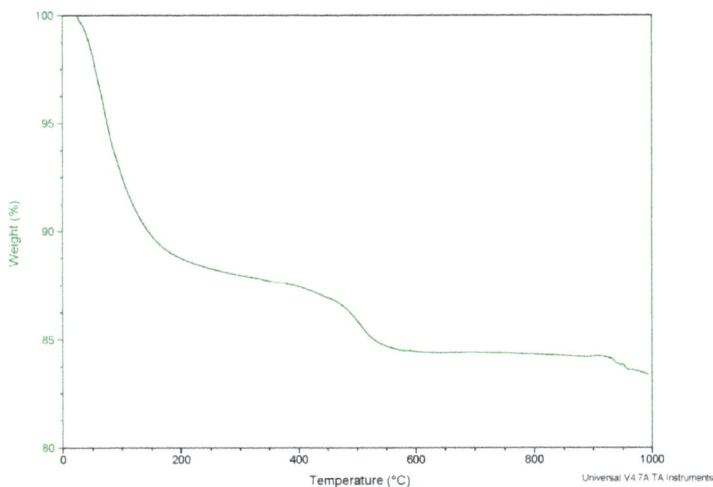

Figure 3. Thermogram of **CsNa-1** from room temperature to 1000 °C.

The synthetic procedure of **1** contains several crucial parameters. An excess of yttrium ions over and above the stoichiometric ratio was needed for optimal yield. Furthermore, the type and pH of the buffer are essential for obtaining the desired product, as well as the presence of cesium ions. In fact, **1** could not be isolated in the absence of cesium ions, and a sufficient amount was needed to induce precipitation, before filtration and crystallization of the product salt. On the other hand, **CsNa-1** proved to be only slightly soluble in D₂O, and hence our solution studies by ^{183}W and ^{89}Y NMR spectroscopy were unsuccessful. NMR measurements on a freshly prepared reaction solution also remained inconclusive. Attempts to prepare isostructural lanthanide derivatives of **1** resulted in reported structures [20]. Finally, to the best of our knowledge, **1** incorporates the largest number of yttrium ions, together with Francesconi's $[(PY_2W_{10}O_{38})_4(W_3O_{14})]^{30-}$ [35]. Boscovic's penta- and tetra-yttrium-containing $[As^{III}_4(Y^{III}W^{VI}_3)W^{VI}_{44}Y^{III}_4O_{159}(Gly)_8(H_2O)_{14}]^{9-}$ and $[As^{III}_4(Mo^V_2Mo^{VI}_2)W^{VI}_{44}Y^{III}_4O_{160}(Nle)_9(H_2O)_{11}]^{18-}$, respectively, should also be mentioned here [38].

Infrared Spectroscopy. The Fourier transform infrared (FTIR) spectrum of **CsNa-1** (see Figure 4) displays a fingerprint region characteristic for the tungsten-oxo framework, indicating the presence of $\{AsW_9\}$ units. The bands in the range of 940–945 cm^{-1} are associated with the antisymmetric stretching vibrations of the terminal $W=O$ bonds, whereas the bands in the range of 850–900 cm^{-1} can be mainly attributed to the antisymmetric stretching vibrations of the As–O(W). The two bands at about 780 cm^{-1} and 700 cm^{-1} arise from antisymmetric stretching of the W–O(W) bridges, whereas the bands below 650 cm^{-1} are due to bending vibrations of the As–O(W) and the W–O(W) bridges [50]. Furthermore, the bands in the range of 1350 – 1620 cm^{-1} can be assigned to vibrations of the bridging acetate group in the polyanion. The broad band at 3440 cm^{-1} and the strong one at 1620 cm^{-1} correspond to crystal waters.

Thermogravimetric Analysis. Thermogravimetric analysis of **CsNa-1** was performed between 25 and 1000 °C under a nitrogen atmosphere to determine the number of crystal waters (see Figure 3). The weight loss of about 92% between 25 and 230 °C can be assigned to the loss of 230 crystal waters per formula unit. In addition, the second continuous weight loss step from 250 to 550 °C corresponds to the removal of 18 coordinated water molecules and decomposition of the acetate group.

3. Experimental Section

General Methods and Materials. All reagents were used as purchased without further purification. The trilacunary POM precursor $Na_9[B\text{-}\alpha\text{-}AsW_9O_{33}]\cdot27H_2O$ was prepared according to the published procedure, and its purity was confirmed by infrared spectroscopy [51].

Synthesis of $Cs_{8.5}Na_{34.5}[Y_8(CH_3COO)(H_2O)_{18}(As_2W_{19}O_{68})_4(W_2O_6)_2(WO_4)]\cdot230H_2O$ (CsNa-1): 0.50 g (0.20 mmol) $Na_9[B\text{-}\alpha\text{-}AsW_9O_{33}]\cdot27H_2O$ and 0.18 g (0.60 mmol) YCl_3 were added to 20 mL of 1 M NaOAc/HOAc buffer at pH 4.8. The solution was heated to 80 °C for 60 min, cooled to room temperature, and then filtered. Addition of five drops 1.0 M CsCl solution to the filtrate resulted in a precipitate. The reaction mixture was kept at 30 °C for 5 min, and then filtered. The clear colorless filtrate was kept in an open vial at room temperature for slow evaporation. After one week, a colorless crystalline product started to appear. Evaporation was allowed to continue until about half the solvent had evaporated. The solid product was then collected by filtration and air-dried. Yield: 0.25 g (38%). IR (2% KBr pellet, v/cm^{-1}): 1623(s), 949 (m), 864(s), 788(m), 707(m), 454 (w) (see Figure 4). Elemental analysis for $Cs_{8.5}Na_{34.5}[Y_8(CH_3COO)(H_2O)_{18}(As_2W_{19}O_{68})_4(W_2O_6)_2(WO_4)]\cdot230H_2O$ (**CsNa-1**), calcd. (found): Na 2.91% (2.72%), Cs 4.14% (3.90%) As 2.20% (2.09%), Y 2.61% (2.87%), W 54.7% (55.2%).

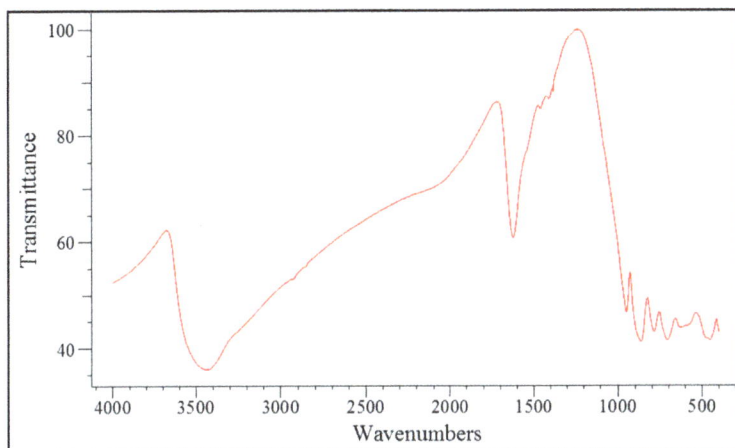

Figure 4. Infrared spectrum of **CsNa-1**.

Instrumentation (IR, TGA). Infrared spectra were recorded on a Nicolet Avatar 370 FT-IR spectrophotometer using KBr pellets. The following abbreviations were used to assign the peak intensities: w = weak, m = medium and s = strong. Thermogravimetric analyses were carried out on a TA Instruments SDT Q600 thermobalance with a 100 mL/min flow of nitrogen; the temperature was ramped from 20 to 1000 °C at a rate of 5 °C /min. Elemental analysis for **CsNa-1** was performed by CNRS, Service Central d'Analyze, Solaize, France.

X-ray Crystallography. A single crystal of **CsNa-1** was mounted on a Hampton cryoloop in light oil for data collection at 173 K. Indexing and data collection were performed on a Bruker D8 SMART APEX II CCD diffractometer with kappa geometry and Mo-Kα radiation (graphite monochromator, λ = 0.71073 Å). Data integration was performed using SAINT [52]. Routine Lorentz and polarization corrections were applied. Multiscan absorption corrections were performed using SADABS [53]. Direct methods (SHELXS97) successfully located the tungsten atoms, and successive Fourier syntheses (SHELXL2014) revealed the remaining atoms [54]. Refinements were full-matrix least-squares against $|F^2|$ using all data. In the final refinement, all non-disordered heavy atoms (As, W, Y, Cs, Na) were refined anisotropically; oxygen atoms and disordered counter cations were refined isotropically. No hydrogen atoms were included in the models. Crystallographic data are summarized in Table 1. We observed an extra W atom (W82) with an occupancy of 16.67%, but could not model its coordination environment due to serious disorder (see CIF file for details). This result implies that there is a small amount of $\{Y_8As_8W_{82}\}$ polyanion impurity present, which we could not avoid during synthesis or eliminate afterwards, in spite of many attempts.

Table 1. Crystal Data for **CsNa-1**.

Formula	$Cs_{8.5}Na_{34.5}[Y_8(CH_3COO)(H_2O)_{18}(As_2W_{19}O_{68})_4(W_2O_6)_2(WO_4)] \cdot 230H_2O$
Formula weight, g/mol	27260.36
Crystal system	Triclinic
Space group	$P\bar{1}$
a, Å	21.7759(5)
b, Å	32.0368(7)
c, Å	33.0799(8)
α, °	94.1720(10)
β, °	107.5370(10)
γ, °	90.5610(10)
Volume, Å³	21934.4(9)
Z	2
D_{calc}, g/cm³	4.127
Absorption coefficient	23.661
F(000)	24464
Crystal size, mm	$0.281 \times 0.16 \times 0.126$
Theta range for data collection, °	3.406–23.257
Completeness to Θ_{max}, %	99.6
Index ranges	$-24 \leqslant h \leqslant 24$ $-35 \leqslant k \leqslant 33$ $-36 \leqslant l \leqslant 36$
Total Reflections	651748
Independent Reflections	62815
Calculated Reflections ($I > 2\sigma$)	49552
R(int)	0.0411
Data / restraints / parameters	62815/0/2553
Goodness-of-fit on F2	1.005
R_1 [a]	0.0643
wR_2 [b]	0.1719
Highest / deepest electron density	5.239/−5.002

[a] $R_1 = \sum ||F_o| - |F_c|| / \sum |F_o|$. [b] $wR_2 = [\sum w (F_o^2 - F_c^2)^2 / \sum w(F_o^2)^2]^{1/2}$.

4. Conclusions

In conclusion, we have synthesized the 8-yttrium(III)-containing 81-tungsto-8-arsenate(III) $[Y_8(CH_3COO)(H_2O)_{18}(As_2W_{19}O_{68})_4(W_2O_6)_2(WO_4)]^{43-}$ (**1**) using a simple, one-pot procedure by reacting the trilacunary POM precursor $[B-\alpha-AsW_9O_{33}]^{9-}$ with Y^{3+} ions in 1 M NaOAc/AcOH buffer at pH 4.8. Polyanion **1** was characterized in the solid state by IR spectroscopy, single crystal XRD, and thermogravimetric analysis. The title polyanion demonstrates the ability of yttrium(III) to act as an efficient linker of POM subunits resulting in large, discrete assemblies. The yttrium(III) ions in **1** possess labile terminal aqua ligands, which allow for Lewis acid type catalysis.

Acknowledgements

U. K. thanks the German Science Foundation (DFG KO-2288/20-1 and KO-2288/8-1) and Jacobs University for research support, and also acknowledges the COST Actions CM1006 (EUFEN) and CM1203 (PoCheMoN). M. I. thanks DAAD and Higher Education Commission of Pakistan for a doctoral fellowship at Jacobs University, and DFG and Institute of Nanotechnology, Karlsruhe Institute of Technology (KIT) for a postdoctoral fellowship. M. I. also thanks the University of Balochistan, Quetta, Pakistan for allowing her to pursue her doctoral and postdoctoral studies at Jacobs University and KIT, respectively. Figures 1 and 2 were generated by *Diamond*, Version 3.2 (copyright Crystal Impact GbR).

Author Contributions

UK proposed the research project and planned for the publication. The manuscript was written by MI and BSB, and then edited by UK. MI performed the lab experiments and BSB did the X-ray diffraction measurements.

Conflicts of Interest

The authors declare no conflict of interest.

References

1. Pope, M.T. *Heteropoly and Isopoly Oxometalates*; Springer: Berlin, Germany, 1983.
2. Pope, M.T.; Müller, A. Polyoxometalate chemistry: An old field with new dimensions in several disciplines. *Angew. Chem., Int. Ed. Engl.* **1991**, *30*, 34–48.
3. Hasenknopf, B.; Micoine, K.; Lacôte, E.; Thorimbert, S.; Malacria, M.; Thouvenot, R. Chirality in polyoxometalate chemistry. *Eur. J. Inorg. Chem.* **2008**, 5001–5013.
4. Kortz, U.; Müller, A.; van Slageren, J.; Schnack, J.; Dalal, N.S.; Dressel, M. Polyoxometalates: Fascinating structures, unique magnetic properties. *Coord. Chem. Rev.* **2009**, *253*, 2315–2327.
5. Kortz, U. Special Issue: Polyoxometalates; John Wiley & Sons, Inc.: Weinheim, Germany, 2009, Volume 34, 5055–5276.
6. Long, D.L.; Tsunashima, R.; Cronin, L. Polyoxometalates: Building blocks for functional nanoscale systems. *Angew. Chem. Int. Ed.* **2010**, *49*, 1736–1758.
7. Izarova, N.V.; Pope, M.T.; Kortz, U. Noble metals in polyoxometalates. *Angew. Chem. Int. Ed.* **2012**, *51*, 9492–9510.
8. Clemente-Juan, J.M.; Coronado, E.; Gaita-Ariño, A. Magnetic polyoxometalates: From molecular magnetism to molecular spintronics and quantum computing. *Chem. Soc. Rev.* **2012**, *41*, 7464–7478.
9. Lv, H.; Geletii, Y.V.; Zhao, C.; Vickers, J.W.; Zhu, G.; Luo, Z.; Song, J.; Lian, T.; Musaev, D.G.; Hill, C.L. Polyoxometalate water oxidation catalysts and the production of green fuel. *Chem. Soc. Rev.* **2012**, *41*, 7572–7589.
10. Pope, M.T.; Kortz, U. Polyoxometalates. In *Encyclopedia of Inorganic and Bioinorganic Chemistry*; Scott, R.A., Ed.; John Wiley: Chichester, UK, 2012.
11. Fang, X.K.; Kögerler, P. PO_4^{3-}-mediated polyoxometalate supercluster assembly. *Angew. Chem. Int. Ed.* **2008**, *47*, 8123–8126.

12. Bassil, B.S.; Ibrahim, M.; Al-Oweini, R.; Asano, M.; Wang, Z.; van Tol, J.; Dalal, N.S.; Choi, K.-Y.; Biboum, R.N.; Keita, B.; Nadjo, L.; Kortz, U. A Planar {$Mn_{19}(OH)_{12}$}$^{26+}$ Unit Incorporated in a 60-Tungsto-6-Silicate Polyanion. *Angew. Chem. Int. Ed.* **2011**, *50*, 5961–5964.

13. Ibrahim, M.; Lan, Y.; Bassil, B.S.; Xiang, Y.; Suchopar, A.; Powell, A.K.; Kortz, U. Hexadecacobalt(II)-containing polyoxometalate-based single-molecule magnet. *Angew. Chem. Int. Ed.* **2011**, *50*, 4708–4711.

14. Zheng, S.-T.; Yang, G.-Y. Recent advances in paramagnetic-TM-substituted polyoxometalates (TM = Mn, Fe, Co, Ni, Cu). *Chem. Soc. Rev.* **2012**, *41*, 7623–7646.

15. Oms, O.; Dolbecq, A.; Mialane, P. Diversity in structures and properties of 3d-incorporating polyoxotungstates. *Chem. Soc. Rev.* **2012**, *41*, 7497–7536.

16. Ibrahim, M.; Xiang, Y.; Bassil, B.S.; Lan, Y.; Powell, A.K.; de Oliveira, P.; Keita, B.; Kortz, U. Synthesis, Magnetism, and Electrochemistry of the Ni_{14}- and Ni_5-Containing Heteropolytungstates [$Ni_{14}(OH)_6(H_2O)_{10}(HPO_4)_4(P_2W_{15}O_{56})_4$]$^{34-}$ and [$Ni_5(OH)_4(H_2O)_4(\beta$-$GeW_9O_{34})(\beta$-$GeW_8O_{30}(OH))$]$^{13-}$. *Inorg. Chem.* **2013**, *52*, 8399–8408.

17. Ibrahim, M.; Haider, A.; Lan, Y.; Bassil, B.S.; Carey, A.M.; Liu, R.; Zhang, G.; Keita, B.; Li, W.; Kostakis, G.E.; Powell, A.K.; Kortz, U. Multinuclear cobalt(II)-containing heteropolytungstates: structure, magnetism, and electrochemistry. *Inorg. Chem.* **2014**, *53*, 5179–5188.

18. Hussain, F.; Gable, W.R.; Speldrich, M.; Kögerler, P.; Boscovic, C. Polyoxotungstate-encapsulated Gd_6 and Yb_{10} complexes. *Chem. Commun.* **2009**, 328–330.

19. Hussain, F.; Conrad, F.; Patzke, G.R. A Gadolinium-bridged polytungstoarsenate(III) nanocluster: [$Gd_8As_{12}W_{124}O_{432}(H_2O)_{22}$]$^{60-}$. *Angew. Chem. Int. Ed.* **2009**, *48*, 9088–9091.

20. Hussain, F.; Spingler, B.; Conrad, F., Speldrich, M., Kögerler, P.; Boskovic, C.; Patzke, G.R. Cesium-templated lanthanoid-containing polyoxotungstates. *Dalton Trans.* **2009**, 4423–4425.

21. Ritchie, C.; Moore, E.G.; Speldrich, M.; Kögerler, P.; Boskovic, C. Terbium polyoxometalate organic complexes: correlation of structure with luminescence properties. *Angew. Chem. Int. Ed.* **2010**, *49*, 7702–7705.

22. Ritchie, C.; Miller, C.E.; Boskovic, C. The generation of a novel polyoxometalate-based 3D framework following picolinate-chelation of tungsten and potassium centres. *Dalton Trans.* **2011**, *40*, 12037–12039.

23. Ritchie, C.; Baslon, V.; Moore, E.G.; Reber, C.; Boskovic, C. Sensitization of lanthanoid luminescence by organic and inorganic ligands in lanthanoid-organic-polyoxometalates. *Inorg. Chem.* **2012**, *51*, 1142–1151.

24. Cotton, S.A. Lanthanides: Comparison to 3d metals. In *Encyclopedia of inorganic and Bioinorganic Chemistry*; Wiley: Weinheim, Germany, 2012.

25. Wassermann, K.; Dickman, M.H.; Pope, M.T. Self-Assembly of supramolecular polyoxometalates: The compact, water-soluble heteropolytungstate Anion [As^{III}_{12} Ce^{III}_{16} $(H_2O)_{36}W_{148}O_{524}$]$^{76-}$. *Angew. Chem. Int. Ed. Engl.* **1997**, *36*, 1445–1448.

26. Bassil, B.S.; Dickman, M.H.; Römer, I.; von der Kammer, B.; Kortz, U. The tungstogermanate [$Ce_{20}Ge_{10}W_{100}O_{376}(OH)_4(H_2O)_{30}$]$^{56-}$: A polyoxometalate containing 20 cerium(III) Atoms *Angew. Chem. Int. Ed.* **2007**, *46*, 6192–6195.

27. Reinoso, S.; Giménez-Marqués, M., Galán-Mascarós, J.R.; Vitoria, P.; Gutiérrez-Zorrilla, J.M. Giant crown-shaped polytungstate formed by self-assembly of Ce^{III}-stabilized dilacunary Keggin fragments. *Angew. Chem. Int. Ed.* **2010**, *49*, 8384–8388.

28. AlDamen, M.A.; Clemente-Juan, J.M.; Coronado, E.; Martí-Gastaldo, C.; Gaita-Ariňo, A. Mononuclear lanthanide single-molecule magnets based on polyoxometalates. *J. Am. Chem. Soc.* **2008**, *130*, 8874–8875.

29. Boglio, C.; Lemiére, G.; Hasenknopf, B.; Thorimbert, S.; Lacôte, E.; Malacria, M. Lanthanide complexes of the monovacant Dawson polyoxotungstate [α_1-P$_2$W$_{17}$O$_{61}$]$^{10-}$ as selective and recoverable Lewis acid catalysts. *Angew. Chem. Int. Ed.* **2006**, *45*, 3324–3327.

30. Dupre, N.; Remy, P.; Micoine, K.; Boglio, C.; Thorimbert, S.; Lacôte, E.; Hasenknopf, B.; Malacria, M. Chemoselective catalysis with organosoluble Lewis acidic polyoxotungstates. *Chem. Eur. J.* **2010**, *16*, 7256–7264.

31. Kikukawa, Y.; Suzuki, K.; Sugawa, M.; Hirano, T.; Kamata, K.; Yamaguchi, K.; Mizuno, N. Cyanosilylation of carbonyl compounds with trimethylsilyl cyanide catalyzed by an Yttrium-pillared silicotungstate dimer. *Angew. Chem. Int. Ed.* **2012**, *51*, 3686–3690.

32. Suzuki, K.; Sugawa, M.; Kikukawa, Y.; Kamata, K.; Yamaguchi, K.; Mizuno, N. Strategic design and refinement of Lewis acid–base catalysis by rare-earth-metal-containing polyoxometalates. *Inorg. Chem.* **2012**, *51*, 6953–6961.

33. Peacock, R.D.; Weakley, T.J. R. Heteropolytungstate complexes of the lanthanide elements. Part I. Preparation and reactions. *J. Chem. Soc. A* **1971**, 1836–1839.

34. Griffith, W.P.; Morley-Smith, N.; Nogueira, H.I. S.; Shoair, A.G. F.; Suriaatmaja, M.; White, A.J.P.; Williams, D.J. Studies on polyoxo and polyperoxo-metalates: Part 7. Lanthano- and thoriopolyoxotungstates as catalytic oxidants with H$_2$O$_2$ and the X-ray crystal structure of Na$_8$[ThW$_{10}$O$_{36}$]·28H$_2$O. *J. Organomet. Chem.* **2000**, *607*, 146–155.

35. Howell, R.C.; Perez, F.G.; Horrocks, W.D.; Jain, S.; Rheingold, A.L.; Francesconi, L.C. A new type of heteropolyoxometalates formed from lacunary polyoxotungstate ions and europium or yttrium cations. *Angew. Chem. Int. Ed.* **2001**, *40*, 4031–4034.

36. Fang, X.; Anderson, M.T.; Neiwert, W.A.; Hill, C.L. Yttrium polyoxometalates. synthesis and characterization of a carbonate-encapsulated sandwich-type complex. *Inorg. Chem.* **2003**, *42*, 8600–8602.

37. Fang, X.; Anderson, T.M.; Benelli, C.; Hill, C.L. Polyoxometalate-supported Y– and Yb^{III}– hydroxo/oxo clusters from carbonate-assisted hydrolysis. *Chem. Eur. J.* **2005**, *11*, 712–718.

38. Vonci, M. Akhlaghi Bagherjeri, F.; Hall, P.D.; Gable, R.W.; Zavras, A.; O'Hair, R.A. J.; Liu, Y.; Zhang, J.; Field, M.R.; Taylor, M.B.; Du Plessis, J.; Bryant, G.; Riley, M.; Sorace, L.; Aparicio, P.A.; López, X.; Poblet, J.M.; Ritchie, C.; Boskovic, C. Modular molecules: Site-selective metal substitution, photoreduction, and chirality in polyoxometalate hybrids. *Chem. Eur. J.* **2014**, *20*, 14102–14111.

39. Xue, G.; Liu, B.; Hua, H.; Yang, J.; Wang, J.; Fu, F. Large heteropolymetalate complexes formed from lanthanide (Y, Ce, Pr, Nd, Sm, Eu, Gd), nickel cations and cryptate [As$_4$W$_{40}$O$_{140}$]$^{28-}$ synthesis and structure characterization. *J. Mol. Str.* **2004**, *690*, 95–103.

40. Wang, J.P.; Duan, X.Y.; Du, X.D.; Niu, J.Y. Novel rare earth germanotungstates and organic hybrid derivatives: Synthesis and structures of M/[α-GeW$_{11}$O$_{39}$] (M = Nd, Sm, Y, Yb) and Sm/[α-GeW$_{11}$O$_{39}$](DMSO). *Cryst. Growth Des.* **2006**, *6*, 2266–2270.

41. Zhang, S.W.; Wang, Y.; Zhao, J.W.; Ma, P.T.; Wang, J.P.; Niu, J.Y. Two types of oxalate-bridging rare-earth-substituted Keggin-type phosphotungstates {[(α-PW$_{11}$O$_{39}$)RE(H$_2$O)]$_2$(C$_2$O$_4$)}$^{10-}$ and {(α-x-PW$_{10}$O$_{38}$)RE$_2$(C$_2$O$_4$)(H$_2$O)$_2$}$^{3-}$. *Dalton Trans.* **2012**, *41*, 3764–3772.

42. Wang, Y.; Sun, X.P.; Li, S.Z.; Ma, P.T.; Wang, J.P.; Niu, J.Y. Synthesis and magnetic properties of tartrate-bridging rare-earth-containing polytungstoarsenate aggregates from an adaptive precursor [As$_2$W$_{19}$O$_{67}$(H$_2$O)]$^{14-}$. *Dalton Trans.* **2015**, *44*, 733–738.

43. Barsukova, M.; Dickman, M.H.; Visser, E.; Mal, S.S.; Kortz, U. Synthesis and structural characterization of the yttrium containing isopolytungstate [YW$_{10}$O$_{36}$]$^{9-}$. *Z. Anorg. Allg. Chem.* **2008**, *634*, 2423–2427.

44. Ismail, A.H.; Dickman, M.H.; Kortz, U. 22-Isopolytungstate fragment [H$_2$W$_{22}$O$_{74}$]$^{14-}$ coordinated to lanthanide ions. *Inorg. Chem.* **2009**, *48*, 1559–1565.

45. Hussain, F.; Degonda, A.; Sandriesser, S.; Fox, T.; Mal, S.S.; Kortz, U.; Patzke, G.R. Yttrium containing head-on complexes of silico- and germanotungstate: Synthesis, structure and solution properties. *Inorg. Chim. Acta.* **2010**, *363*, 4324–4328.

46. Ibrahim, M.; Mal, S.S.; Bassil, B.S.; Banerjee, A.; Kortz, U. Yttrium(III)-containing tungstoantimonate(III) stabilized by capping, tetrahedral WO$_4^{2-}$ unit, [{Y(α-SbW$_9$O$_{31}$(OH)$_2$)(CH$_3$COO)(H$_2$O)}$_3$(WO$_4$)]$^{17-}$. *Inorg. Chem.* **2011**, *50*, 956–960.

47. Barsukova, M.; Izarova, N.V.; Ngo Biboum, R.; Keita, B.; Nadjo, L.; Ramachandran, V.; Dalal, N.S.; Antonova, N.S.; Carbó, J.J.; Poblet, J.M.; Kortz, U. Polyoxopalladates Encapsulating Yttrium and Lanthanide Ions, [X^{III}Pd$^{II}_{12}$(AsPh)$_8$O$_{32}$]$^{5-}$ (X = Y, Pr, Nd, Sm, Eu, Gd, Tb, Dy, Ho, Er, Tm, Yb, Lu). *Chem.-Eur. J.* **2010**, *16*, 9076–9085.

48. Kortz, U.; Savelieff, M.G.; Bassil, B.S.; and Dickman, M.H. A Large, Novel Polyoxotungstate [As$^{III}_6$W$_{65}$O$_{217}$(H$_2$O)$_7$]$^{26-}$. *Angew. Chem. Int. Ed. Engl.* **2001**, *40*, 3384–3386.

49. Brown, I.D.; Altermatt, D. Bond-valence parameters obtained from a systematic analysis of the Inorganic Crystal Structure Database. *Acta Crystallogr.* **1985**, *B41*, 244–247.

50. Thouvenot, R.; Fournier, M.; Franck, R.; Rocchiccioli-Deltcheff, C. Vibrational investigations of polyoxometalates. 3. Isomerism in molybdenum(VI) and tungsten(VI) compounds related to the Keggin structure. *Inorg. Chem.* **1984**, *23*, 598–605.

51. Tourné, C.; Revel, A.; Tourné, G.; Vendrell, M. Heteropolytungstates containing elements of phosphorus family with degree of oxidation (iii) or (v)—identification of species having composition X$_2$W$_{19}$ and XW$_9$ (X = P, As, Sb, Bi) and relation to those with composition XW$_{11}$. *C. R. Acad. Sci. Paris, Ser. C* **1973**, *277*, 643–645.

52. *SAINT*, Version 2.1-4; Bruker AXS Inc.: Madison, WI, USA, 2007.

53. Sheldrick, G.M. *SADABS*; Program for empirical absorption correction of area detector data; University of Göttingen: Göttingen, Germany, 1996

54. Sheldrick, G.M. *SHELX-97/2013*; Program for solution of crystal structures; University of Göttingen: Göttingen, Germany, 2013.

Fully Oxidized and Mixed-Valent Polyoxomolybdates Structured by Bisphosphonates with Pendant Pyridine Groups: Synthesis, Structure and Photochromic Properties

Olivier Oms, Tarik Benali, Jérome Marrot, Pierre Mialane, Marin Puget,
Hélène Serier-Brault, Philippe Deniard, Rémi Dessapt and Anne Dolbecq

Abstract: Hybrid organic-inorganic polyoxometalates (POMs) were synthesized in water by the reaction of a Mo^{VI} precursor with bisphosphonate ligands functionalized by pyridine groups. The fully oxidized POM $[(Mo^{VI}_3O_8)_2(O)(O_3PC(O)(C_3H_6NH_2CH_2C_5H_4NH)PO_3)_2]^{4-}$ has been isolated as water insoluble pure Na salt (**NaMo₆(Ale-4Py)₂**) or mixed Na/K salt (**NaKMo₆(Ale-4Py)₂**) and their structure solved using single-crystal X-ray diffraction. The mixed-valent complex $[(Mo^V_2O_4)(Mo^{VI}_2O_6)_2\{O_3PC(O)(C_3H_6N(CH_2C_5H_4N)_2(Mo^{VI}O_3))PO_3\}_2]^{8-}$ was obtained as an ammonium salt (**NH₄Mo₆(AlePy₂Mo)₂**), in the presence of a reducing agent (hydrazine). ^{31}P NMR spectroscopic studies in aqueous media have allowed determining the pH stability domain of **NH₄Mo₆(AlePy₂Mo)₂**. **NaMo₆(Ale-4Py)₂** and **NaKMo₆(Ale-4Py)₂** exhibit remarkable solid-state photochromic properties in ambient conditions. Under UV excitation, they develop a very fast color-change from white to deep purple and proved to be the fastest photochromic organoammonium/POM systems. The coloration kinetics has been fully quantified for both salts and is discussed in light of the hydrogen-bonding networks.

Reprinted from *Inorganics*. Cite as: Oms, O.; Benali, T.; Marrot, J.; Mialane, P.; Puget, M.; Serier-Brault, H.; Deniard, P.; Dessapt, R.; Dolbecq, A. Fully Oxidized and Mixed-Valent Polyoxomolybdates Structured by Bisphosphonates with Pendant Pyridine Groups: Synthesis, Structure and Photochromic Properties. *Inorganics* **2015**, *3*, 279-294.

1. Introduction

Polyoxometalates (POMs) are discrete soluble metal oxide fragments of d-block transition metals in high oxidation states ($W^{V,VI}$, $Mo^{V,VI}$, $V^{IV,V}$) with a wide range of properties going from magnetism [1] to biology [2] and catalysis [3]. Among these properties, photochromism attracts increasing interest due to the potential applications of POM-based photoresponsive molecular materials as reversible memory photodevices [4]. Reversible color-change effects under UV irradiation have been first evidenced on alkylammonium polyoxomolybdates [5,6]. The mechanism has been rationalized considering the transfer of an electron from an O atom of the POM unit to a Mo^{VI} center, coupled with the transfer of a hydrogen atom (see below). These materials are ionic crystals combining anionic POMs and cationic alkylammonium counter-ions. However, an alternative and attractive approach to get functionalized POM-based materials consists in the covalent attachment of organic ligands directly onto the POM core [7]. It has indeed been demonstrated that the presence of covalently bonded organic ligands can bring new functionalities and allow the elaboration of unique materials and devices [8], as demonstrated by the characterization of surfaces patterned with POMs [9,10], the synthesis of fluorescent [11] and catalytic [12] POMs, or the elaboration of extended frameworks based on POM building units [13]. We have developed for several years the functionalization of POMs by bisphosphonates (BPs) with the general formula $H_2O_3PC(R_1)(R_2)PO_3H_2$ (R_1 = H, OH) [14]. Once deprotonated, BPs act as multidentate ligands and can form stable POM complexes [15]. We initiated a few years ago a study on the synthesis of polyoxomolybdates with grafted BPs, starting with Mo^V/BP complexes. Alendronic acid (Ale, Scheme 1) and its aminophenol derivative were reacted with Mo^V ions, leading to low nuclearity complexes [16]. The reaction of Mo^{VI} ions with Ale was then investigated and we showed that the synthetic pH governs the formation of one complex over the other. The mononuclear POM $[(MoO_2)(Ale)_2]^{6-}$ ($Mo(Ale)_2$) [17] was isolated at pH 7.5, the hexanuclear complex $[(Mo_3O_8)_2O(Ale)_2]^{6-}$ ($Mo_6(Ale)_2$) [18] at pH = 4.5 and the dodecanuclear complex $[(Mo_3O_8)_4(Ale)_4]^{8-}$ ($Mo_{12}(Ale)_4$) at pH = 3.0 [16]. The hexa and dodecanuclear hybrid POMs possess highly efficient solid-state photochromic properties which are brought by the grafting of the organic ligand [18]. We then decided to post-functionalize the Ale ligand with pyridine groups with the aim to explore its chelating properties towards additional metal ions and study their influence on the structure and properties of the molybdobisphosphonate complexes. We thus report here our results with two BP ligands bearing one or two pyridine groups, abbreviated Ale-4Py and AlePy$_2$, respectively (Scheme 1). The photochromic properties of the Mo^{VI} derivatives with Ale-4Py have been studied and are compared with those of related complexes.

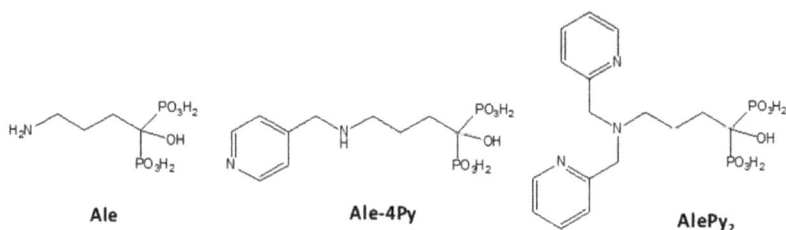

Ale   **Ale-4Py**   **AlePy$_2$**

Scheme 1. Structure and abbreviations of the ligands used in this work.

2. Results and Discussion

2.1. Synthesis and Characterization

The new BP Ale-4Py ligand has been obtained by coupling Ale and 4-pyridinecarboxaldehyde, followed by the reduction of the resulting imine derivative. We explored the synthesis of hybrid polyoxomolybdates functionalized by BP ligands, namely Ale-4Py and AlePy$_2$, using two different synthetic protocols (Scheme 2). The first one consists in reacting an aqueous solution of Mo^{VI} oxoanions with Ale-4Py. The pH is set at 3.1 by addition of 2 M HCl and the solution is heated to 80 °C for 30 min. White crystals of the pure Na (**NaMo$_6$(Ale-4Py)$_2$**) and of the mixed Na/K salt (**NaKMo$_6$(Ale-4Py)$_2$**) of the hybrid POMs were isolated after slow evaporation and characterized by single crystal X-ray diffraction and elemental analysis (see below). Surprisingly, the X-ray study revealed the presence of the hexanuclear anion $[(Mo^{VI}_3O_8)_2(O)(O_3PC(O)$ $(C_3H_6NH_2CH_2C_5H_4NH)PO_3)_2]^{4-}$ (**Mo$_6$(Ale-4Py)$_2$**). Indeed, the crystallization of a dodecanuclear POM of general formula $Mo_{12}(BP)_4$ was expected as these species were shown to be stable at pH 3 while the formation of the hexanuclear POMs $Mo_6(BP)_2$ (BP = Ale, $O_3PC(C_3H_6NRR'R")(O)PO_3$ with R, R', R" = H or CH_3) occurs at pH ~ 6 [18].

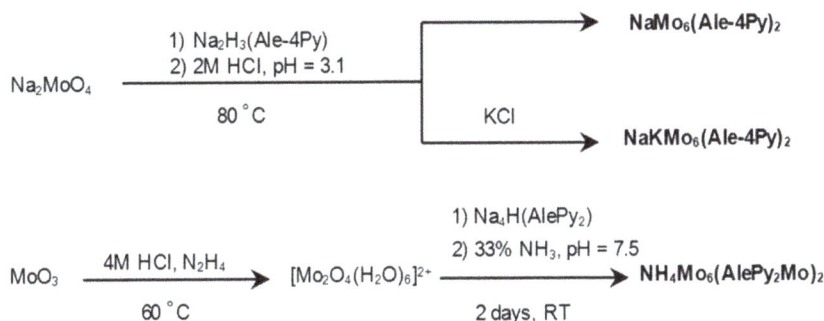

Scheme 2. Synthetic pathways.

The second synthetic pathway consists in using a solution of $[Mo^V_2O_4(H_2O)_6]^{2+}$ ions prepared by reduction of MoO_3 by hydrazine in 4 M HCl. $AlePy_2$ is added in the solution and the pH is adjusted to 7.5 by addition of aqueous ammonium hydroxide. An abundant precipitate forms and progressively redissolves while stirring for two days at room temperature. Red crystals (Figure 1) are isolated after slow evaporation of the resulting solution. The formula of the hybrid POM, $[(Mo^V_2O_4)(Mo^{VI}_2O_6)_2\{O_3PC(O)(C_3H_6N(CH_2C_5H_4N)_2(Mo^{VI}O_3))PO_3\}_2]^{8-}$ **(Mo_6(AlePy$_2$Mo)$_2$)** has been deduced from single crystal X-ray analysis (see below) and confirmed by elemental analysis. The mixed-valent anion contains two Mo^V and six Mo^{VI} ions, which are formed by slow reoxidation in air of the Mo^V ions of the reacting solution. Such reoxidation has been previously observed [16,19,20]. The red color of the complex is correlated to the presence of the dimeric unit $\{Mo^V_2O_4\}$, which can be found in oxomolybdenum(V) clusters isolated in organic solvents [21] as well as in cyclic POMs and ε-Keggin type species synthesized in aqueous solutions [14], and is the indication of electron localization while the blue color of mixed-valent Mo species is the sign of electron delocalization [22].

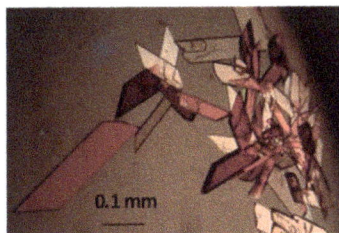

Figure 1. Picture of the crystals of **NH$_4$Mo$_6$(AlePy$_2$Mo)$_2$** in the synthetic medium.

NaMo$_6$(Ale-4Py)$_2$ and **NaKMo$_6$(Ale-4Py)$_2$** are insoluble in all common solvents. Therefore, it was not possible to investigate their stability in solution by ^{31}P NMR spectroscopy. Nevertheless, it has been possible to record the ^{31}P NMR spectra of the synthetic solutions (Figure 2a). Three peaks around 22 ppm are observed which can be tentatively attributed to the presence of conformers in equilibrium in solution [18].

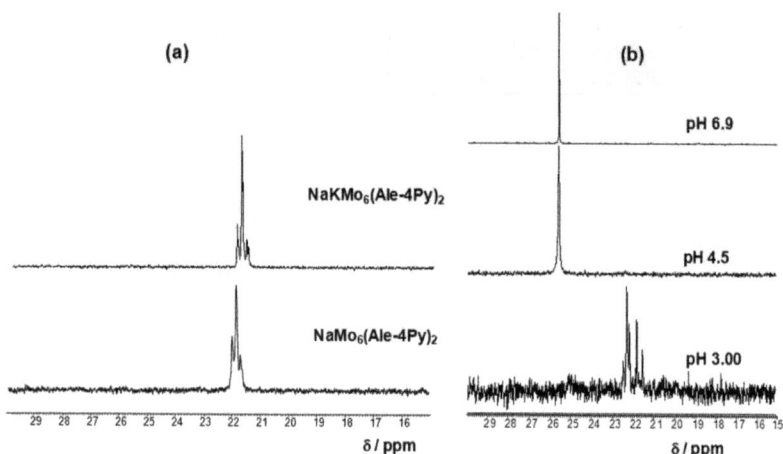

Figure 2. ^{31}P NMR spectra of a) the synthetic medium of **NaMo$_6$(Ale-4Py)$_2$** and **NaKMo$_6$(Ale-4Py)$_2$** and b) an aqueous solution of **NH$_4$Mo$_6$(AlePy$_2$Mo)$_2$** at various pH.

Crystals of **NH$_4$Mo$_6$(AlePy$_2$Mo)$_2$** have been dissolved in D$_2$O and the pH adjusted to 7. A single peak at 25.6 ppm is observed by ^{31}P{^{1}H} NMR (Figure 2b). This indicates that the two P atoms in each BP ligand are magnetically equivalent although they appear inequivalent in the crystal structure (Figure S1). This is in contrast also with the observation of doublets at 30.1 and 26.2 ppm on the ^{31}P{^{1}H} NMR spectrum of a POM possessing the {(Mo$^V_2$O$_4$)(Mo$^{VI}_2$O$_6$)$_2$(BP)$_2$} (BP = O$_3$PC(O)(CH$_2$-3-C$_5$NH$_4$)PO$_3$ = Ris) core [20]. The ^{31}P NMR spectrum of **NH$_4$Mo$_6$(AlePy$_2$Mo)$_2$** at pH 4.5 is identical, showing the stability of the POM at this pH. In contrast, when the pH decreases to 3, the evolution of the ^{31}P NMR spectrum together with the appearance of a blue color clearly indicate the decomposition of the POM.

2.2. Structures

The hexanuclear anion [(Mo$^{VI}_3$O$_8$)$_2$(O)(O$_3$PC(O)(C$_3$H$_6$NH$_2$CH$_2$C$_5$H$_4$NH)PO$_3$)$_2$]$^{4-}$ (**Mo$_6$(Ale-4Py)$_2$**), common in the structures of **NaMo$_6$(Ale-4Py)$_2$** and **NaKMo$_6$(Ale-4Py)$_2$**, contains two trinuclear {Mo$_3$O$_8$} fragments connected by a central oxo bridge (Figure 3), encountered also in structures reported with etidronate [23], alendronate [17,24] and its methyl derivatives [18]. Each trinuclear unit is built of two face-sharing octahedra which share also a corner with a third Mo^{VI} octahedron. The three octahedra are linked to one bisphosphonate ligand via four Mo-O-P bonds with the phosphonate groups and one Mo-O-C bond with the deprotonated hydroxo group. Two conformations have been previously observed for the hexanuclear complexes of general formula [(Mo$^{VI}_3$O$_8$)$_2$(O)(BP)$_2$]$^{n-}$ [14]. In the A conformer (Figure 3a), the six Mo^{VI} ions are approximately coplanar. This conformer is the most frequently encountered, and is found in particular in

$Rb_{0.25}(NH_4)_{5.75}[(Mo^{VI}_3O_8)_2(O)(Ale)_2]$ [18]. The *B* conformer (Figure 3b), observed for example in $Na_6[(Mo^{VI}_3O_8)_2(O)(Ale)_2]$ [24], results from a rotation around the central oxygen atom of a trimeric unit, so that the three Mo^{VI} ions of a trimeric unit are located in a plane perpendicular to the three Mo^{VI} ions of the other trimeric unit. The conformers are very close in energy, and the stabilization of one conformer or another depends on the inter- and intra-molecular interactions governed by the crystal packing. A comparison of the structures of the anions in **NaMo₆(Ale-4Py)₂** and **NaKMo₆(Ale-4Py)₂** with those of the alendronate derivatives shows that the **Mo₆(Ale-4Py)₂** entities in **NaMo₆(Ale-4Py)₂** and **NaKMo₆(Ale-4Py)₂** adopt the *B* conformation. The orientation of the organic ligand is slightly different from one compound to the other.

In the unit cell, the anions alternate with alkali counter-ions (4 Na^+ ion in **NaMo₆(Ale-4Py)₂** and 3.5 K^+ and 0.5 Na^+ in **NaKMo₆(Ale-4Py)₂**) and the crystal packing thus appear quite different in the two structures (Figure S2). Despite this, the hydrogen bond networks involving the POM units appear remarkably similar. Both intra and intermolecular hydrogen bonds are observed between the protonated amino and pyridine groups of the Ale-4Py ligands and bridging oxygen atoms of the POM cores or crystallized water molecules (Figure 4 and Table S1). The structure of **NaMo₆(Ale-4Py)₂** contains one crystallographically independent **Mo₆(Ale-4Py)₂** anion. As shown in Figure 4a, two of them are connected together into a supramolecular dimer {**Mo₆(Ale-4Py)₂**}₂ via two H-bonds involving two of the four pyridinium groups (the other two being connected to water molecules). Each dimer also develops eight additional N-H⋯O interactions: four of them involve the protonated amino groups of the four Ale-4Py ligands and the oxygen atoms of the two Mo₆(Ale-4Py)₂ units, while the other four connect the protonated amino groups with water molecules.

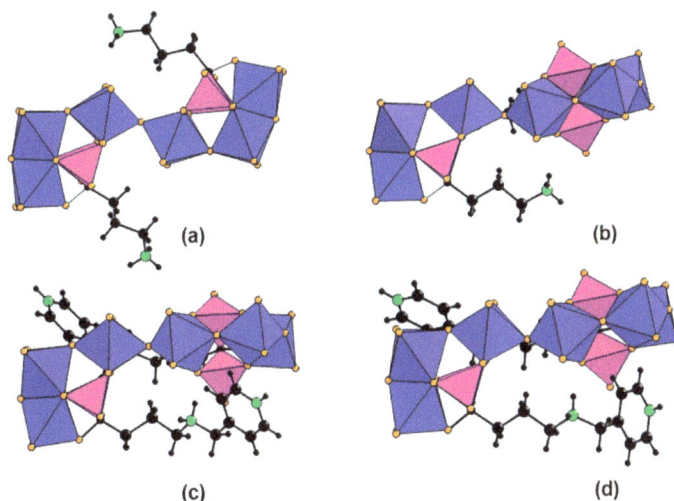

Figure 3. Mixed polyhedral/ball-and-stick representation of the hexanuclear POM of general formula $[(Mo^{VI}_3O_8)_2(O)(BP)_2]^{n-}$ in (a) $Rb_{0.25}(NH_4)_{5.75}[(Mo^{VI}_3O_8)_2(O)(Ale)_2]$[18] (*A* conformation), (b) $Na_6[(Mo^{VI}_3O_8)_2(O)(Ale)_2]$[24] (*B* conformation), (c) **NaMo₆(Ale-4Py)₂** and (d) **NaKMo₆(Ale-4Py)₂**; blue octahedra: $Mo^{VI}O_6$, pink tetrahedra: PO_4, orange spheres: O, black spheres: C, green spheres: N.

The structure of **NaKMo₆(Ale-4Py)₂** is built upon two crystallographically independent **Mo₆(Ale-4Py)₂** units which are assembled into two supramolecular dimers (Figure 4b). The relative arrangement of anions within the dimers is as in **NaMo₆(Ale-4Py)₂**, and therefore, very similar N-H⋯O contacts are developed.

The **Mo₆(AlePy₂Mo)₂** POM contains a mixed-valent $\{(Mo^V_2O_4)(Mo^{VI}_2O_6)_2(BP)_2\}$ core (Figure 5). Two face-sharing Mo^{VI} octahedra are connected symmetrically to a $\{Mo^V_2O_4\}$ dimeric unit via a common oxygen atom and two pentadentate AlePy₂ ligands. We can note that AlePy₂ and Ale-4Py behave very differently. Indeed, while in Ale-4Py the aminopyridine fragments cannot act as chelating groups, the two pyridine and the alkylammino groups of AlePy₂ can strongly bind to an additional metal ion. This is highlighted by the fact that in **NaKMo₆(Ale-4Py)₂** and **NaMo₆(Ale-4Py)₂** the pyridine and alkylammonium groups do not act as a ligand. In contrast in **Mo₆(AlePy₂Mo)₂**, the nitrogen atoms of the two pyridine and of the amino groups of the AlePy₂ ligand are bound to an isolated Mo^{VI} ion in octahedral coordination. The coordination sphere of this ion is completed by three terminal oxo groups. Valence bond calculations confirm the oxidation states of the Mo ions in this octanuclear complex (Figure S1). It can be noticed that the oxygen atoms of the POM develop hydrogen bonding interactions exclusively with water molecules (Figure S3a). In the solid, there is a segregation of the organic and inorganic parts with the formation of double layers of Mo₆ clusters at $z = \frac{1}{2}$ and Mo-organic regions at $z = 0$ (Figure S3b).

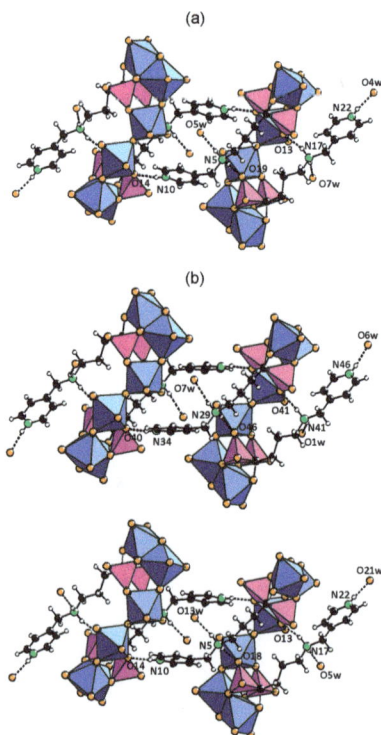

Figure 4. Hydrogen bonding schemes involving the supramolecular dimers {Mo₆(Ale-4Py)₂}₂ in **(a)** **NaMo₆(Ale-4Py)₂** and **(b)** **NaKMo₆(Ale-4Py)₂**.

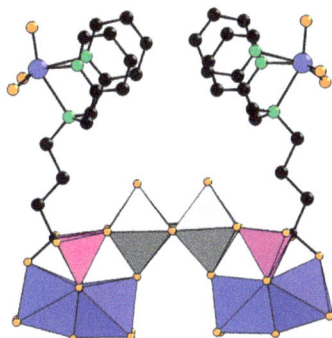

Figure 5. Mixed polyhedral/ball-and-stick representation of the mixed-valent POM [(Mo$^V_2$O$_4$)(Mo$^{VI}_2$O$_6$)$_2$\{O$_3$PC(O)(C$_3$H$_6$N(CH$_2$C$_5$H$_4$N)$_2$(Mo^{VI}O$_3$))PO$_3$\}$_2$]$^{8-}$ in **NH$_4$Mo$_6$(AlePy$_2$Mo)$_2$**; blue octahedra: Mo^{VI}O$_6$, grey octahedra: Mo^{V}O$_6$, blue spheres: Mo^{VI}, pink tetrahedra: P, orange spheres: O, black spheres: C, green spheres: N; hydrogen atoms have been omitted for clarity.

2.3. Solid-State Photochromic Properties

The solid-state photochromic properties of **NaMo$_6$(Ale-4Py)$_2$** and **NaKMo$_6$(Ale-4Py)$_2$** have been thoroughly investigated in ambient conditions by diffuse reflectance spectroscopy. Under low-power UV excitation at λ_{ex} = 365 nm (3.34 eV), the two compounds exhibit strong photochromic responses. The color of the white microcrystalline powders gradually shifts to deep purple with a remarkable coloration contrast, and finally reaches an eye-detected saturation level after only 20 min (Figure 6a and Figure S4a for **NaKMo$_6$(Ale-4Py)$_2$** and **NaMo$_6$(Ale-4Py)$_2$**, respectively). As displayed in Figure 6b, the color change of **NaKMo$_6$(Ale-4Py)$_2$** deals with the growth of a broad absorption band in the visible range at λ_{max} = 508 nm (2.44 eV) that mainly dictates the photogenerated hue, and a second very less-intense absorption band rising up at ~750 nm (1.65 eV). Similar absorption bands are also observed for **NaMo$_6$(Ale-4Py)$_2$** (Figure S4). It is worth noting that these bands are quite comparable with those of other photochromic Mo$_6$(BP)$_2$ compounds [18], and then, they well characterize the photoreduced Mo$_6$ core. In such photochromic organoammonium/POM systems, the UV excitation transfers an electron from an O atom of the POM unit to the adjacent Mo^{6+} site [5,25]. This electron/hole pair segregation is coupled with the transfer of a hydrogen atom from the ammonium group to an adjacent \{MoO$_6$\} octahedron. The H atom moves along the hydrogen bond creating a \{Mo^{5+}(OH)O$_5$\} site. The coloration is then due to the photoreduction of Mo^{6+} (4d^0) to Mo^{5+} (4d^1) cations and occurs via d–d transitions and/or Mo^{6+}/Mo^{5+} intervalence transfers. The photocoloration rate is strongly correlated to the strength of the N$^+$-H bonds considering that, the lower the dissociation energy of the N$^+$-H bond, the faster the coloration speed [6].

The coloration kinetics of **NaMo$_6$(Ale-4Py)$_2$** and **NaKMo$_6$(Ale-4Py)$_2$** has been quantified by monitoring the evolution of their reflectivity values *versus* the UV irradiation time (R(*t*)). According to the photocoloration kinetics model of organoammonium/POM systems, the decrease of $R^{\lambda max}$(*t*) (*i.e.*, the reflectivity at the wavelength of the maximum photogenerated absorption in the visible) with the UV irradiation time is correlated to the decrease of the concentration of reducible Mo^{6+} cations, according to a pseudo second-order kinetic law [6]. Herein, the $R^{508}(t)$ *vs.* *t* curves for **NaMo$_6$(Ale-4Py)$_2$** and **NaKMo$_6$(Ale-4Py)$_2$** are perfectly fitted as $R^{508}(t) = a/(bt+1) + R^{508}(\infty)$ (Figure 6c). $R^{508}(\infty)$ is the reflectivity value at the end of the photochromic process, that is at $t = \infty$. The a parameter is defined as a = $R^{508}(0)$-$R^{508}(\infty)$, *i.e.* the difference between the reflectivity values just before UV illumination ($t = 0$) and at $t = \infty$. The b parameter is defined as b = $k^c \times C_{6+,r}(0)$, where k^c is the coloration rate constant, and $C_{6+,r}(0)$ is the initial concentration of photoreducible Mo^{6+} centers per unit volume. The details of the coloration kinetic parameters are given in Table S2. **NaMo$_6$(Ale-4Py)$_2$** and **NaKMo$_6$(Ale-4Py)$_2$** exhibit a remarkably fast photocoloration rates well characterized by a very short coloration half-time ($t_{1/2} = 0.37$ min for both compounds). The very fast photoreduction rate of the Mo^{6+} ions is also well assessed by monitoring the evolution of the photoreduction degree Y(t) which reaches 98% for **NaMo$_6$(Ale-4Py)$_2$** and 97% for **NaKMo$_6$(Ale-4Py)$_2$** after only 20 min of UV irradiation (Figure S5).

In addition, for two hybrid materials, *i* and *j* built upon the same POM unit (then which develop the same photogenerated absorption bands), the relative coloration rate constants k^c_i/k^c_j can be extracted from the ratio b$_i$a$_j$/b$_j$a$_i$, to compare their coloration rates [6]. Here, $k^c_{NaMo6(Ale-4Py)2}$/ $k^c_{NaKMo6(Ale-4Py)2}$ = 1.1, and both compounds exhibit the same photocoloration rates. This quite evidences that in these two hybrid systems that contain the same **Mo$_6$(Ale-4Py)$_2$** anion and very similar H-bonding networks, the intrinsic photochromic process is not influenced by the packing with the other alkali countercations. Noticeably, **NaMo$_6$(Ale-4Py)$_2$** and **NaKMo$_6$(Ale-4Py)$_2$** exhibit remarkably improved solid-state photochromic performances compared to those of **Mo$_6$-Ale** that was, to date, the fastest photochromic members of the Mo$_6$(BP)$_2$ series [18,26] (Table S2) (for example, the coloration rate constant ratio $k^c_{NaKMo6(Ale-4Py)4}$/$k^c_{Mo6-Ale}$ reaches 10.6). Consequently, these two new compounds are the most efficient known photochromic organoammonium/POM systems (Table 1). It should be reasonably assumed that the very high photochromic performances of **NaMo$_6$(Ale-4Py)$_2$** and **NaKMo$_6$(Ale-4Py)$_2$** should originate from a beneficial combination of three main factors that are: (i) the presence of the easily photoreducible Mo$_6$ core [18], (ii) a great number of N-H···O contacts between the **Mo$_6$(Ale-4Py)$_2$** units and the organoammonium groups of the Ale-4Py ligands, and (iii) the implication in the H-bonding interactions of the diprotonated amino group which possesses a low dissociation energy of the N^+-H bond [6].

NaMo$_6$(Ale-4Py)$_2$ and **NaKMo$_6$(Ale-4Py)$_2$** show reversible color-change effects. The bleaching is correlated to the back oxidation of the Mo^{5+} cations by O$_2$ in ambient conditions [27]. After switching off the UV irradiation, the bleaching occurs slowly in air and at room temperature. No color change occurs when the colored samples are kept in O$_2$ free atmosphere, and at the opposite the bleaching is much faster when samples are exposed to O$_2$ flux. The fading process is also strongly accelerated by keeping the samples in air under moderate heating at 40 °C (the deep purple color disappears after a few minutes). To date, about 15 coloration/fading cycles at room temperature as well as at 40 °C can be performed without detecting any fatigue resistance with naked eyes.

Table 1. Comparison of the $t_{1/2}$ values for various photochromic organoammonium/POM systems.

Compound	λ_{max} (nm) [a]	$t_{1/2}$ (min) [b]	Ref.
NaMo$_6$(Ale-4Py)$_2$	508	0.37	this work
NaKMo$_6$(Ale-4Py)$_2$	508	0.37	this work
Mo$_6$(Ale)$_2$	508	2.87	[18]
Mo$_6$(Ale-1C$_a$)$_2$	508	31.25	[18]
Mo$_6$(Ale-1C$_b$)$_2$	508	4.78	[18]
Mo$_6$(Ale-2C)$_2$	508	12.82	[18]
Mo$_6$(1C-Ale)$_2$ Mo$_6$(Ris)$_2$	508	5.71	[18]
Mo$_6$(Ris)$_2$	508	14.89	[26]
Mo$_{12}$(Ale)$_4$	464	8.20	[18]
Mo$_{12}$(Ale-1C$_a$)$_4$	464	5.44	[18]
Mo$_{12}$(Ale-2C)$_4$	464	51.81	[18]
(H$_2$DABCO)$_2$(NH$_4$)$_2$[Mo$_8$O$_{27}$]	525	37.0	[6]
(H$_2$DABCO)$_2$(H$_2$pipz)[Mo$_8$O$_{27}$]	525	5.6	[6]
(H$_2$pipz)$_3$[Mo$_8$O$_{27}$]	525	7.8	[6]
(H$_2$DABCO)$_2$(HDMA)$_{0.5}$Na$_{0.75}$[Mo$_8$O$_{27}$]	525	0.8	[6]

[a] Photoinduced absorption band wavelength. [b] Coloration kinetic half-life time (min).

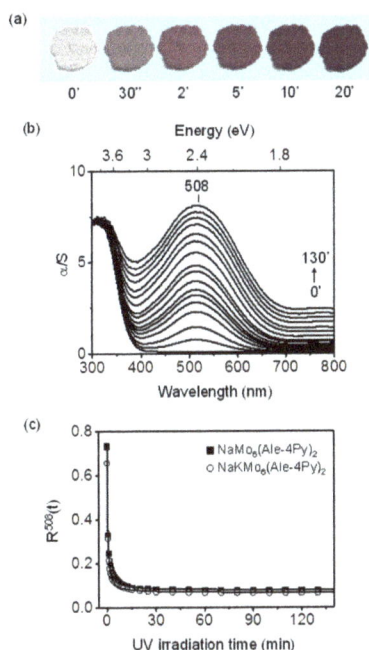

Figure 6. (a) Photographs of the powder of **NaKMo$_6$(Ale-4Py)$_2$** at different UV irradiation time (in min). (b) Evolution of the photo-generated absorption in **NaKMo$_6$(Ale-4Py)$_2$** after 0, 0.5, 1, 2, 3, 4, 5, 7, 10, 15, 20, 30, 40, 60, 80, 100 and 130 min of UV irradiation (λ_{ex} = 365 nm). (c) Comparison of the $R^{508}(t)$ vs. t plots for **NaMo$_6$(Ale-4Py)$_2$** (■), and **NaKMo$_6$(Ale-4Py)$_2$** (○). The lines show the fits of the plots according to rate law $R^{508}(t) = a/(bt + 1) + (R^{508}(0) - a)$.

3. Experimental Section

3.1. General

The alendronic acid $H_2O_3PC(OH)(C_3H_6NH_2)PO_3H_2$ (Ale) [27] and its derivative $Na_4[O_3PC(OH)(C_3H_6N(CH_2C_5H_4N)_2)PO_3]$ ($Na_4HAlePy_2$) [28] were synthesized according to reported procedures. All other chemicals were used as purchased without purification. Ale-4Py and AlePy$_2$ were reacted with Mo^{VI} and Mo^V precursors. However, only the reactions of Ale-4Py with Mo^{VI} and AlePy$_2$ with Mo^V led to crystalline compounds which could be characterized by single-crystal X-ray diffraction analysis.

3.2. Synthesis of $Na_2[(HO_3P)_2(OH)C(C_3H_6NHCH_2(C_5H_4N)]$ (Na_2H_3Ale-4Py)

Alendronic acid (1.5 g, 6 mmol) and triethylamine (1.6 mL, 12 mmol) were stirred at room temperature in 36 mL of MeOH until complete dissolution. Then, 4-pyridinecarboxaldehyde (0.62 mL, 6.6 mmol) was added to the mixture. The solution was stirred at reflux for 2 h. and then allowed to cool down to room temperature. After slow addition of (TBA)BH$_4$ (1.549 g, 6 mmol), the resulting mixture was refluxed overnight. The solution was then concentrated to about 10 mL and NaPF$_6$ (2.022 g, 12 mmol) was added. A white precipitate appeared instantly. The solid was filtered under vacuum and washed with MeOH and Et$_2$O. The ligand (1.612 g, yield 70%) was used without further purification. ^{31}P NMR (200 MHz, D$_2$O): δ 17.7. ^{1}H NMR (200 MHz, D$_2$O): δ 8.47 (d, 2H, ^3J = 5.41 Hz), 7.40 (d, 2H, ^3J = 5.41 Hz), 4.18 (s, 2H), 3.05 (m, 2H, N-CH_2-CH$_2$), 1.91 (m, 4H, N-CH$_2$-CH_2-CH_2). IR (KBr pellets): ν(cm^{-1}) = 3192 (w), 2968 (w), 2950 (w), 1610 (m), 1501 (m), 1424 (m), 1381 (w), 1224 (m), 1191 (m), 1141 (s), 1122 (s), 1071 (vs), 1036 (vs), 959 (s), 930 (s), 882 (s), 844 (s), 810 (m), 787 (m), 673 (s), 605 (m), 589 (m), 556 (m), 541 (m).

3.3. Synthesis of $Na_4[(Mo^{VI}_3O_8)_2(O)(O_3PC(O)(C_3H_6NH_2CH_2C_5H_4NH)PO_3)_2].14H_2O$ ($NaMo_6(Ale$-4Py))

To a solution of Na$_2$MoO$_4$ (0.291 g, 1.2 mmol) in 10 mL of water was added Na$_2$H$_3$(Ale-4Py) (0.154 g, 0.4 mmol). The pH was adjusted to 3.1 with 2 M HCl and the mixture was stirred at 80 °C for 30 min. A fine powder was then eliminated by centrifugation (0.134 g) and the filtrate was left to evaporate at room temperature. The IR spectrum of the powder shows the presence of Mo-O and P-O vibrations close to the vibrations found in Mo$_{12}$(BP)$_4$ species. After one month, white crystals (yield 0.037 g; 10% based on Mo) suitable for X-ray diffraction studies were collected by filtration. The formation of the white insoluble powder explains the low yield of the synthesis.

IR (KBr pellets): ν(cm^{-1}) = 3361 (br), 3082 (s), 2832 (m), 2788 (m), 1644 (s), 1624 (s), 1508 (m), 1467 (m), 1252 (w), 1137 (s), 1067 (s), 1015 (s), 907 (vs), 872 (vs), 815 (s), 675 (s), 548 (s). Anal. calcd. for C$_{20}$H$_{58}$N$_4$Mo$_6$Na$_4$O$_{45}$P$_4$ (1866.16 g.mol^{-1}) (found): C 12.87 (12.60), H 3.13 (2.95), N 3.00 (2.94). EDX analyses confirm the values of the Mo/P and Mo/Na ratios.

3.4. Synthesis of $Na_{0.5}K_{3.5}[(Mo^{VI}_3O_8)_2(O)(O_3PC(O)(C_3H_6NH_2CH_2C_5H_4NH)PO_3)_2].13H_2O$
$(NaKMo_6(Ale-4Py)_2)$

To a solution of Na_2MoO_4 (0.291 g, 1.2 mmol) in 10 mL of water was added Na_2H_3(Ale-4Py) (0.154 g, 0.4 mmol). The pH was adjusted to 3.1 with 2 M HCl and the mixture was stirred at 80 °C for 30 minutes. A fine powder was then eliminated by centrifugation. KCl (0.100 g) was added to the filtrate and the solution was left to evaporate at room temperature. After one month, white crystals (yield 0.056 g; 15% based on Mo) suitable for X-ray diffraction studies were collected by filtration.

IR (KBr pellets): $v(cm^{-1})$ = 3365 (br), 3082 (s), 2832 (m), 2789 (m), 1644 (s), 1626 (s), 1509 (m), 1468 (m), 1256 (w), 1137 (s), 1065 (s), 1016 (s), 907 (vs), 871 (vs), 811 (s), 678 (s), 551 (s). Anal. calcd. for $C_{20}H_{56}N_4K_{3.5}Mo_6Na_{0.5}O_{44}P_4$ (1904.53 g.mol^{-1}) (found): C 12.61 (12.50), H 2.96 (2.82), N 2.94 (2.85). EDX analyses confirm the values of the Mo/P, Mo/K and Mo/Na ratios.

3.5. Synthesis of $(NH_4)_8[(Mo^V_2O_4)(Mo^{VI}_2O_6)_2\{O_3PC(O)(C_3H_6N(CH_2C_5H_4N)_2(Mo^{VI}O_3))PO_3\}_2]\cdot20H_2O$
$(NH_4Mo_6(AlePy_2Mo)_2)$

A 0.1 M solution of $[Mo^V_2O_4(H_2O)_6]^{2+}$ was prepared as previously reported [19] by reduction of MoO_3 by N_2H_4 in 4M HCl. $Na_4HAlePy_2$ (0.325 g, 0.21 mmol) was added in aqueous $[Mo^V_2O_4(H_2O)_6]^{2+}$ (0.1 M, 6.25 mL, 0.625 mmol). A green precipitate appeared. Aqueous ammonium hydroxide (33%) was added dropwise to pH 7.5. An orange precipitate was formed. The beaker was covered with a watch glass and the mixture was stirred for 2 days until the solution progressively turned limpid. The deep red solution was filtered, and left to evaporate at room temperature after addition of NH_4Br (1.2 g, 12.24 mmol). Red crystals were collected by filtration two days later (0.095 g, yield 24% based on Mo). Elemental analysis calcd. (found) for $C_{32}H_{108}N_{14}Mo_8O_{56}P_4$ (2477 g mol^{-1}): C 15.52 (15.20), H 4.39 (4.04), N 7.92 (8.42), Mo 30.98 (30.01), P 5.00 (4.91). IR (v/cm^{-1}): 1638 (m), 1606 (m), 1419 (s), 1314 (w), 1283 (w), 1142 (sh), 1123 (s), 1075 (sh), 1052 (s), 1013 (s), 913 (m), 896 (sh), 856 (s), 760 (w), 702 (m), 682 (w), 645 (m), 479 (m).

3.6. X-ray Crystallography

Data collection was carried out by using a Bruker Nonius X8 APEX 2 diffractometer equipped with a CCD bi-dimensional detector using the monochromatized wavelength λ(Mo Kα) = 0.71073 Å. Absorption correction was based on multiple and symmetry-equivalent reflections in the data set using the SADABS program [29] based on the method of Blessing [30].The structures were solved by direct methods and refined by full-matrix least-squares using the SHELX-TL package [31]. In all structures, there are small discrepancies between the formulae determined by elemental analysis and those deduced from the crystallographic atom list because of the difficulty in locating all disordered water molecules (and also the counter-cations in some cases). In the structures of **NH₄Mo₆(AlePy₂Mo)₂**, NH_4^+ and H_2O could not be distinguished based on the observed electron densities; therefore, all the positions were labeled O and assigned the oxygen atomic scattering factor. Crystallographic data are given in Table 2. Partial atomic labeling schemes, selected bond distances and bond valence sum calculations [32] are reported in Figure S1.

Table 2. Crystallographic data.

Parameters	NaMo$_6$(Ale-4Py)$_2$	NaKMo$_6$(Ale-4Py)$_2$	NH$_4$Mo$_6$(AlePy$_2$Mo)$_2$
Formula	C$_{20}$H$_{58}$N$_4$Mo$_6$Na$_4$O$_{45}$P$_4$	C$_{20}$H$_{56}$N$_4$K$_{3.5}$Mo$_6$Na$_{0.5}$O$_{44}$P$_4$	C$_{32}$H$_{108}$N$_{14}$Mo$_8$O$_{56}$P$_4$
Formula weight/g mol^{-1}	1866.16	1904.53	2476.67
Crystal system	triclinic	triclinic	triclinic
Space group	P-1	P-1	P-1
a/Å	9.0240(3)	17.9501(5)	17.602(2)
b/Å	19.4282(6)	19.4745(4)	22.134(2)
c/Å	21.1568(7)	19.8487(5)	28.323(3)
α/°	103.992(2)	76.188(1)	97.443(5)
β/°	95.5612(2)	69.590(1)	95.464(5)
γ/°	103.407(1)	76.669(1)	107.430(5)
V/Å^3	3454.8(2)	6231.6(3)	10334(2)
Z	2	4	4
D_{calc}/g cm^{-3}	1.813	2.028	1.560
μ/mm^{-1}	1.273	1.616	1.087
Data/parameters	12208/865	36431/1558	36283 / 2119
R_{int}	0.0301	0.0429	0.0467
GOF	1.023	1.070	1.035
R_1 ($I > 2\sigma$ (I))	0.0710	0.0568	0.1051
wR_2	0.1960	0.1649	0.2719

3.7. Physical Methods

Infrared (IR) spectra were recorded on a Nicolet 30 ATR 6700 FT spectrometer. EDX measurements were performed on a JEOL JSM 5800LV apparatus. The ^{31}P NMR spectra were recorded on a Brüker AC-300 spectrometer operating at 121.5 MHz in 5-mm tubes with ^{1}H decoupling. ^{31}P NMR chemical shifts were referenced to the external usual standard 85% H$_3$PO$_4$. Diffuse reflectance spectra were collected at room temperature on a finely ground sample with a Cary 5G spectrometer (Varian) equipped with a 60 mm diameter integrating sphere and computer control using the "Scan" software. Diffuse reflectance was measured from 250–1000 nm with a 2 nm step using Halon powder (from Varian) as reference (100% reflectance). The reflectance data were treated by a Kubelka-Munk transformation [33] to better locate the absorption thresholds. The samples were irradiated with a Fisher Bioblock labosi UV lamp (λ_{exc} = 365 nm, 6W) at a distance of 50 mm.

4. Conclusions

In summary, the synthesis of two novel molybdobisphosphonate complexes with BPs functionalized by pyridine groups has been performed by a one step or two step procedure in water. These POMs have been characterized in the solid state by single crystal X-ray diffraction and in solution by ^{31}P NMR spectroscopy. **NH$_4$Mo$_6$(AlePy$_2$Mo)$_2$** is the ammonium salt of the mixed-valent anion [(Mo$^V_2$O$_4$)(Mo$^{VI}_2$O$_6$)$_2$(AlePy$_2$)$_2$(Mo^{VI}O$_3$)$_2$]$^{8-}$ in which the two pyridine arms of each BP ligand chelate an additional Mo^{VI} ion. [(Mo$^{VI}_3$O$_8$)$_2$(O)(Ale-4Py)$_2$]$^{4-}$ (**Mo$_6$(Ale-4Py)$_2$**) crystallizes as a pure Na salt and as a mixed Na/K salt. Although the nature of the alkali counter-ions is different in the two salts and thus the 3D structure, the hydrogen bonding networks involving the **Mo₆(Ale-4Py)₂** units are quite similar. This strongly impacts the photochromic properties of both compounds. Noticeably, **NaMo₆(Ale-4Py)₂** and **NaKMo₆(Ale-4Py)₂** exhibit remarkably fast and quite comparable color-change effects from white to deep purple under UV excitation, and to date they are the most efficient known photochromic organoammonium/POM systems. Their highly solid-state photochromic performances in ambient conditions can be correlated to a great number of N-H···O contacts between the oxygen atoms of the hexanuclear units and the organoammonium groups of the Ale-4Py ligands as well as the presence of the diprotonated amino groups.

Acknowledgments

This work was supported by CNRS, UVSQ and the French ANR (grant ANR-11-BS07-011-01-BIOOPOM). Guillaume Rousseau, Bertrand Gaulupeau and Noura Saada are gratefully acknowledged for their participation to the synthesis of the compounds.

Author Contributions

The synthesis and characterizations of the compounds were performed by Tarik Benali and Olivier Oms. Characterization by single crystal X-ray diffraction was performed by Jérôme Marrot. Photochromic properties were studied by Marin Puget, Hélène Serier-Brault, Philippe Deniard, and Rémi Dessapt. Full research and methodology were provided by Olivier Oms, Pierre Mialane and Anne Dolbecq. The preparation of the manuscript was made by Olivier Oms, Anne Dolbecq and Rémi Dessapt.

Conflicts of Interest

The authors declare no conflict of interest.

References

1. Clemente-Juan, J.M.; Coronado, E.; Gaita-Ariño, A. Magnetic Polyoxometalates: from Molecular Magnetism to Molecular Spintronics and Quantum Computing. *Chem. Soc. Rev.* **2012**, *41*, 7464–7478.
2. Hasenknopf, B. Polyoxometalates: Introduction to a Class of Inorganic Compounds and their Biomedical Applications. *Front. Biosci.* **2005**, *10*, 275–287.
3. Song, Y.-F.; Tsunashima, R. Recent Advances on Polyoxometalate-based Molecular and Composite Materials. *Chem. Soc. Rev.* **2012**, *41*, 7384–7402.
4. Kawata, S.; Kawata, Y. Three-Dimensional Optical Data Storage Using Photochromic Materials. *Chem. Rev.* **2000**, *100*, 1777–1788.
5. Yamase, T. Photo- and Electrochromism of Polyoxometalates and Related Materials. *Chem. Rev.* **1998**, *98*, 307–325.
6. Dessapt, R.; Collet, M.; Coué, V.; Bujoli-Doeuff, M.; Jobic, S.; Lee, C.; Whangbo, M.-H. Kinetics of Coloration in Photochromic Organoammonium Polyoxomolybdates. *Inorg. Chem.* **2009**, *48*, 574–580.

144

7. Dolbecq, A.; Dumas, E.; Mayer, C.R.; Mialane, P. Hybrid Organic-Inorganic Polyoxometalate Compounds : From Structural Diversity to Applications. *Chem. Rev.* **2010**, *110*, 6009.

8. Proust, A.; Matt, B.; Villanneau, R.; Guillemot, G.; Gouzerh, P.; Izzet, G. Functionalization and Post-functionalization : a Step Towards Polyoxometalate-based Materials. *Chem. Soc. Rev.* **2012**, *41*, 7605–7622.

9. Musumeci, C.; Luzio, A.; Pradeep, C.P.; Miras, H.N.; Rosnes, M.H.; Song, Y.-F.; Long, D.-L.; Cronin, L.; Pignataro, B. Programmable Surface Architectures Derives from Hybrid Polyoxometalate-Based Clusters. *J. Phys. Chem. C* **2011**, *115*, 4446–4455.

10. Errington, R.J.; Petkar, S.S.; Horrocks, B.R.; Houlton, A.; Lie, L.H.; Patole, S.N. Covalent Immobilization of a TiW5 Polyoxometalate on Derivatized Silicon Surfaces. *Angew. Chem. Int. Ed.* **2005**, *44*, 1254–1257.

11. Geisberger, G.; Gyenge, E.B.; Hinger, D.; Bösiger, P.; Maake, C.; Patzke, G.R. Synthesis, Characterization and Bioimaging of Fluorescent Labeled Polyoxometalates. *Dalton Trans.* **2013**, *42*, 9914–9920.

12. Li, J.; Huth, I.; Chamoreau, L.M.; Hasenknopf, B.; Lacôte, E.; Thorimbert, S.; Malacria, M. Insertion of Amides into a Polyoxometalate. *Angew. Chem. Int. Ed.* **2009**, *48*, 2035.

13. Du, D.-Y.; Qin, J.-S.; Li, S.-L.; Su, Z.-M.; Lan, Y.-Q. Recent Advances in Porous Polyoxometalate-based Metal-Organic Framework Materials. *Chem. Soc. Rev.* **2014**, *43*, 4615–4632.

14. Dolbecq, A.; Mialane, P.; Sécheresse, F.; Keita, B.; Nadjo, L. Functionalized Polyoxometalates with covalently linked bisphosphonate, N-donor or Carboxylate Ligands: From Electrocatalytic to Optical Properties. *Chem. Commun.* **2012**, *48*, 8299–8316.

15. Banerjee, A.; Bassil, B.S.; Röschenthaler, G.-V.; Kortz, U. Diphosphates and Diphosphonates in Polyoxometalate Chemistry. *Chem. Soc. Rev.* **2012**, *41*, 7590–7604.

16. Compain, J.-D.; Mialane, P.; Marrot, J.; Sécheresse, F.; Zhu, W.; Oldfield, E.; Dolbecq, A. Tetra- to Dodecanuclear Oxomolybdate Complexes with Functionalized Bisphosphonate Ligands: Activity in Killing Tumor Cells. *Chem. Eur. J.* **2010**, *16*, 13741–13748.

17. El Moll, H.; Zhu, W.; Oldfield, E.; Rodriguez-Albelo, L.M.; Mialane, P.; Marrot, J.; Vila, N.; Mbomekallé, I.-M.; Rivière, E.; Duboc, C.; Dolbecq, A. Polyoxometalates Functionalized by Bisphosphonate Ligands: Synthesis, Structural, Magnetic, and Spectroscopic Characterizations and Activity on Tumor Cell Lines. *Inorg. Chem.* **2012**, *51*, 7921–7931.

18. El Moll, H.; Dolbecq, A.; Mbomekallé, I.-M.; Marrot, J.; Deniard, P.; Dessapt, R.; Mialane, P. Tuning the Photochromic Properties of Molybdenum Bisphosphonate Polyoxometalates. *Inorg. Chem.* **2012**, *51*, 2291–2302.

19. Dolbecq, A.; Compain, J.-D.; Mialane , P.; Marrot, J.; Sécheresse, F.; Keita, B.; Brudna Holzle, L.R.; Miserque, F.; Nadjo, L. Hexa- and Dodecanuclear Polyoxomolybdate Cyclic Compounds: Application toward the Facile Synthesis of Nanoparticles and Film Electrodeposition. *Chem. Eur. J.* **2009**, *15*, 733–741.

20. Banerjee, A.; Raad, F.S.; Vankova, N.; Bassil, B.S.; Heine, T.; Kortz, U. Polyoxomolybdodiphosphonates : Examples Incorporating Ethylenepyridines. *Inorg. Chem.* **2011**, *50*, 11667–11675.

21. Modec, B.; Brenčič, J.V. From Small $\{Mo_2O_4\}^{2+}$ Aggregates to Infinite Solids. *J. Clust. Sci.* **2002**, *13*, 279–302.

22. Pope, M.T.; Müller, A. Polyoxometalate Chemistry: An Old Field with New Dimensions in Several Disciplines. *Angew. Chem. Int. Ed. Engl.* **1991**, *30*, 34–48

23. Sergienko, V.S. Structural Chemistry of 1-Hydroxyethylidenediphosphonic Acid Complexes. *Russ. J. Coord. Chem.* **2001**, *27*, 681–710.

24. Tan, H.; Chen, W.; Liu, D.; Feng, X.; Li, Y.; Yan, A.; Wang, E. Two Diphosphonate-Functionalized Asymmetric Polyoxomolybdates with Catalytic Activity for Oxidation of Benzyl Alcohol to Benzaldehyde. *Dalton Trans.* **2011**, *40*, 8414–8418.

25. Coué, V.; Dessapt, R.; Bujoli-Doeuff, M.; Evain, M.; Jobic, S. Synthesis, characterizations and photochromic properties of hybrid organic-inorganic materials based on molybdate, DABCO and piperazine. *Inorg. Chem.* **2007**, *46*, 2824–2835.

26. Yang, L.; Zhou, Z.; Ma, P.; Wang, J.; Niu, J. Assembly of Dimeric and Tetrameric Complexes of Polyoxomolybdobisphosphonates Built from $[(Mo_3O_8)\{O_3PC(O)(CH_2-3-C_5NH_5)PO_3\}]^{2-}$ Subunits. *Cryst. Growth Des.* **2013**, *13*, 2540–2547.

27. Kubíček, V.; Kotek, J.; Hermann, P.; Lukeš, I. Aminoalkylbis(phosphonates): Their Complexation Properties in Solution and in the Solid State. *Eur. J. Inorg. Chem.* **2007**, 333–344.

28. De Rosales, R.T.M.; Finucane, C.; Mather, S.J.; Blower, P.J. Bifunctional Bisphosphonate Complexes for the Diagnosis and Therapy of Bone Metastases. *Chem. Commun.* **2009**, 4847–4849.

29. Sheldrick, G.M. *SADABS*; Program for Scaling and Correction of Area detector data; University of Göttingen: Germany, 1997.

30. Blessing, R. An Empirical Correction for Absorption Anisotropy *Acta Crystallogr.* **1995**, *A51*, 33–38.

31. Sheldrick, G.M. *SHELX-TL*, version 5.03; Software Package for the Crystal Structure Determination; Siemens Analytical X-ray Instrument Division: Madison, WI, USA, 1994.

32. Brese, N.E.; O'Keeffe, M. Bond Valence Parameters for Solids. *Acta Crystallogr. Sect. B* **1991**, *47*, 192–197.

33. Kubelka, P.; Munk, F. An Article on Optics of Paint Layers. *Z. Techn. Physik* **1931**, *12*, 593–601.

Structure Transformation among Deca-, Dodeca- and Tridecavanadates and Their Properties for Thioanisole Oxidation

Yuji Kikukawa, Kazuhiro Ogihara and Yoshihito Hayashi

Abstract: The transformation of three types of polyoxovanadates, $\{(n\text{-}C_4H_9)_4N\}_3[H_3V_{10}O_{28}]$, $\{(n\text{-}C_4H_9)_4N\}_4[V_{12}O_{32}]$ and $\{(n\text{-}C_4H_9)_4N\}_3[V_{13}O_{34}]$ have been investigated according to the rational chemical equations, and the best transformation conditions were reported. By the reaction of $[H_3V_{10}O_{28}]^{3-}$ with 0.33 equivalents of $\{(n\text{-}C_4H_9)_4N\}OH$ in acetonitrile at 80 °C, $[V_{12}O_{32}]^{4-}$ was formed with 92% yield. The reaction in nitroethane with 0.69 equivalents of p-toluenesulfonic acid gave $[V_{13}O_{34}]^{3-}$ with 91% yield. The ^{51}V NMR observation of each reaction suggests the complete transformations of $[H_3V_{10}O_{28}]^{3-}$ to $[V_{12}O_{32}]^{4-}$ and to $[V_{13}O_{34}]^{3-}$ proceeded without the formation of any byproducts and it provides the reliable synthetic route. Decavanadates were produced by the hydrolysis of $[V_{12}O_{32}]^{4-}$ or $[V_{13}O_{34}]^{3-}$. While the direct transformation of $[V_{13}O_{34}]^{3-}$ to $[V_{12}O_{32}]^{4-}$ partly proceeded, the reverse one could not be observed. For the thioanisole oxidation, $[V_{13}O_{34}]^{3-}$ showed the highest activity of the three.

Reprinted from *Inorganics*. Cite as: Kikukawa, Y.; Ogihara, K.; Hayashi, Y. Structure Transformation among Deca-, Dodeca- and Tridecavanadates and Their Properties for Thioanisole Oxidation. *Inorganics* **2015**, *3*, 295-308.

1. Introduction

Polyoxometalates have been attracting much attention in broad fields such as catalyst chemistry, magnetochemistry, and pharmaceutical chemistry [1–3]. Polyoxovanadates have a unique structural chemistry due to the availability of various coordination environments of tetrahedral $\{VO_4\}$, square-pyramidal $\{VO_5\}$ and octahedral $\{VO_6\}$ units [4]. Due to the high reactivity, the chemistry of polyoxovanadates is limited in decavanadate species. The only stable example of isopolyoxovanadates in aqueous solution is decavanadates, which consist of $\{VO_6\}$ units, in addition to the metavanadate species, $[VO_3]_n^{n-}$ with $\{VO_4\}$ units [5]. The employment of non-aqueous solvent increases the diversity of the polyoxovanadates. Up to now, various isopolyoxovanadates with only fully oxidized V^{5+}, such as $[V_3O_9]^{3-}$, $[V_4O_{12}]^{4-}$ and $[V_5O_{14}]^{3-}$ with $\{VO_4\}$ units; $[V_{11}O_{29}F_2]^{4-}$, $[HV_{12}O_{32}(Cl)]^{4-}$, $[V_{12}O_{32}]^{4-}$, $[V_{12}O_{32}(Cl)]^{5-}$ and $[H_2V_{14}O_{38}(Cl)]^{5-}$ with $\{VO_5\}$ units; $[V_7O_{19}F]^{4-}$ with both $\{VO_4\}$ and $\{VO_5\}$ units; and $[V_6O_{19}]^{8-}$ derivatives, $[H_nV_{10}O_{28}]^{(6-n)-}$ and $[V_{13}O_{34}]^{3-}$ with $\{VO_6\}$ units have been synthesized [6–19]. Since the properties such as redox and optics of polyoxovanadates depend on the anion structures, it is crucially important to control those cluster frameworks [14,19].

The half spherical dodecavanadates $[V_{12}O_{32}]^{4-}$ consist of twelve $\{VO_5\}$ units and have an open cavity into which several molecules and anions, such as nitriles, dichloromethane, NO^- and Cl^-, are incorporated and the host-guest interactions have been theoretically investigated

(Figure 1) [9,10,20–23]. To clarify the role of an electron-rich guest and the anion, further investigations including systematic synthesis are required.

Tridecavanadate $[V_{13}O_{34}]^{3-}$ was reported as an oxidation catalyst for a number of organic substrates (Figure 1) [19]. After the pioneering work on the chemistry of $[V_{13}O_{34}]^{3-}$, few further investigations were reported. Our group successfully synthesized the four electron reduced tridecavanadate $[H_4V_{13}O_{34}]^{3-}$ [24]. During our investigation, the electrochemical oxidation of $[H_4V_{13}O_{34}]^{3-}$ gave $[V_{12}O_{32}]^{4-}$ instead of $[V_{13}O_{34}]^{3-}$. The protonation may prevent the oxidation of $[H_4V_{13}O_{34}]^{3-}$ to form $[V_{13}O_{34}]^{3-}$. Cronin's group synthesized the manganese-substituted polyoxovanadate with the same anion structure of $[V_{13}O_{34}]^{3-}$ [25]. The hetero-metal containing polyoxometalates have the potential to show unique catalytic and magnetic properties [26–29]. As like synthesis of Cronin's cluster from the related manganese-containing polyoxovanadate, the control of the structure transformation is important.

In this work, we report on the structure transformation among $[H_3V_{10}O_{28}]^{3-}$, $[V_{12}O_{32}]^{4-}$ and $[V_{13}O_{34}]^{3-}$, including the improved synthesis of $\{(n\text{-}C_4H_9)_4N\}_3[V_{13}O_{34}]$. In addition, oxidation of thioanisole using these polyoxovanadates is discussed.

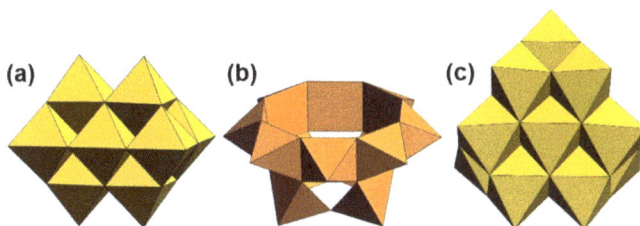

Figure 1. Anion structures of (**a**) $[H_3V_{10}O_{28}]^{3-}$, (**b**) $[V_{12}O_{32}]^{4-}$ and (**c**) $[V_{13}O_{34}]^{3-}$. Yellow octahedra and orange square-pyramids represent $\{VO_6\}$ and $\{VO_5\}$ units, respectively.

2. Results and Discussion

2.1. Structure Transformation among Deca-, Dodeca- and Tridecavanadates

The structure of decavanadates, $[H_nV_{10}O_{28}]^{(6-n)-}$, consists of ten octahedral $\{VO_6\}$ units stacked together in one molecule (Figure 1). While tridecavanadate $[V_{13}O_{34}]^{3-}$ also has a similar structure with three additional $\{VO_6\}$ octahedra in a triangular arrangement as add-on units on one of the surfaces of the decavanadate. Dodecavanadate $[V_{12}O_{32}]^{4-}$ has a totally different structure which is based on twelve square-pyramidal $\{VO_5\}$ units in a cage arrangement that has a cavity to incorporate a small molecule like acetonitrile at the center of the anion (Figure 1). These three complexes have distinct structures, yet the size of the anions are similar, and we tried to find the reaction conditions which can transform one molecule into another by adjusting the stoichiometry according to the molecular formula. The transformation process is monitored through ^{51}V NMR since all chemical shifts for each of those complexes were already reported [17,19,30]. In the transformation process, the employment of decavanadates as a starting material is especially beneficial because they are one of the most widely available species and thoroughly investigated in water as well as acetonitrile [8,31].

2.1.1. Transformation between Deca- and Dodecavanadates

The synthesis of the dodecavanadates was achieved in two different routes, even if the precursors are different in oxidation states, formulas and total charges on the cluster. In the first report by Klemperer's group, $[V_{12}O_{32}]^{4-}$ was prepared by refluxing an acetonitrile solution of the decavanadate with only two protonation sites, $\{(n\text{-}C_4H_9)_4N\}_4[H_2V_{10}O_{28}]$, for 1–2 min with >80% yield [10]. Later, by our group, $[V_{12}O_{32}]^{4-}$ was prepared by the oxidation of another type of decavanadate $\{(n\text{-}C_4H_9)_4N\}_4[V_{10}O_{26}]$ for 1 h at room temperature with 90% yield [9]. While $\{(n\text{-}C_4H_9)_4N\}_4[H_2V_{10}O_{28}]$ consists ten $\{V^{5+}O_6\}$ units, $\{(n\text{-}C_4H_9)_4N\}_4[V_{10}O_{26}]$ consists eight $\{V^{5+}O_4\}$ and two $\{V^{4+}O_5\}$ units. These differences show that the precursors are not important and provide us an opportunity to investigate the transformation of polyoxovanadates in accordance with the reaction equations. The triprotonated decavanadate $[H_3V_{10}O_{28}]^{3-}$ was selected as a precursor for investigation since the solution state of $[H_3V_{10}O_{28}]^{3-}$ in acetonitrile and water is well discussed before [17]. The conventional equation for the synthesis of $\{(n\text{-}C_4H_9)_4N\}_4[V_{12}O_{32}]$ from $\{(n\text{-}C_4H_9)_4N\}_3[H_3V_{10}O_{28}]$ can be written by adjusting the stoichiometry as expressed in Equation (1).

$$6\{(n\text{-}C_4H_9)_4N\}_3[H_3V_{10}O_{28}] + 2\{(n\text{-}C_4H_9)_4N\}OH \rightleftarrows 5\{(n\text{-}C_4H_9)_4N\}_4[V_{12}O_{32}] + 10H_2O \qquad (1)$$

By the reaction of $\{(n\text{-}C_4H_9)_4N\}_3[H_3V_{10}O_{28}]$ with 0.33 equivalents of $\{(n\text{-}C_4H_9)_4N\}OH$ in acetonitrile at room temperature for 24 h, ^{51}V NMR spectra showed signals due to $[H_3V_{10}O_{28}]^{3-}$ and $[H_2V_{10}O_{28}]^{4-}$, and $[V_{12}O_{32}]^{4-}$ as a minor product (Figure A1). The transfer reaction was slow at room temperature, and to accelerate the reaction, heating the solution was necessary. After refluxing for 1 h, only signals due to $[V_{12}O_{32}]^{4-}$ were observed with >99% chemoselectivity, and the isolated yield was 92% (Figure 2). IR spectrum of the isolated product show a perfect match with that of $\{(n\text{-}C_4H_9)_4N\}_4[V_{12}O_{32}]$ (Figure 3).

In the presence of a stoichiometric amount of water, $[V_{12}O_{32}]^{4-}$ was stable with retention of the structure. An excess amount of water leads to the transformation of $[V_{12}O_{32}]^{4-}$ to decavanadates. This is consistent with the established distribution study in aqueous media that dodecavanadates are not involved in the complicated equilibrium of polyoxovanadates [5]. In a mixed solvent of acetonitrile and water (3:1, v/v), the mixture of $[H_3V_{10}O_{28}]^{3-}$ and $[H_2V_{10}O_{28}]^{4-}$ was obtained (Figure A2). By addition of 0.4 equivalents of p-toluenesulfonic acid (TsOH) with respect to $[V_{12}O_{32}]^{4-}$, only $[H_3V_{10}O_{28}]^{3-}$ was detected by ^{51}V NMR and the isolated yield was 83% (Figure 2). IR spectrum of the isolated product shows convincing agreement with that of $\{(n\text{-}C_4H_9)_4N\}_3[H_3V_{10}O_{28}]$ (Figure 3).

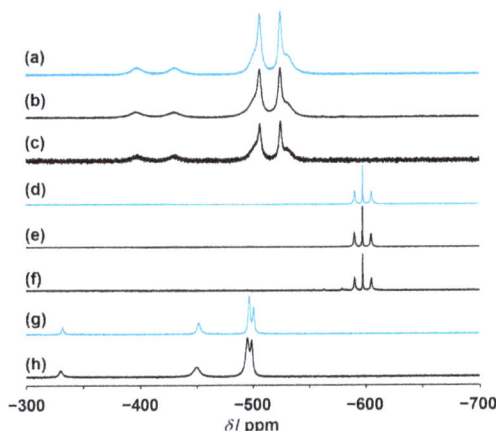

Figure 2. ^{51}V NMR spectra of (**a**) $[H_3V_{10}O_{28}]^{3-}$, (**d**) $[V_{12}O_{32}]^{4-}$ and (**g**) $[V_{13}O_{34}]^{3-}$ crystals dissolved in acetonitrile and of the reaction solution after the transformations of (**b**) $[V_{12}O_{32}]^{4-}$ to $[H_3V_{10}O_{28}]^{3-}$, (**c**) $[V_{13}O_{34}]^{3-}$ to $[H_3V_{10}O_{28}]^{3-}$, (**e**) $[H_3V_{10}O_{28}]^{3-}$ to $[V_{12}O_{32}]^{4-}$, (**f**) $[V_{13}O_{34}]^{3-}$ to $[V_{12}O_{32}]^{4-}$ and (**h**) $[H_3V_{10}O_{28}]^{3-}$ to $[V_{13}O_{34}]^{3-}$.

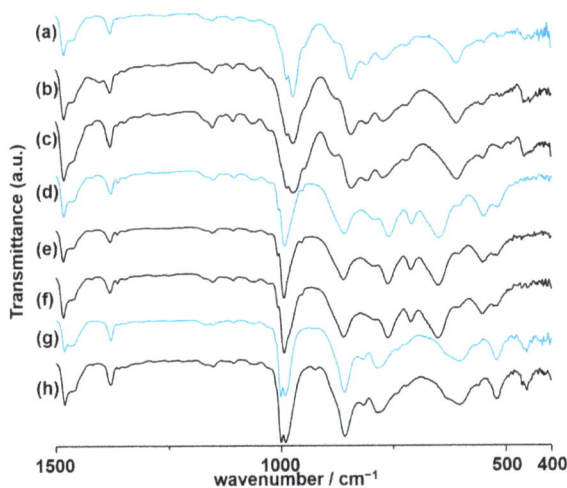

Figure 3. IR spectra of (**a**) $[H_3V_{10}O_{28}]^{3-}$, (**d**) $[V_{12}O_{32}]^{4-}$ and (**g**) $[V_{13}O_{34}]^{3-}$ crystals and of the powder obtained after the transformations of (**b**) $[V_{12}O_{32}]^{4-}$ to $[H_3V_{10}O_{28}]^{3-}$, (**c**) $[V_{13}O_{34}]^{3-}$ to $[H_3V_{10}O_{28}]^{3-}$, (**e**) $[H_3V_{10}O_{28}]^{3-}$ to $[V_{12}O_{32}]^{4-}$, (**f**) $[V_{13}O_{34}]^{3-}$ to $[V_{12}O_{32}]^{4-}$ and (**h**) $[H_3V_{10}O_{28}]^{3-}$ to $[V_{13}O_{34}]^{3-}$.

2.1.2. Transformation between Deca- and Tridecavanadates

In the reported procedure, $\{(n\text{-}C_4H_9)_4N\}_3[V_{13}O_{34}]$ was obtained by refluxing 2.6 M $\{(n\text{-}C_4H_9)_4N\}_3[H_3V_{10}O_{28}]$ acetonitrile solution for 7 h under dry nitrogen [19]. In our investigation under the same condition, $[V_{13}O_{34}]^{3-}$ was formed as a minor product. The time dependent observation of ^{51}V NMR spectra at 80 °C reveals that the formation of $[V_{12}O_{32}]^{4-}$ is faster than that of $[V_{13}O_{34}]^{3-}$ (Figure A3). Once $[V_{12}O_{32}]^{4-}$ is formed, $[V_{13}O_{34}]^{3-}$ is no longer obtained in acetonitrile (see below). Although the solubility of $[V_{13}O_{34}]^{3-}$ is lower than $[V_{12}O_{32}]^{4-}$ and $[V_{13}O_{34}]^{3-}$ is selectively obtained by crystallization, the synthesis of $[V_{13}O_{34}]^{3-}$ is difficult in our hands. Therefore, the improved synthesis of $[V_{13}O_{34}]^{3-}$ is suggested in this paper. Since acetonitrile acts as a template to stabilize the host anion $[V_{12}O_{32}]^{4-}$, the choice of the appropriate solvent is important for the rational synthesis of $[V_{13}O_{34}]^{3-}$. $\{(n\text{-}C_4H_9)_4N\}_3[H_3V_{10}O_{28}]$ was dissolved in nitroethane and stirred for 3 h at 80 °C. Despite the indication that $[V_{13}O_{34}]^{3-}$ is selectively formed by ^{51}V NMR, the solution color turned to green during the reaction, suggesting that some of the vanadium atoms are reduced by hot nitroethane. The reduced byproducts are no longer detectable by ^{51}V NMR. After the removal of $[V_{13}O_{34}]^{3-}$ as crystals, addition of diethyl ether gave green precipitates. IR spectrum of this precipitate was in agreement with that of reported $[V_{18}O_{46}(NO_3)]^{5-}$ (Figure A4) [32]. The transformation of $\{(n\text{-}C_4H_9)_4N\}_3[H_3V_{10}O_{28}]$ into $\{(n\text{-}C_4H_9)_4N\}_3[V_{13}O_{34}]$ may be accomplished according to Equation (2).

$$13\{(n\text{-}C_4H_9)_4N\}_3[H_3V_{10}O_{28}] \rightleftarrows 10\{(n\text{-}C_4H_9)_4N\}_3[V_{13}O_{34}] + 9\{(n\text{-}C_4H_9)_4N\}OH + 15H_2O \qquad (2)$$

The formation of a base according to Equation (2) prompted the reductive condition for those polyoxovanadates. In the presence of 0.69 equivalents of TsOH with respect to $[H_3V_{10}O_{28}]^{3-}$ to neutralize $\{(n\text{-}C_4H_9)_4N\}OH$, the solution color remained brown. ^{51}V NMR spectrum of the reaction solution showed pure signals due to $[V_{13}O_{34}]^{3-}$ (Figure 2). IR spectrum of the brown precipitates formed by addition of an excess amount of diethyl ether into the reaction solution, was the same as that of $[V_{13}O_{34}]^{3-}$ crystals, suggesting that the complete transformation of $[H_3V_{10}O_{28}]^{3-}$ to $[V_{13}O_{34}]^{3-}$ is achieved with 91% isolated yield (Figure 3). We find this reaction useful for further investigation of $[V_{13}O_{34}]^{3-}$ chemistry.

By the reaction of $[V_{13}O_{34}]^{3-}$ with 0.9 equivalents of $\{(n\text{-}C_4H_9)_4N\}OH$ in a mixed solvent of acetonitrile and water (2:1, v/v), quantitative formation of $[H_3V_{10}O_{28}]^{3-}$ was detected by ^{51}V NMR and IR spectra (Figures 2 and 3). The transformation of $[V_{13}O_{34}]^{3-}$ to $[H_3V_{10}O_{28}]^{3-}$ was $ca.$ 10 times faster than that of $[V_{12}O_{32}]^{4-}$ partly due to the structural resemblance between $[H_3V_{10}O_{28}]^{3-}$ and $[V_{13}O_{34}]^{3-}$. Both of the structures of $[H_3V_{10}O_{28}]^{3-}$ and $[V_{13}O_{34}]^{3-}$ consist of $\{VO_6\}$ units sharing the edge of the octahedra. Only difference in the $[V_{13}O_{34}]^{3-}$ structure from $[H_3V_{10}O_{28}]^{3-}$ is the successive capping of the additional three $\{VO_6\}$ octahedra in a triangular arrangement as add-on units on top of the same decavanadate framework.

2.1.3. Transformation between Dodeca- and Tridecavanadates

In addition to the transformation between $\{(n\text{-}C_4H_9)_4N\}_3[H_3V_{10}O_{28}]$ and $\{(n\text{-}C_4H_9)_4N\}_4[V_{12}O_{32}]$ or $\{(n\text{-}C_4H_9)_4N\}_3[V_{13}O_{34}]$, transformation between $\{(n\text{-}C_4H_9)_4N\}_4[V_{12}O_{32}]$ and $\{(n\text{-}C_4H_9)_4N\}_3[V_{13}O_{34}]$ was investigated based on the Equation (3).

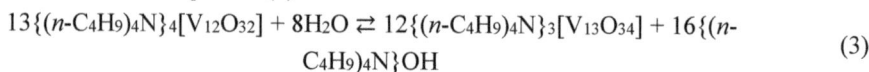

$$13\{(n\text{-}C_4H_9)_4N\}_4[V_{12}O_{32}] + 8H_2O \rightleftarrows 12\{(n\text{-}C_4H_9)_4N\}_3[V_{13}O_{34}] + 16\{(n\text{-}C_4H_9)_4N\}OH \qquad (3)$$

In acetonitrile, $[V_{12}O_{32}]^{4-}$ was not converted to $[V_{13}O_{34}]^{3-}$ even by addition of water and TsOH with heating. The resemblance of stoichiometry between Equations (1) and (3) also make it difficult to control the selective conversion of $[V_{12}O_{32}]^{4-}$ to $[V_{13}O_{34}]^{3-}$. Decavanadates are thermodynamically stable in the presence of water and $[V_{12}O_{32}]^{4-}$ is hardly transformed to other species in acetonitrile without water even when a stoichiometric amount of acid was added. To stimulate the formation of $[V_{13}O_{34}]^{3-}$, we tried to use a different solvent from acetonitrile and nitroethane was successful in converting $[V_{12}O_{32}]^{4-}$ into other forms. However, only the decomposition reaction was observed under the reaction condition with 1.2 equivalents of TsOH in nitroethane at 80 °C, and the formation of $[V_{13}O_{34}]^{3-}$ could not be detected by ^{51}V NMR.

Without any additives, $[V_{13}O_{34}]^{3-}$ was stable even in the refluxing condition in acetonitrile. According to Equation (3), the addition of 1.3 equivalents of $\{(n\text{-}C_4H_9)_4N\}OH$ to the solution of $\{(n\text{-}C_4H_9)_4N\}_3[V_{13}O_{34}]$ was investigated. From ^{51}V NMR spectrum, decavanadates and $[V_{12}O_{32}]^{4-}$ were formed in 2 h at room temperature (Figure 4). The formation of decavanadates could be proceeded via Equation (2). Without heating, $[V_{12}O_{32}]^{4-}$ was hardly obtained via Equation (1) for 2 h (Figure A1). Therefore, $[V_{13}O_{34}]^{3-}$ was directly converted to $[V_{12}O_{32}]^{4-}$ via Equation (3). To achieve high yield of $[V_{12}O_{32}]^{4-}$, the reaction solution was refluxed and the formation of $[V_{12}O_{32}]^{4-}$ was detected by ^{51}V NMR with >95% chemoselectivity (Figure 2). The isolated yield was 86%. IR spectrum of the isolated product was identical to that of $\{(n\text{-}C_4H_9)_4N\}_4[V_{12}O_{32}]$ (Figure 3).

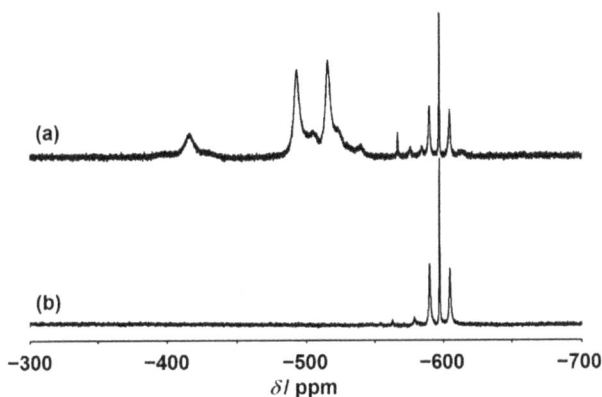

Figure 4. ^{51}V NMR spectra of the reaction solution of $\{(n\text{-}C_4H_9)_4N\}_3[V_{13}O_{34}]$ with 1.3 equivalents of $\{(n\text{-}C_4H_9)_4N\}OH$ after 2 h (**a**) at room temperature and (**b**) in reflux condition.

2.2. Oxidation of Thioanisole

Next, to clarify the differences among $[H_3V_{10}O_{28}]^{3-}$, $[V_{12}O_{32}]^{4-}$ and $[V_{13}O_{34}]^{3-}$ in the activity for oxidation, the thioanisole oxidation with t-butyl hydroperoxide (TBHP) was carried out (Table 1). The oxidation of sulfides to sulfoxides and sulfones has been a widely researched subject due to the importance of products as intermediates in organic synthesis [33]. This method is also useful for oxidative desulfurization of oil [34]. TBHP and H_2O_2 are usual oxidants for the oxidation of sulfides. It is reported that the redox properties of $[H_3V_{10}O_{28}]^{3-}$, $[V_{12}O_{32}]^{4-}$ and $[V_{13}O_{34}]^{3-}$ are different from one another [19]. Although there are some reports on oxidation catalysis for several organic substrates with $[H_3V_{10}O_{28}]^{3-}$ and $[V_{13}O_{34}]^{3-}$, the comparison of the activity in these compounds is not investigated [19,35,36].

Table 1. Oxidation of thioanisole with vanadium species [a].

Entry	Vanadium Species (μmol)	Total Yield (%)	Ratio of Sulfoxide:Sulfone
1	$\{(n\text{-}C_4H_9)_4N\}_3[V_{13}O_{34}]$ (7.7)	91	93:7
2	$\{(n\text{-}C_4H_9)_4N\}_4[V_{12}O_{32}]$ (8.4)	79	79:21
3	$\{(n\text{-}C_4H_9)_4N\}_3[H_3V_{10}O_{28}]$ (10)	33	94:6
4 [b]	V_2O_5 (20)	67	97:3
5	-	3	-

[a] Reaction condition: thioanisole (1 mmol), TBHP (1 mmol), acetonitrile (2 mL), 25 °C, 2.5 h. Yields were determined by GC with an internal standard. Total yield (%) = {sulfoxide (mol) + sulfone (mol) × 2}/initial TBHP (mol) × 100; [b] V_2O_5 was suspended.

Each polyoxovanadate retains their anion structures in acetonitrile at 25 °C. In the presence of 20 equivalents of TBHP with respect to polyoxovanadates, their anion structures were maintained (Figure A5). In the presence of $[V_{13}O_{34}]^{3-}$, 91% total yield of methyl phenyl sulfoxide and methyl phenyl sulfone was achieved in 2.5 h (Table 1). The ratio of sulfoxide and sulfone was 93:7. Under the same reaction conditions, oxidation reaction hardly proceeded in the absence of vanadium species. $[H_3V_{10}O_{28}]^{3-}$ showed a lower activity for this reaction (33% yield in 2.5 h). The reactivity of $[V_{12}O_{32}]^{4-}$ was different from that of $[V_{13}O_{34}]^{3-}$. The yield with $[V_{12}O_{32}]^{4-}$ in 0.5 h (59%) was higher than that with $[V_{13}O_{34}]^{3-}$ (25%), and the reaction ended in 90 min with *ca.* 80% yield. Successive oxidation of sulfoxide to sulfone proceeded from the initial stage of the reaction in the presence of $[V_{12}O_{32}]^{4-}$. While the $[V_{12}O_{32}]^{4-}$ framework is closely related to layered V_2O_5, in which vanadium atoms adopt the square-pyramidal coordination geometry, $[V_{12}O_{32}]^{4-}$ showed the higher activity for the formation of sulfone than V_2O_5. The reactivity depends on the electrophilicity of the active oxidant [37].

After the reaction, the precipitates of polyoxovanadates were formed by addition of a large amount of ethyl acetate and the powder was collected by filtration. From IR and ^{51}V NMR spectra, $[H_3V_{10}O_{28}]^{3-}$ and $[V_{12}O_{32}]^{4-}$ retained their structures and $[V_{13}O_{34}]^{3-}$ was partly decomposed to decavanadates (Figures A6 and A7). Since decavanadates are less active for the oxidation of thioanisole, $[V_{13}O_{34}]^{3-}$ is contributed to the high activity.

3. Experimental Section

3.1. Chemicals and Instruments

All solvents were purchased from Wako Pure Chemical Industries (Osaka, Japan) and used as received. *p*-Toluenesulfonic acid monohydrate (TsOH) and 40% tetra-*n*-butylammonium hydroxide aqueous solution were purchased from Wako Pure Chemical Industries and Sigma-Aldrich (Tokyo, Japan) and used after the dilution to 0.1 M solution with acetone (for the transformations to $[V_{13}O_{34}]^{3-}$) or acetonitrile (for the other transformations). Thioanisole, naphthalene, methyl phenyl sulfoxide, methyl phenyl sulfone and 5.5 M *t*-butyl hydroxide in decane were obtained from

Wako Pure Chemical Industries, Tokyo Chemical Industry (Tokyo, Japan) and Sigma-Aldrich and used as received. $\{(n\text{-}C_4H_9)_4N\}_3[H_3V_{10}O_{28}]$ was synthesized according to the reported procedure [17].

IR spectra were measured on Jasco FT/IR-4200 (Hachioji, Japan) using KBr disks. NMR spectra were recorded with JEOL JNM-LA400 (Akishima, Japan). ^{51}V NMR spectra were measured at 105.15 MHz at 25 °C unless otherwise noted. The chemical shift reference standard for ^{51}V NMR spectroscopy is VOCl$_3$. Elemental analyses of C, H and N were performed by the Research Institute for Instrumental Analysis at Kanazawa University (Kanazawa, Japan). GC analyses were performed on Shimadzu GC-2014 (Kyoto, Japan) with a flame ionization detector (FID) equipped with a NEUTRABOND-1 capillary column (internal diameter = 0.25 mm, length = 30 m).

3.2. Transformation of $\{(n\text{-}C_4H_9)_4N\}_3[H_3V_{10}O_{28}]$ to $\{(n\text{-}C_4H_9)_4N\}_4[V_{12}O_{32}]$

$\{(n\text{-}C_4H_9)_4N\}_3[H_3V_{10}O_{28}]$ (100 mg, 0.06 mmol) was dissolved in acetonitrile (3 mL), followed by addition of 0.1 M $\{(n\text{-}C_4H_9)_4N\}OH$ (0.2 mL, 0.02 mmol) and refluxed for 1 h. After cooling, the solution was added to diethyl ether (40 mL) and stirred for 10 min. Then, the precipitates formed were collected and dried to afford 98 mg of $\{(n\text{-}C_4H_9)_4N\}_4[V_{12}O_{32}]$ (92% yield based on vanadium atoms). ^{51}V NMR of the reaction solution: δ −590, −597, −605 ppm. Anal. Calcd. for $\{(n\text{-}C_4H_9)_4N\}_4[V_{12}O_{32}]$: C,36.72; H, 6.93; N, 2.68; found: C, 35.33; H, 6.62; N, 2.67. IR (KBr pellet; 1500–400 cm^{-1}): 1483, 1379, 994, 860, 792, 761, 711, 649, 609, 549, 519 cm^{-1}.

3.3. Transformation of $\{(n\text{-}C_4H_9)_4N\}_3[H_3V_{10}O_{28}]$ to $\{(n\text{-}C_4H_9)_4N\}_3[V_{13}O_{34}]$

$\{(n\text{-}C_4H_9)_4N\}_3[H_3V_{10}O_{28}]$ (100 mg, 0.06 mmol) was dissolved in nitroethane (3 mL), followed by addition of 0.1 M TsOH (0.4 mL, 0.04 mmol) and stirred for 3 h at 80 °C. After cooling, the solution was added to diethyl ether (40 mL) and stirred for 10 min. Then, the precipitates formed were collected and dried to afford 81 mg of $\{(n\text{-}C_4H_9)_4N\}_3[V_{13}O_{34}]$ (91% yield based on vanadium atom). ^{51}V NMR of the reaction solution: δ −331, −451, −496, −500 ppm. Anal. Calcd. for $\{(n\text{-}C_4H_9)_4N\}_3[V_{13}O_{34}]$: C, 29.82; H, 5.63; N, 2.17; found: C, 29.64; H, 5.59; N, 2.34. IR (KBr pellet; 1500–400 cm^{-1}): 1482, 1379, 1001, 991, 859, 816, 786, 605, 523, 457 cm^{-1}.

3.4. Transformation of $\{(n\text{-}C_4H_9)_4N\}_4[V_{12}O_{32}]$ to $\{(n\text{-}C_4H_9)_4N\}_3[H_3V_{10}O_{28}]$

$\{(n\text{-}C_4H_9)_4N\}_4[V_{12}O_{32}]$ (106 mg, 0.05 mmol) was dissolved in a mixed solvent of acetonitrile and water (4 mL, 3:1, v/v), followed by addition of 0.1 M TsOH (0.2 mL, 0.02 mmol) and stirred for 10 h at 60 °C. After cooling, the solution was added to the mixed solution of diethyl ether and tetrahydrofuran (45 mL, 8:1, v/v) and stirred for 10 min. Then, the precipitates formed were collected and dried to afford 84 mg of $\{(n\text{-}C_4H_9)_4N\}_3[H_3V_{10}O_{28}]$ (83% yield based on vanadium atom). ^{51}V NMR of the reaction solution: δ −397, −431, −506, −524 ppm. IR (KBr pellet; 1500–400 cm^{-1}): 1482, 1379, 987, 972, 876, 843, 808, 771, 721, 609, 550, 506, 461 cm^{-1}.

3.5. Transformation of {(n-C₄H₉)₄N}₃[V₁₃O₃₄] to {(n-C₄H₉)₄N}₃[H₃V₁₀O₂₈]

3.5. Transformation of $\{(n\text{-}C_4H_9)_4N\}_3[V_{13}O_{34}]$ to $\{(n\text{-}C_4H_9)_4N\}_3[H_3V_{10}O_{28}]$

$\{(n\text{-}C_4H_9)_4N\}_3[V_{13}O_{34}]$ (19 mg, 0.01 mmol) was dissolved in a mixed solvent of acetonitrile and water (3 mL, 2:1, v/v), followed by addition of 0.1 M $\{(n\text{-}C_4H_9)_4N\}OH$ (0.09 mL, 0.009 mmol) and stirred for 1 h at 60 °C. After cooling, the solution was added to the mixed solution of diethyl ether and tetrahydrofuran (45 mL, 8:1, v/v) and stirred for 10 min. Then, the precipitates formed were collected and dried to afford 19 mg of $\{(n\text{-}C_4H_9)_4N\}_3[H_3V_{10}O_{28}]$ (86% yield based on vanadium atom). ^{51}V NMR of the reaction solution: δ −397, −430, −506, −525 ppm. IR (KBr pellet; 1500–400 cm^{-1}): 1482, 1379, 987, 973, 875, 843, 809, 770, 719, 609, 550, 507, 461 cm^{-1}.

3.6. Transformation of {(n-C₄H₉)₄N}₃[V₁₃O₃₄] to {(n-C₄H₉)₄N}₄[V₁₂O₃₂]

3.6. Transformation of $\{(n\text{-}C_4H_9)_4N\}_3[V_{13}O_{34}]$ to $\{(n\text{-}C_4H_9)_4N\}_4[V_{12}O_{32}]$

$\{(n\text{-}C_4H_9)_4N\}_3[V_{13}O_{34}]$ (39 mg, 0.02 mmol) was dissolved in acetonitrile (3 mL), followed by addition of 0.1 M $\{(n\text{-}C_4H_9)_4N\}OH$ (0.26 mL, 0.026 mmol) and refluxed for 1 h. After cooling, the solution was added to diethyl ether (40 mL) and stirred for 10 min. Then, the precipitates formed were collected and dried to afford 40 mg of $\{(n\text{-}C_4H_9)_4N\}_4[V_{12}O_{32}]$ (86% yield based on vanadium atom). ^{51}V NMR of the reaction solution: δ −590, −597, −605 ppm. IR (KBr pellet; 1500–400 cm^{-1}): 1483, 1380, 993, 860, 789, 762, 711, 649, 605, 553, 519 cm^{-1}.

3.7. Oxidation of Thioanisole

The oxidations of thioanisole were carried out in an 18 mL glass tube reactor containing a magnetic stir bar. A procedure was as follows: Vanadium species, thioanisole, naphthalene as internal standard, acetonitrile and TBHP were successively placed into the glass tube reactor. The reaction mixture was stirred at 25 °C. The yields of the products were periodically determined by GC analysis using an internal standard technique. All products are known and confirmed by comparison of their GC retention times with the authentic samples.

4. Conclusions

The quantitative conversions among deca-, dodeca- and tridecavanadates were established by monitoring through ^{51}V NMR to determine the optimized transformation conditions. The inorganic host cage complex $[V_{12}O_{32}]^{4-}$ is converted from $[H_3V_{10}O_{28}]^{3-}$ by adjusting the amount of base. Decavanadate is self-condensed into $[V_{13}O_{34}]^{3-}$ in nitroethane by utilizing proton on the decavanadate for the dehydration condensation and the further addition of the controlled amount of acid helps to prevent the byproduct formation. The quantitative formation of $[V_{13}O_{34}]^{3-}$ was achieved with one of the most available $[H_3V_{10}O_{28}]^{3-}$ as a starting material. Transformation of $[V_{12}O_{32}]^{4-}$ or $[V_{13}O_{34}]^{3-}$ to decavanadates proceeded by the hydrolysis reaction. While the direct transformation of $[V_{13}O_{34}]^{3-}$ to $[V_{12}O_{32}]^{4-}$ proceeded by addition of base, conversion of $[V_{12}O_{32}]^{4-}$ to $[V_{13}O_{34}]^{3-}$ could not be observed during our investigation. These transformation reactions proceeded according to the equations and provide the reliable synthetic pathway for the synthesis of $[V_{13}O_{34}]^{3-}$. The studies shows that a careful control of the amount of acid or base in appropriate solvents is able to achieve the transformation among those isopolyoxovanadates, which is fundamentally important species in polyoxovanadate chemistry.

We also demonstrated that $[V_{13}O_{34}]^{3-}$ shows the highest activity for the oxidation of thioanisole among these polyoxovanadates.

Acknowledgments

This work was supported in part by a Grant-in-Aid for Young Scientists (26820349).

Author Contributions

This work was conceived by Y. Hayashi and Y. Kikukawa. The rational structural transformation experiments and oxidation of sulfide were undertaken by K. Ogihara and Y. Kikukawa, respectively. The manuscript was written with contributions from all authors. Y. Kikukawa and K. Ogihara contributed equally.

Conflicts of Interest

The authors declare no conflict of interest.

Appendix

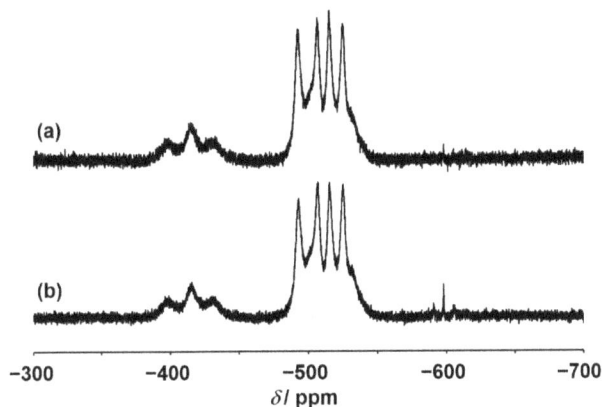

Figure A1. ^{51}V NMR spectra of the forward reaction solution in Equation (1) for the conversion of $\{(n\text{-}C_4H_9)_4N\}_3[H_3V_{10}O_{28}]$ to $\{(n\text{-}C_4H_9)_4N\}_4[V_{12}O_{32}]$ in acetonitrile at room temperature after (**a**) 12 h; (**b**) 24 h.

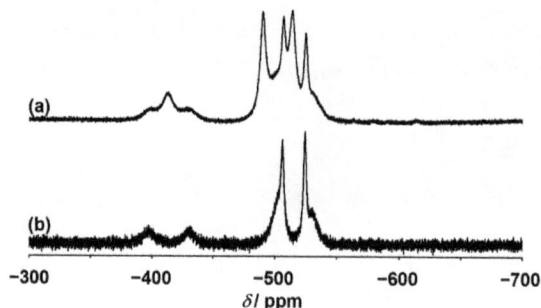

Figure A2. ^{51}V NMR spectra of the reverse reaction solution in Equation (1) for the conversion of $\{(n\text{-}C_4H_9)_4N\}_4[V_{12}O_{32}]$ to $\{(n\text{-}C_4H_9)_4N\}_3[H_3V_{10}O_{28}]$ in a mixed solvent of acetonitrile and water (3:1, v/v) (**a**) without TsOH; (**b**) with 0.4 equivalents of TsOH.

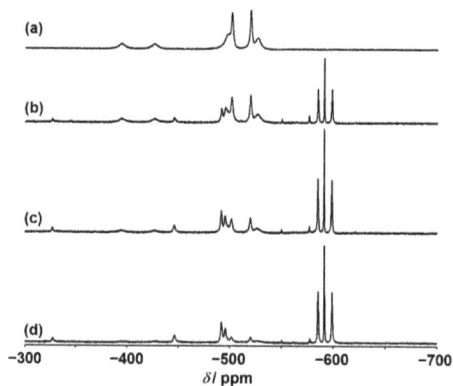

Figure A3. ^{51}V NMR spectra of the forward reaction solution in Equation (2) for the conversion of $\{(n\text{-}C_4H_9)_4N\}_3[H_3V_{10}O_{28}]$ to $\{(n\text{-}C_4H_9)_4N\}_3[V_{13}O_{34}]$ in acetonitrile at 80 °C under dry nitrogen atmosphere after (**a**) 0 h; (**b**) 1 h; (**c**) 3 h; (**d**) 7 h.

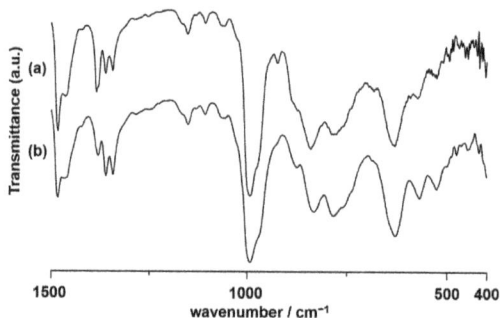

Figure A4. (**a**) IR spectrum of the pale-green precipitated samples after the separation of $\{(n\text{-}C_4H_9)_4N\}_3[V_{13}O_{34}]$. The powder was obtained from the forward reaction solution in Equation (2) for the conversion of $\{(n\text{-}C_4H_9)_4N\}_3[H_3V_{10}O_{28}]$ to $\{(n\text{-}C_4H_9)_4N\}_3[V_{13}O_{34}]$ without addition of acid in nitroethane at 80 °C. (**b**) IR spectrum of the authentic samples of $\{(n\text{-}C_4H_9)_4N\}_5[V_{18}O_{46}(NO_3)]$ [32].

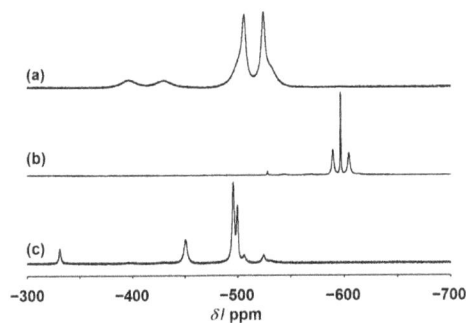

Figure A5. ^{51}V NMR spectra of (**a**) $\{(n\text{-}C_4H_9)_4N\}_3[H_3V_{10}O_{28}]$; (**b**) $\{(n\text{-}C_4H_9)_4N\}_4[V_{12}O_{32}]$; (**c**) $\{(n\text{-}C_4H_9)_4N\}_3[V_{13}O_{34}]$ in the presence of 20 equivalents of TBHP. No spectral change was observed after 23 h.

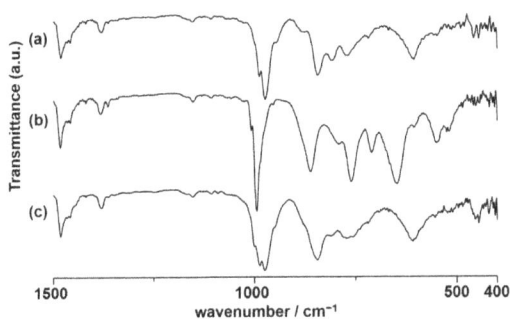

Figure A6. IR spectra of the samples obtained after the oxidation reaction with (**a**) $\{(n\text{-}C_4H_9)_4N\}_3[H_3V_{10}O_{28}]$; (**b**) $\{(n\text{-}C_4H_9)_4N\}_4[V_{12}O_{32}]$; (**c**) $\{(n\text{-}C_4H_9)_4N\}_3[V_{13}O_{34}]$.

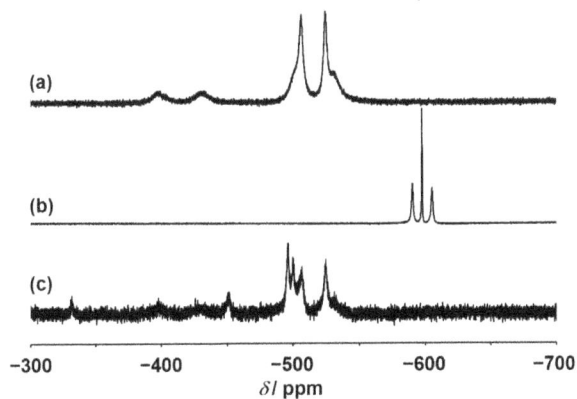

Figure A7. ^{51}V NMR spectra of the samples obtained after the oxidation reaction with (**a**) $\{(n\text{-}C_4H_9)_4N\}_3[H_3V_{10}O_{28}]$; (**b**) $\{(n\text{-}C_4H_9)_4N\}_4[V_{12}O_{32}]$; (**c**) $\{(n\text{-}C_4H_9)_4N\}_3[V_{13}O_{34}]$. Samples were dissolved in acetonitrile.

References

1. Hill, C.L. Thematic issue on polyoxometalates. *Chem. Rev.* **1998**, *98*, 1–390.
2. Kozhevnikov, I.V. *Catalysis by Polyoxometalates*; John Wiley & Sons, Ltd.: Chichester, UK, 2002.
3. Cronin, L.; Müller, A. Thematic issue on polyoxometalates. *Chem. Soc. Rev.* **2012**, *41*, 7325–7648.
4. Khan, M.I. Novel Extended Solids Composed of Transition Metal Oxide Clusters. *J. Solid Stat. Chem.* **2000**, *152*, 105–112.
5. Baes, C.F., Jr.; Mesmer, R.E. *The Hydrolysis of Cations*; Wiley & Sons: New York, NY, USA, 1976.
6. Hamilton, E.E.; Fanwick, P.E.; Wilker, J.J. The Elusive Vanadate $(V_3O_9)^{3-}$: Isolation, Crystal Structure, and Nonaqueous Solution Behavior. *J. Am. Chem. Soc.* **2002**, *124*, 78–82.
7. Román, P.; José, A.S.; Luque, A.; Gutiérrez-Zorrilla, J.M. Observation of a Novel Cyclic Tetrametavanadate Anion Isolated from Aqueous Solution. *Inorg. Chem.* **1993**, *32*, 775–776.
8. Day, V.W.; Klemperer, W.G.; Yaghi, O.M. A New Structure Type in Polyoxoanion Chemistry: Synthesis and Structure of the $V_5O_{14}^{3-}$ Anion. *J. Am. Chem. Soc.* **1989**, *111*, 4518–4519.
9. Okaya, K.; Kobayashi, T.; Koyama, Y.; Hayashi, Y.; Isobe, K. Formation of V^V Lacunary Polyoxovanadates and Interconversion Reactions of Dodecavanadate Species. *Eur. J. Inorg. Chem.* **2009**, *2009*, 5156–5163.
10. Day, V.W.; Klemperer, W.G.; Yaghi, O.M. Synthesis and Characterization of a Soluble Oxide Inclusion Complex, $[CH_3CN \subset (V_{12}O_{32}^{4-})]$. *J. Am. Chem. Soc.* **1989**, *111*, 5959–5961.
11. Kastner, K.; Margraf, J.T.; Clark, T.; Streb, C. A Molecular Placeholder Strategy to Access a Family of Transition-Metal-Functionalized Vanadium Oxide Clusters. *Chem. Eur. J.* **2014**, *20*, 12269–12273.
12. Kobayashi, T.; Kuwajima, S.; Kurata, T.; Hayashi, Y. Structural Conversion from Bowl- to Ball-Type Polyoxovanadates: Synthesis of a Spherical Tetradecavanadate through a Chloride-Incorporated Bowl-Type Dodecavanadate. *Inorg. Chim. Acta* **2014**, *420*, 69–74.
13. Omri, I.; Graia, M.; Mhiri, T. Synthesis, Crystal Structure, Vibrational and Optical Properties of $(Hdea)_4(V_7O_{19}F) \cdot 0.42H_2O$, an Original $(V_7O_{19}F)^{4-}$ Cluster Oxyfluoride. *J. Clust. Sci.* **2015**, *26*, 815–825.
14. Kikukawa, Y.; Yokoyama, T.; Kashio, S.; Hayashi, Y. Synthesis and Characterization of Fluoride-Incorporated Polyoxovanadates. *J. Inorg. Biochem.* **2015**, *147*, 221–226.
15. Chen, Q.; Zubieta, J. Synthesis and Structural Characterization of a Polyoxovanadate Coordination Complex with a Hexametalate Core: $[(n\text{-}C_4H_9)_4N]_2[V_6O_{13}\{O_2NC(CH_2O)_3\}_2]$. *Inorg. Chem.* **1990**, *29*, 1456–1458.
16. Domae, K.; Uchimura, D.; Koyama, Y.; Inami, S.; Hayashi, Y.; Isobe, K.; Kameda, H.; Shimoda, T. Synthesis of a Bowl-Type Dodecavanadate by the Coupling Reaction of Alkoxohexavanadate and Discovery of a Chiral Octadecavanadate. *Pure Appl. Chem.* **2009**, *81*, 1323–1330.

17. Day, V.W.; Klemperer, W.G.; Maltbie, D.J. Where Are the Protons in $H_3V_{10}O_{28}{}^{3-}$? *J. Am. Chem. Soc.* **1987**, *109*, 2991–3002.

18. Nakamura, S.; Ozeki, T. Hydrogen-bonded Aggregates of Protonated Decavanadate Anions in Their Tetraalkylammonium Salts. *J. Chem. Soc. Dalton Trans.* **2001**, doi:10.1039/B008128K.

19. Hou, D.; Hagen, K.S.; Hill, C.L. Tridecavanadate, $[V_{13}O_{34}]^{3-}$, a New High-Potential Isopolyvanadate. *J. Am. Chem. Soc.* **1992**, *114*, 5864–5866.

20. Klemperer, W.G.; Marquart, T.A.; Yaghi, O.M. Shape-Selective Binding of Nitriles to the Inorganic Cavitand, $V_{12}O_{32}{}^{4-}$. *Mater. Chem. Phys.* **1991**, *29*, 97–104.

21. Kawanami, N.; Ozeki, T.; Yagasaki, A. NO^- Anion Trapped in a Molecular Oxide Bowl. *J. Am. Chem. Soc.* **2000**, *122*, 1239–1240.

22. Kurata, T.; Hayashi, Y.; Isobe, K. Synthesis and Characterization of Chloride-Incorporated Dodecavanadate from Dicopper Complex of Macrocyclic Octadecavanadate. *Chem. Lett.* **2010**, *39*, 708–709.

23. Rohmer, M.-M.; Devémy, J.; Wiest, R. Bénard, M. *Ab initio* Modeling of the Endohedral Reactivity of Polyoxometallates: 1. Host-Guest Interactions in $[RCN \subset (V_{12}O_{32})^{4-}]$ (R = H, CH_3, C_6H_5). *J. Am. Chem. Soc.* **1996**, *118*, 13007–13014.

24. Kurata, T.; Hayashi, Y.; Uehara, A.; Isobe, K. Synthesis of a Reduced Tridecavanadate Dimer Linked by Eight Hydrogen Bonds. *Chem. Lett.* **2003**, *32*, 1040–1041.

25. Li, F.; Long, D.-L.; Cameron, J.M.; Miras, H.N.; Pradeep, C.P.; Xu, L.; Cronin, L. Cation Induced Structural Transformation and Mass Spectrometric Observation of the Missing Dodecavanadomanganate(IV). *Dalton Trans.* **2012**, *41*, 9859–9862.

26. Yin, Q.; Tan, J.M.; Besson, C.; Geletti, Y.V.; Musaev, D.G.; Kuznetsov, A.E.; Luo, Z.; Hardcastle, K.I.; Hill, C.L. A Fast Soluble Carbon-Free Molecular Water Oxidation Catalyst Based on Abundant Metals. *Science* **2010**, *328*, 342–345.

27. Kikukawa, Y.; Yamaguchi, K.; Mizuno, N. Zinc(II) Containing γ-Keggin Sandwich-Type Silicotungstate: Synthesis in Organic Media and Oxidation Catalysis. *Angew. Chem. Int. Ed.* **2010**, *49*, 6096–6100.

28. Ibrahim, M.; Lan, Y.; Bassil, B.S.; Xiang, Y.; Suchopar, A.; Powell, A.K.; Kortz, U. Hexadecacobalt(II)-Containing Polyoxometalate-Based Single-Molecule Magnet. *Angew. Chem. Int. Ed.* **2011**, *50*, 4708–4711.

29. Suzuki, K.; Sato, R.; Mizuno, N. Reversible Switching of Single-Molecule Magnet Behaviors by Transformation of Dinuclear Dysprosium Cores in Polyoxometalates. *Chem. Sci.* **2013**, *4*, 596–600.

30. Wagner, G.W. Two-Dimensional $^{17}O-^{51}V$ Heteronuclear Shift Correlation NMR Spectroscopy of the ^{17}O-Enriched Inclusion Complex $[CH_3CN \subset (V_{12}O_{32}{}^{4-})]$. Relationship of Cross-Peak Intensity to Bond Order. *Inorg. Chem.* **1991**, *30*, 1960–1962.

31. Hayashi, Y.; Fukuyama, K.; Takatera, T.; Uehara, A. Synthesis and Structure of a New Reduced Isopolyvanadate, $[V_{17}O_{42}]^{4-}$. *Chem. Lett.* **2000**, *29*, 770–771.

32. Koyama, Y.; Hayashi, Y.; Isobe, K. Self-Assembled All-Inorganic Chiral Polyoxovanadate: Spontaneous Resolution of Nitrate-Incorporated Octadecavanadate. *Chem. Lett.* **2008**, *37*, 578–579.

33. Carreño, M.C. Applications of Sulfoxides to Asymmetric Synthesis of Biologically Active Compounds. *Chem. Rev.* **1995**, *95*, 1717–1760.

34. Zannikos, F.; Lois, E.; Stournas, S. Desulfurization of Petroleum Fractions by Oxidation and Solvent Extraction. *Fuel Process. Tech.* **1995**, *42*, 35–45.

35. Csányi, L.J.; Jáky, K.; Dombi, G.; Evanics, F.; Dezsö, G.; Kóta, Z. Onium-Decavanadate Ion-Pair Complexes as Catalysts in the Oxidation of Hydrocarbons by O_2. *J. Mol. Catal. A* **2003**, *195*, 101–111.

36. Coletti, A.; Whiteoak, C.J.; Conte, V.; Kleij, A.W. Vanadium Catalyzed Synthesis of Cyclic Organic Carbonates. *ChemCatChem* **2012**, *4*, 1190–1196.

37. Adam, W.; Malisch, W.; Roschmann, K.J.; Saha-Möller, C.R.; Schenk, W.A. Catalytic Oxidations by Peroxy, Peroxo and Oxo Metal Complexes: An Interdisciplinary Account with a Personal View. *J. Organomet. Chem.* **2002**, *661*, 3–16.

Structural and Electronic Properties of Polyoxovanadoborates Containing the [V$_{12}$B$_{18}$O$_{60}$] Core in Different Mixed Valence States

Patricio Hermosilla-Ibáñez, Karina Muñoz-Becerra, Verónica Paredes-García, Eric Le Fur, Evgenia Spodine and Diego Venegas-Yazigi

Abstract: This review summarizes all published data until April 2015 related to crystalline lattices formed by the [V$_{12}$B$_{18}$O$_{60}$] core, which generates polyanionic clusters with different degrees of protonation and mixed-valence ratios. The negative charge of this cluster is counterbalanced by different cations such as protonated amines, hydronium, and alkaline, and transition metal ions. The cluster is shown to form extended 1D, 2D, or 3D frameworks by forming covalent bonds or presenting hydrogen bond interactions with the present secondary cations. These cations have little influence on the solid state reflectance UV-visible spectra of the polyanionic cluster, but are shown to modify the FT-IR spectra and the magnetic behavior of the different reported species.

Reprinted from *Inorganics*. Cite as: Hermosilla-Ibáñez, P.; Muñoz-Becerra, K.; Paredes-García, V.; Le Fur, E.; Spodine, E.; Venegas-Yazigi, D. Structural and Electronic Properties of Polyoxovanadoborates Containing the [V$_{12}$B$_{18}$O$_{60}$] Core in Different Mixed Valence States. *Inorganics* **2015**, *3*, 309-331.

1. Introduction

Polyoxometalates (POMs) have been systematically studied during the last decades due to their structural variety and physico-chemical properties [1–13]. The structural variety of these polyanionic clusters is generated by their interaction with or bonding to different cationic species such as protonated amines, ammonium and hydronium ions, alkaline, transition metal, and lanthanide ions. These cationic species can exist in the lattice as charge-compensating agents, or linked to the external oxygen atoms of the polyanionic species, and thus serve to increase the dimensionality of these systems [14–19]. The crystalline packing is stabilized by hydrogen bonds, in the case of ammonium cations, or by the formation of covalent bonds through the interaction of the external oxygen atoms of the polyanionic species with different metal cations, such as alkaline or transition metal ions.

The polyoxovanadates (VO) constitute an appealing family of polyanions, due principally to the structural plasticity that vanadium atoms present in their different oxidation states. The variable topology of the polynuclear VO systems is given by the {V$_x$O$_y$} entities with diverse connectivities. The literature reports polyoxovanadates containing from five to 34 metal centers: [V$_5$O$_{14}$]$^{3-}$ [20], [V$_{10}$O$_{28}$]$^{6-}$ [21], [V$_{12}$O$_{32}$]$^{4-}$ [22], [V$_{13}$O$_{34}$]$^{3-}$ [23], [V$_{15}$O$_{42}$]$^{9-}$ [24], [V$_{15}$O$_{36}$]$^{5-}$ [25], [V$_{18}$O$_{42}$]$^{12-}$ [26], [V$_{19}$O$_{49}$]$^{9-}$ [27], [V$_{22}$O$_{54}$]$^{6-}$ [28], and [V$_{34}$O$_{82}$]$^{10-}$ [29]. When a heteroatom such as boron is condensed in these systems, the structural richness of the VO cluster is increased; this new family of polyoxometalates is termed polyoxovanadoborates (VBO). To date the known polyanions included in the VBO family are [V$_6$B$_{20}$] [30–33], [V$_6$B$_{22}$] [34], [V$_{10}$B$_{28}$] [35–37], [V$_{12}$B$_{16}$] [38–41], [V$_{12}$B$_{17}$] [42], [V$_{12}$B$_{18}$] [31,39–59], and [V$_{12}$B$_{32}$] [60,61]. The first VBO polyanions were reported at the end of the

1990s by Rijssenbeek *et al.* [42], and in 2000 Williams *et al.* published the first review on vanadoborate clusters as a chapter in *Contemporary Boron Chemistry* [62].

As a result of V^V and V^{IV} sharing similar geometries, the redox processes between both states can be accessed by keeping the initial structure of the polyanion constant, thus generating several mixed valence systems. This feature is directly related to the magnetic and electronic properties shown by these compounds. However, these properties have been studied less extensively than the structural ones. In this review we report and discuss the structural and electronic properties of several polyanions derived from the $[V_{12}B_{18}O_{60}]$ core and their corresponding crystalline lattices.

2. Polyoxovanadoborates: Vanadate and Borate Fragments

The structures of all the VBO compounds reported in the literature are formed by the condensation of vanadate and borate fragments. The vanadate fragments always contain five-coordinated vanadium atoms (VO_5) in a [4+1] square base pyramidal coordination geometry, while the borate fragments include trigonal (BO_3) and tetrahedral (BO_4) borate units. The different connectivity between the (VO_5), (BO_3), and (BO_4) units leads to varied VBO cores with open barrel-like structures ($[V_{10}B_{28}]$ and $[V_{12}B_{32}]$), and closed spherical-like structures ($[V_6B_{20}]$, $[V_6B_{22}]$, $[V_{12}B_{16}]$, $[V_{12}B_{17}]$, and $[V_{12}B_{18}]$). Among all the abovementioned clusters, $[V_{12}B_{18}]$ is the one most studied from a structural point of view.

2.1. Vanadate Fragments

Some selected examples of clusters have been analyzed in order to summarize the bond lengths given for the following fragments.

$[V_6O_{18}]$-type A: The $[V_6O_{18}]$-type A hexanuclear fragment shown in Figure 1a consists of six (VO_5) units connected by two equatorial oxygen atoms sharing two opposite edges to form a ring-like structure. In this moiety, the vanadium atoms are coplanar and the V=O groups are on the periphery. This hexanuclear ring is found in the structures formed by the $[V_6B_{20}]$ and $[V_6B_{22}]$ cores. The V=O bond lengths range from 1.605 to 1.625 Å, while the V-O distances are between 1.944 and 1.970 Å [30,33].

$[V_6O_{18}]$-type B: The topology of this fragment changes from the above mentioned hexanuclear ring to a triangular array (Figure 1b), due to the different condensation of the (VO_5) units, which are arranged in an alternated way of adjacent and opposite edges. This fragment can be found solely in the $[V_{12}B_{18}]$ core. The V=O and V-O bonds lengths range from 1.599 to 1.638 Å, and between 1.906 and 2.023 Å, respectively [43,52].

$[V_{10}O_{30}]$: This decanuclear moiety is similar to the $[V_6O_{18}]$-type A, where the (VO_5) polyhedra share their opposite edges, forming a coplanar toroidal ring (Figure 1c). The compounds constituted by the $[V_{10}B_{28}]$ core present this fragment in their structures. The V=O bond distances range from 1.599 to 1.640 Å; the V-O bond distances range from 1.894 to 1.989 Å [35,63].

$[V_{12}O_{36}]$-type A: This dodecanuclear vanadate fragment can be seen as the union of two mutually perpendicular semicircles, each one formed by five (VO_5) polyhedra linked by opposite edges. Both semicircles are connected by two additional (VO_5) polyhedral, thus forming a continuous ring as shown in Figure 1d. This moiety is found in the compounds containing the $[V_{12}B_{16}]$ and $[V_{12}B_{17}]$ cores. The V=O bond distances are between 1.609 and 1.632 Å, while the V-O bond distances range from 1.919 to 2.019 Å [40,41].

$[V_{12}O_{36}]$-type B: This dodecanuclear fragment is described as a planar ring, whose topology is similar to the above-mentioned $[V_6O_{18}]$-type A and $[V_{10}O_{30}]$ rings (Figure 1e). This fragment can be found in the structures of compounds with the $[V_{12}B_{32}]$ core. The V=O and V-O bonds lengths range from 1.569 to 1.649 Å, and between 1.890 and 1.990 Å, respectively [60,61].

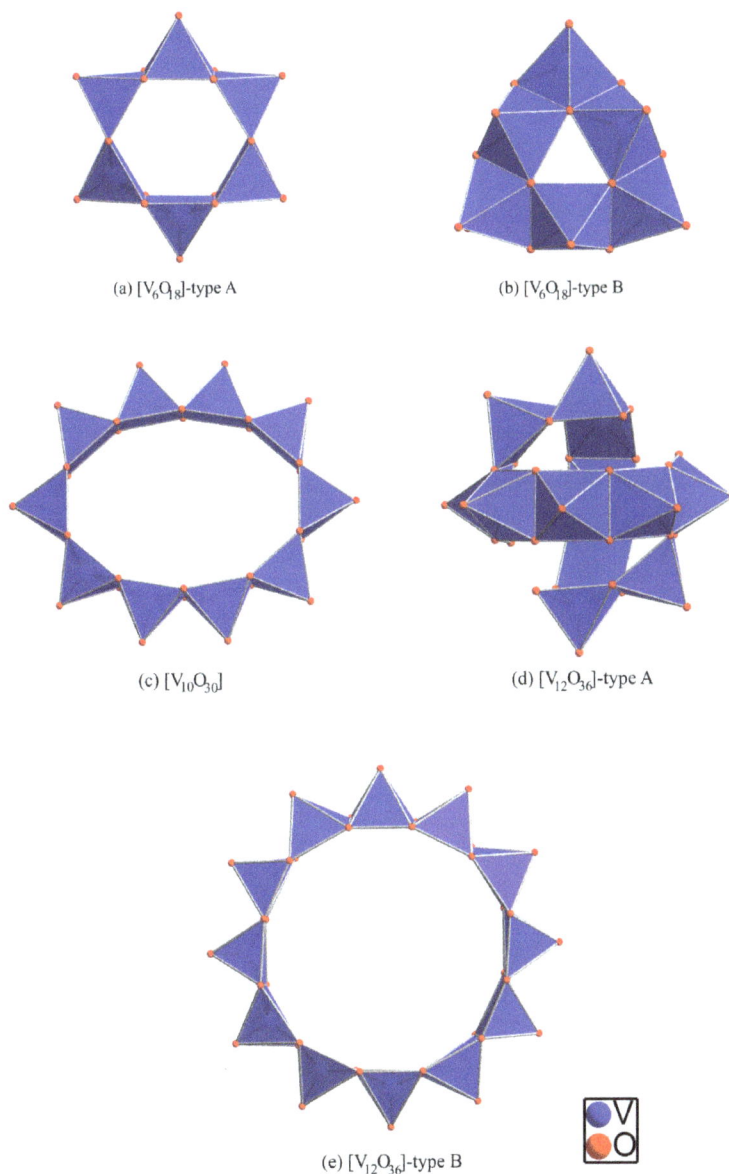

(a) $[V_6O_{18}]$-type A

(b) $[V_6O_{18}]$-type B

(c) $[V_{10}O_{30}]$

(d) $[V_{12}O_{36}]$-type A

(e) $[V_{12}O_{36}]$-type B

Figure 1. Polyhedral representation of the vanadate fragments.

As expected, the shorter distances are observed for the vanadyl groups (V=O), as compared to the single bonds (V-O). However, the analysis of the different vanadium-oxygen distances for all the vanadate fragments will depend on the mixed valence ratio, which is not the case for the borate fragments. Therefore a comparative analysis is not possible.

2.2. Borate Fragments

In order to describe the different borate units, the topological classification proposed by Christ and Clark will be used [64]. Some selected examples of clusters have been considered in order to analyze the bond lengths given for the following fragments.

$[B_{10}O_{22}]^{14-}$: This unit is composed by three $(B_3O_8)^{7-}$ building-blocks connected to an additional (BO_4) entity, located in the middle of the fragment, as shown in Figure 2a. The triangular decaborate unit has three trigonal boron atoms at the corner of the triangle, and seven tetrahedral boron atoms 10[3: Δ+7T]. This decaborate is present in compounds with the $[V_6B_{20}]$ core. The trigonal B-O bonds range from 1.349-1.391 Å, while the tetrahedral ones are between 1.425 and 1.545 Å [30,33].

$[B_{11}O_{24}]^{15-}$: This undecaborate can be described as constituted by the abovementioned $[B_{10}O_{22}]^{14-}$ fragment with an extra (BO_3) unit linked to the apical oxygen atom of the central tetrahedral (BO_4) group (Figure 2b) 11[4:Δ+7T]. This polyborate fragment has been described in compounds formed by the $[V_6B_{22}]$ core. The trigonal and tetrahedral B-O bond distances range from 1.359 to 1.376 Å, and 1.430 to 1.546 Å, respectively [34].

$[B_{14}O_{32}]^{22-}$: The structure of this ring-like fragment is composed of eight tetrahedral and six trigonal boron atoms. Four (BO_4) units form two pairs and these are linked by two (BO_3) group forming a ring (Figure 2c); four additional (BO_3) entities are terminal groups bridging each pair of (BO_4) groups 14:2[2(3:Δ+2T)+(3:Δ)]. The compounds containing the $[V_{10}B_{28}]$ core have this tetradecaborate unit. The B-O bond distances of the trigonal units range from 1.338 and 1.391 Å, while the B-O bond distances of the tetrahedral units are between 1.422 and 1.500 Å [35,63].

$[B_8O_{21}]^{21-}$: This polyanionic entity is the lowest borate nuclearity fragment described to date. This structure is formed by a chain of six tetrahedral (BO_4) units. Two trigonal (BO_3) units bridge the second/third and fourth/fifth tetrahedral units of the chain (Figure 2d) 8:2[4:Δ+3T]. This fragment is found in clusters with the $[V_{12}B_{16}]$ core. The B-O bond distances of the trigonal and tetrahedral units are between 1.348 and 1.389 Å, and 1.416 and 1.530 Å, respectively [40,41].

$[B_{18}O_{42}]^{30-}$: This polyborate ring forms part of the compounds containing the $[V_{12}B_{18}]$ core. It is made of six $(B_3O_7)^{5-}$ units, each one constituted by one trigonal and two tetrahedral boron atoms. Each $(B_3O_7)^{5-}$ building-block contains one terminal (BO_3) group, and therefore each $[B_{18}O_{42}]^{30-}$ fragment has six peripheral (BO_3) units, 18[6: Δ+12T] (Figure 2e). The trigonal B-O bonds range from 1.340 to 1.382 Å, while the tetrahedral ones are between 1.420 and 1.525 Å [43,52].

$[B_{16}O_{36}]^{24-}$: This polyborate ring is formed of four pairs of (BO_4) units linked by four (BO_3) trigonal entities; each pair is additionally condensed to a terminal trigonal BO_3, 16[8:Δ+8T] (Figure 2f). This fragment is part of the $[V_{12}B_{32}]$ family. The B-O bond distances are between 1.310 and 1.450 Å for the trigonal units, and between 1.410 and 1.530 Å for the tetrahedral units [60,61].

Based on the abovementioned data, it is possible to infer that all the trigonal B-O bond distances are shorter than those corresponding to the tetrahedral ones. An exception is that of the $[B_{16}O_{36}]^{24-}$ fragment, which has been reported in only two studies [60].

(a) $[B_{10}O_{23}]^{14-}$

(b) $[B_{11}O_{24}]^{15-}$

(c) $[B_{14}O_{32}]^{22-}$

(d) $[B_{8}O_{21}]^{21-}$

(e) $[B_{18}O_{42}]^{30-}$

(f) $[B_{16}O_{36}]^{24-}$

Figure 2. Polyhedral representation of the borate fragments.

Figure 3 shows the different polyoxovanadoborate clusters $[V_xB_yO_z]$ generated by the condensation of the $[V_iO_j]$ and $[B_hO_k]$ fragments described above.

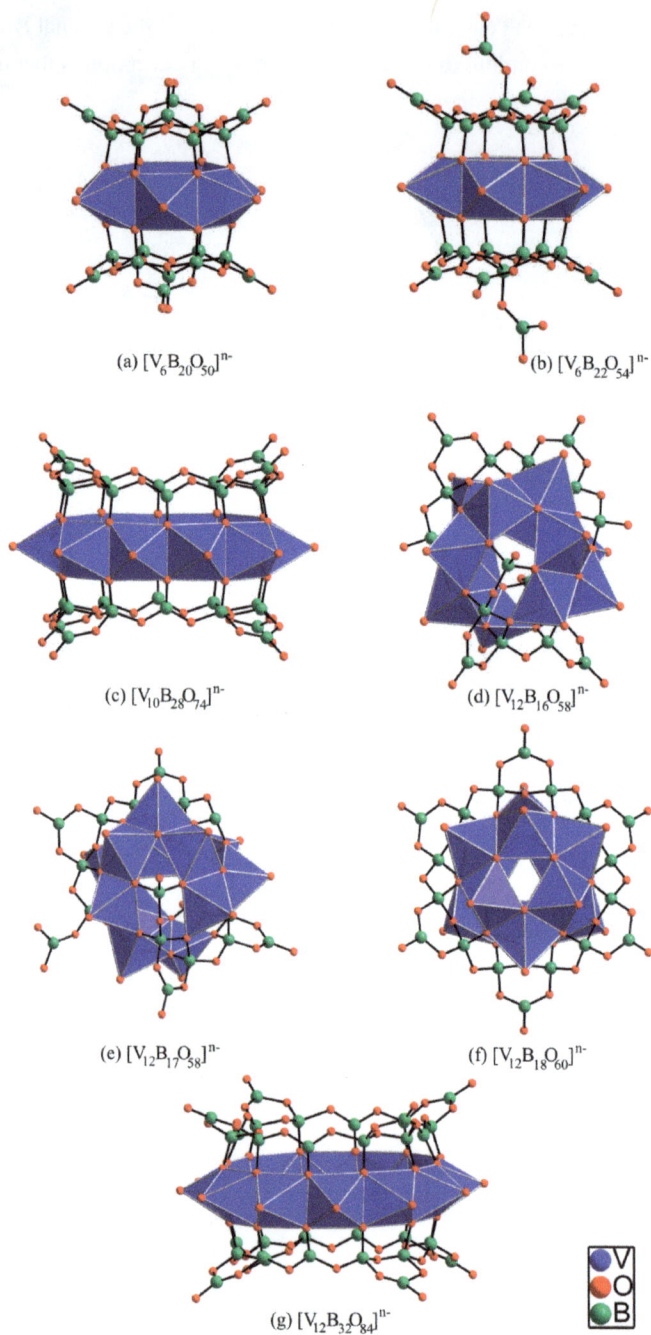

(a) $[V_6B_{20}O_{50}]^{n-}$

(b) $[V_6B_{22}O_{54}]^{n-}$

(c) $[V_{10}B_{28}O_{74}]^{n-}$

(d) $[V_{12}B_{16}O_{58}]^{n-}$

(e) $[V_{12}B_{17}O_{58}]^{n-}$

(f) $[V_{12}B_{18}O_{60}]^{n-}$

(g) $[V_{12}B_{32}O_{84}]^{n-}$

V
O
B

Figure 3. Structural representation of the different polyoxovanadoborate cores.

3. Structural Description of the $[V_{12}B_{18}O_{60}]$ Core

The $[V_{12}B_{18}O_{60}]$ polyoxovanadoborate core consists of two vanadate $[V_6O_{18}]$-type B and one $[B_{18}O_{42}]^{30-}$ borate fragments. Each vanadate fragment has six five-coordinated (VO_5) vanadium centers adopting a [4+1] square base pyramidal coordination geometry. The vanadium atom is displaced from the best mean plane formed by the four equatorial oxygen atoms towards the axial vanadyl group, by ca. 0.7 Å. The angles formed between the V=O bond and the four equatorial V-O bonds are from 100 to 110°. All the V=O distances are in the range of 1.57 Å to 1.68 Å [65]. The borate fragment $[B_{18}O_{42}]^{30-}$ is condensed to the two $[V_6O_{18}]$-type B moieties, thus remaining in a sandwich-type configuration in the middle of the $[V_{12}B_{18}O_{60}]$ polyanion (Figure 3f). A water molecule is always found (with partial occupancy most of the time) within the cavity of the $[V_{12}B_{18}O_{60}]$ polyanion.

4. $[V_{12}B_{18}O_{60}]$ Cores with Protonated Amines as Counterbalancing Ions

In this section lattices based on the $[V_{12}B_{18}O_{60}]$ core including protonated diamines, ammonium, and hydronium as counterbalancing ions will be described, and are listed in Table 1. Rijssenbeek et al. [42], in 1997, obtained by hydrothermal synthesis the first two VBO crystalline systems based on the $[V_{12}B_{18}O_{60}]$ core, with 1,2-ethylenediammonium $(enH_2)^{2+}$ (1) and 1,3-propanediammonium $(1,3-diapH_2)^{2+}$ (2) ions to compensate for the negative charge. The authors mentioned that the organic molecule is included in the synthetic medium as a structure-directing agent of the framework. In both (1) and (2), the diammonium ions occupy the intercluster space along with water solvation molecules. In 2011 Liu et al. [49] reported another lattice having 1,2-propanediammonium $(1,2-dapH_2)^{2+}$ or $(H_2dap)^{2+}$ (3), along with hydronium ions as charge counterbalance cations. (3) was also obtained by hydrothermal synthesis including additionally $Cu(CH_3COO)_2 2H_2O$ in the reaction mixture, a species that was not included in the final crystalline packing. However, the authors did not mention if the same lattice is formed leaving out the copper source. Bigger ammonium cations derived from triethylenetetramine $(H_3teta)^{3+}$ together with hydronium ions were found in the crystalline system (4) based on the $[V_{12}B_{18}O_{60}]$ core, studied by Liu et al. in 2013 [34]. The hydrothermal synthetic procedure used also included a secondary metal source, metallic cobalt powder. In this case, the authors pointed out that the presence of the transition metal is indispensable to obtain (4). In 2014 we reported a new crystalline system (5) containing $(1,3-diapH_2)^{2+}$ and ammonium as counterbalancing ions. The synthesis was also carried out using the hydrothermal method, in which the ammonium ions were included in the lattice by adding $(NH_4)_2HPO_4$ to the reaction mixture [57].

All the abovementioned compounds were described as having the same degree of protonation of the $[V_{12}B_{18}O_{60}]$ core, thus being based on the $[V_{12}B_{18}O_{60}H_6]^{10-}$ polyanion. The five studied clusters have the same mixed valence ratio of V^{IV} to V^V of 10/2.

Lattices (1), (3), (4), and (5) present one crystallographic site for the protonated diamine, while (2) presents three crystallographically different sites for the $(1,3-diapH_2)^{2+}$ cations. The $(1,3-diapH_2)^{2+}$ ions in lattices (2) and (5) adopt a "W"-type conformation [66]. In lattices (2) and (3), the diammonium cations do not adopt a preferential order in the lattice, whereas the $(H_3teta)^{3+}$

molecules in framework (4) are defined by the authors as forming pseudo-hexagonal channels [34]. In framework (5), the (1,3-diapH$_2$)$^{2+}$ cations are described as all oriented along the c axis [57]. Within all the crystalline lattices, each protonated diamine connects four [V$_{12}$B$_{18}$O$_{60}$H$_6$]$^{10-}$ clusters through unidirectional, bifurcated, and trifurcated hydrogen bonds, thus generating supramolecular structures [59].

Table 1. List of the lattices with protonated amines as counterbalancing ions. The mixed valence V^{IV}/V^V ratio is indicated for each lattice.

Compound	Formula	V^{IV}/V^V Ratio	Ref.
1	(enH$_2$)$_5${(VO)$_{12}$O$_6$[B$_3$O$_6$(OH)]$_6$} H$_2$O	10/2	[42]
2	(1,3-diapH$_2$)$_5${(VO)$_{12}$O$_6$[B$_3$O$_6$(OH)]$_6$} 6H$_2$O	10/2	[42]
3	(H$_2$dap)$_2$H$_6${(VO)$_{12}$O$_6$[B$_3$O$_6$(OH)]$_6$(H2O)} 13H$_2$O	10/2	[49]
4	[H$_3$teta]$_3$[V$_{12}$B$_{18}$O$_{54}$(OH)$_6$(H$_2$O)] (H$_3$O) 5H$_2$O	10/2	[34]
5	(NH$_4$)$_8$(1,3-diapH$_2$)[V$_{12}$B$_{18}$O$_{60}$H$_6$] 5H$_2$O	10/2	[57]

5. [V$_{12}$B$_{18}$O$_{60}$] Cores with Transition Metal Ions and Coordination Compounds as Counterbalancing Cations

The lattices that include transition metal ions and coordination compounds, together with the [V$_{12}$B$_{18}$O$_{60}$] core in their crystallographic packing, are listed in Table 2. In some cases the metal cations are found coordinated to the clusters through the oxygen atoms of the polyanions and to water molecules, while in other cases coordination complexes with organic molecules are bonded to the VBO clusters. Thus, this class of systems can be considered as functionalized polyoxovanadoborates.

Compound (12) has a pure inorganic framework that contains six-coordinated Cd^{II} ions and crystallizes in the cubic centrosymmetric space group Pn-3. The asymmetric unit consists of a half of one Cd^{II} ion, two vanadium atoms, and three boron atoms. The divalent cations are connected to the polyanions sharing μ_3-bridge-oxygen atoms from the [B$_{18}$O$_{42}$]$^{30-}$ and [V$_6$O$_{18}$]$^{n-}$ fragments of the [V$_{12}$B$_{18}$O$_{60}$] clusters, leading to a porous 3D lattice. The coordination sphere of the Cd^{II} ions is completed with oxygen atoms of water molecules. The Cd^{II}-(μ_3-O)-B2 distances range from 2.184(3) to 2.552(3) Å. Despite the fact that diethylenetriamine (dien) was added to the reaction mixture, this amine is not present in the crystalline system (12).

Very appealing crystalline structures are formed when metal cations do not act as bridges between clusters, but are only bonded to one [V$_{12}$B$_{18}$O$_{60}$] cluster, thus decorating it as coordination complexes. The first system of this kind is compound (6), reported by Zhang et al. in 1999 [43]. This compound includes five-coordinated Zn(en)$_2$$^{2+}$ cations, whose coordination sphere is completed by coordination to the VBO polyanion through one oxygen atom from the polyborate fragment, forming a Zn-(μ_3-O)-B2 covalent bond (Zn-O; 2.042(2) Å) (Figure 4a). Lin et al. [45] reported compound (8), which has six-coordinated Ni(en)$_2$$^{2+}$ complexes. These complete their coordination sphere with one VBO cluster bonded through two oxygen atoms, thus forming two Ni-(μ_3-O)-B2 bonds (Ni-O; 2.086(4) to 2.224(4) Å) (Figure 4b). Zn(teta)$_2$$^{2+}$ complexes are part of the crystalline system of (9) (Figure 4c) reported by Liu and Zhou [48], while Zn(dien)$_2$$^{2+}$ and [Zn(dien)$_2$(H$_2$O)]$^{2+}$ complexes are introduced in the lattice of compound **(10)** (Figure 4d), reported by Liu *et al.* [49]. The M-O distances of the M-(μ_3-O)-B_2 bonds have values of 1.979(2) Å for **(9)** and 2.001(4) to 2.436(5) Å for **(10)**.

Table 2. List of the lattices with transition metal ions and coordination compounds as counterbalancing ions. The mixed valence V^{IV}/V^V ratio is indicated for each lattice.

Compound	Formula	V^{IV}/V^V Ratio	Ref.
6	$[Zn(en)_2]_6[(VO)_{12}O_6B_{18}O_{39}(OH)_3]$ $13H_2O$	9/3	[43]
7	$H_3\{[Cu(en)_2]_5[(VO)_{12}O_6B_{18}O_{42}]\}[B(OH)_3]_2$ $16H_2O$	7/5 †	[44]
8	$[Ni(en)_2]_6H_2[(VO)_{12}O_6B_{18}O_{42}]$ $15H_2O$	8/4	[45]
9	$[Zn(teta)]_6[(VO)_{12}O_6B_{18}O_{36}(OH)_6](H_2O)$ $8H_2O$	12/0	[48]
10	$\{[Zn(dien)]_2[Zn(dien)(H_2O)]_4(VO)_{12}O_6[B_3O_6(OH)]_6(H_2O)\}_2$ $15H_2O$	12/0	[49]
11	$\{[Cu(dien)(H_2O)]_3V_{12}B_{18}O_{54}(OH)_6(H_2O)\}$ $4H_3O$ $5.5H_2O$	10/2	[55]
12	$\{[Cd(H_2O)_2]_3V_{12}B_{18}O_{54}(OH)_6(H_2O)\}$ $4H_3O$ $9.5H_2O$	10/2	[55]
13	$[Zn(H_2teta)_2V_{12}B_{18}O_{54}(OH)_6]$ $4H_3O$	10/2	[63]

† Calculated by us according to the stoichiometric formula given by the authors.

Six $M(en)_2^{2+}$ moieties (M = Zn, Ni) are coordinated to each VBO polyanion in **(6)** and **(8)**, while six $Zn(teta)^{2+}$ units are bonded in **(9)**. In **(10)** there are two $Zn(dien)^{2+}$ and four $[Zn(dien)(H_2O)]^{2+}$ entities coordinated to the same $[V_{12}B_{18}O_{60}]$ core (Figure 4). The fact that **(10)** presents six different crystallographic positions for the zinc complexes is concordant with its low symmetry, the triclinic crystalline system (*P-1*). Crystalline system **(6)** is rhombohedral, whereas **(8)** and **(9)** both crystallize in a trigonal system presenting one crystallographic site for the metal-amine entities.

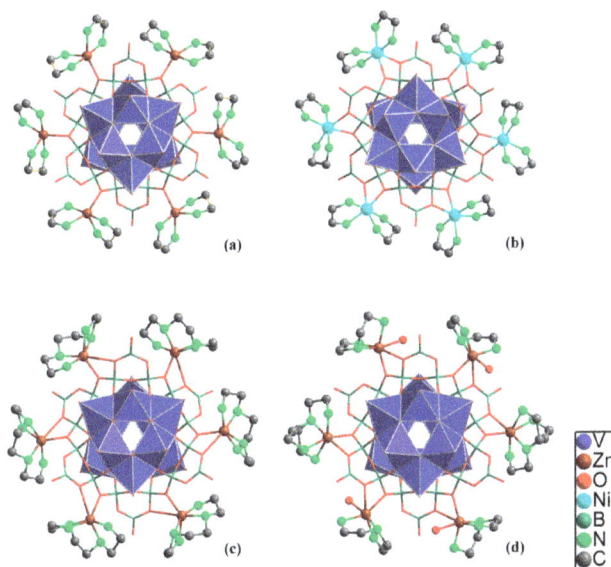

Figure 4. The $[V_{12}B_{18}O_{60}]$ core with the complexes: **(a)** $Zn(en)_2^{2+}$ for **(6)**, **(b)** $Ni(en)_2^{2+}$ for **(8)**, **(c)** $Zn(teta)^{2+}$ for **(9)**, and **(d)** $Zn(dien)^{2+}$ and $[Zn(dien)(H_2O)]^{2+}$ for **(10)**.

Complexes Cu(en)$_2^{2+}$, [Cu(dien)(H$_2$O)]$^{2+}$, and Zn(H$_2$teta)$_2^{2+}$ are part of the crystalline lattices of (7), (11), and (13), respectively, each of them coordinated to the [V$_{12}$B$_{18}$O$_{60}$] clusters. However, in these lattices the metal-complexes are linking adjacent polyoxovanadoborate cores though V-(μ_2-O)-M^{II}-(μ_2-O)-V bonds. In (7) and (11) four and six [V$_{12}$B$_{18}$O$_{60}$] polyanions, respectively, are connected through this type of bonds, while in (13) only two adjacent clusters are linked by the above-mentioned type of bond. The M-O distances range from 2.464(4) to 2.536(4) Å in (7), and 2.033(8) to 2.166(7) Å in (13), while for (11) all distances are 2.292(5) Å.

6. [V$_{12}$B$_{18}$O$_{60}$] Cores with Alkaline Ions as Counterbalancing Cations

Lattices that contain exclusively alkaline cations coordinated to the [V$_{12}$B$_{18}$O$_{60}$] multi-dentate ligands are scarce, in comparison with the lattices that include transition metal ions and organic ammonium ions. To the best of our knowledge, five of the eight reported inorganic frameworks included in this classification, and listed in Table 3, have been published by our research group. Brown *et al.* reported in 2011 the first framework that includes only Na$^+$ acting as charge-compensating ions and coordinated to the [V$_{12}$B$_{18}$O$_{60}$] core (14) [51], while one year later Zhou *et al.* published the same compound (14), as (15) [54]. The crystalline lattices (18), (19), (20), and (21) were obtained using the same hydrothermal synthesis conditions as for (14). In comparison with the aforementioned systems, compounds (15), (16), and (17) were obtained by adding auxiliary reducing agents to the reaction mixtures (Na$_2$SO$_3$, K$_2$SO$_3$, Ni, Co). Despite this fact, all the studied crystalline lattices included in this section present the same 10V^{IV}/2V^V mixed valence ratio. Sodium ions from the isostructural crystalline lattices (14) and (15) are coordinated to six oxygen atoms in a distorted octahedral coordination environment, with Na-O bond lengths ranging from 2.218(5) to 3.040(4) Å for (14) and 2.265(3) to 2.843(1) Å for (15). Zhou *et al.* mentioned that the alternating -Na-O-Na-connectivities in (15) generate 14-ring channels in the (100) direction, where the [V$_{12}$B$_{18}$O$_{60}$] clusters are found, thus permitting a 3D growth of the inorganic framework [54]. Lattices (16) to (19) contain only potassium cations coordinated to oxygen atoms from the VBO polyanionic ligands and/or from water molecules. These ions are found in the form of [KO$_x$] units in different crystallographic sites and with different coordination geometries. [KO$_x$] units with x = 6, 7, 8, 9, and 10 are found in (16) and (17), with K-O distances ranging from 2.610(2) to 3.420(6) Å. In compounds (18) and (19), as reported by Hermosilla-Ibáñez *et al.* [57], the K$^+$ ions are coordinated to six and seven oxygen atoms, with K-O bonds ranging from 2.468(3) to 3.060(2) Å and 2.625(5) to 3.064(3) Å, respectively. (20) has two different alkaline ions (K$^+$ and Cs$^+$) in its crystalline lattice. The potassium cations are found to be six- and seven-coordinated (K-O; 2.662(5) to 3.408(4) Å), while cesium cations are always eight-coordinated (Cs-O; 2.970(3) to 3.516(4) Å).

Hermosilla-Ibáñez *et al.* reported in 2014 the first framework of the VBO family (21) that contains the alkaline ions with the smallest ionic radius (Li$^+$). To the best of our knowledge, this is the sole example of a lattice with lithium counterions [58]. This compound crystallizes in the centrosymmetric cubic space group *Pn-3*, being the first example of such high symmetry. The literature reports examples of crystalline systems with lower symmetries [51,57,67]. Two of the three different crystallographic types of lithium cations are five-coordinated, and one is six-coordinated. The Li-O distances for the five-coordinated ions range from 1.921(2) to 2.976(4) Å, and have values of 3.142(3) Å for the six-coordinated Li^+ centers. Due to the long Li-O distances of the six-coordinated centers, the authors classified the observed distances as pseudo-coordinative interactions [58].

Table 3. List of the lattices with alkaline ions as counterbalancing cations. The mixed valence V^{IV}/V^V ratio is indicated for each lattice.

Compound	Formula	V^{IV}/V^V Ratio	Ref.
14	$(Na)_{10}[(H_2O)V_{12}B_{18}O_{60}H_6]$ 18H_2O	10/2	[51]
15	$\{Na_2B_{18}V_{12}O_{54}(OH)_6(H_2O)[Na_8(H_2O)_{16}]\}$ 2H_2O	10/2	[54]
16	$\{K_2V_{12}B_{18}O_{54}(OH)_6(H_2O)[K_8(H_2O)_{16}]\}$ 3H_2O	10/2	[67]
17	$\{K_{10}V_{12}B_{18}O_{54}(OH)_6(H_2O)\}$ 14H_2O	10/2	[67]
18	$K_8(NH_4)_2[V_{12}B_{18}O_{60}H_6]$ 18H_2O	10/2	[57]
19	$K_{10}[V_{12}B_{18}O_{60}H_6]$ 10H_2O	10/2	[57]
20	$K_8Cs_2[V_{12}B_{18}O_{60}H_6]$ 10H_2O	10/2	[57]
21	$Li_8(NH_4)_2[V_{12}B_{18}O_{60}H_6]$ 8.02H_2O	10/2	[58]

7. $[V_{12}B_{18}O_{60}]$ Cores with Organic Ammonium, Alkaline, and/or Transition Metal Ions as Counterbalancing Cations: The Mixed Family

The systems that include organic ammonium ions along with alkaline and/or transition metal ions in their frameworks are listed in Table 4. (**22**), (**23**), (**27**), (**32**), (**33**), (**34**), (**35**), (**36**), and (**37**) contain alkaline ions and protonated amine ions with different degrees of protonation. Sodium cations are present in the crystalline lattices of (**22**), (**27**), and (**32**) together with protonated ethylenediamine (en), tris(2-aminoethyl)amine (tren), and diethylenetriamine (dien) molecules, respectively. Lin et al. classified framework (**22**) as a two-dimensional lattice built of Na^+ coordinated to the VBO clusters [46], while (**27**) is described by Zhou et al. as a 1D crystalline system considering the Na-POM connectivities [53]. The coordination of Na^+ to oxygen atoms of the polyanonic clusters, together with the hydrogen bond interactions between the protonated amines and the POM, stabilize the three-dimensional lattice (**32**). Additionally to the Na^+, (**23**) includes K^+ in its 3D crystalline lattice along with monoprotonated ethylenediamine (Hen). In this compound, sodium ions are seven-coordinated (Na-O: 2.335 to 3.500 Å), whereas potassium cations are nine-coordinated (K-O: 2.640 to 3.302 Å). Systems (**33**), (**34**), (**35**), (**36**), and (**37**) contain K^+ along with protonated amine ions in their frameworks. Diprotonated dien is found in (**33**) while diprotonated en is present in (**35**) and (**37**); lattice (**36**) contains 1,3-propanediammonium cations. Methylammonium cations are present in framework (**34**) due to the decomposition of the ethylenediamine added to the reaction mixture [68]. Potassium cations are coordinated to different oxygen atoms of the vanadyl groups and borate fragments (terminal and bridging groups), and from water molecules, being six-, seven-, eight-, and nine-coordinated, with K-O distances ranging from 2.507(2) to 3.430(7) Å in (**33**), 2.729(3) to 3.326(3) Å in (**34**), 2.774(5) to 2.865(5) Å for (**35**), 2.552(6) to 3.108(6) Å for (**36**), and 2.140(2) to 3.090(7) Å for (**37**). The crystalline lattice of the abovementioned systems ((**33**)–(**37**)) is stabilized by the coordination of the potassium cations to the polyoxometalate anions, and by the different modes of hydrogen bonds (unidirectional, bifurcated, and trifurcated) between the protonated amine molecules and the VBO clusters. In the crystalline packing of (35), (36), and (37), additional auxiliary cations of hydronium ions also stabilize the lattices. The bond valence sum calculation indicated that the VBO clusters in (33), (34), and (35) have a mixed valence ratio of $10V^{IV}/2V^{V}$, while in (36) and (37) the ratio is $11V^{IV}/1V^{V}$.

Table 4. List of the lattices with organic ammonium, alkaline, and/or transition metal ions as counterbalancing cations. The mixed valence V^{IV}/V^{V} ratio is indicated for each lattice.

Compound	Formula	V^{IV}/V^{V} Ratio	Ref.
22	$(enH_2)_4Na_4H_3[(V_{12}O_6B_{18}O_{42}]\,8H_2O$	9/3	[46]
23	$K_3Na_5(H_2NCH_2CH_2NH_3)_2\{(VO)_{12}O_6[B_3O_6(OH)]_6\}(H_2O)\,12H_2O$	10/2	[47]
24	$Na_8[Cu(en)_2]_2[V_{12}B_{18}O_{60}H_6](NO_3)_2\,14.7H_2O$	8/4	[51]
25	$Na_7[Cu(en)_2]_2[V_{12}B_{18}O_{60}H_6](NO_3)\,15.5H_2O$	8/4	[51]
26	$[Na(H_2O)]_2[Na(H_2O)_2]_2[Cu(en)_2][V_{12}B_{18}O_{54}(OH)_6]\,(H_3O)_2\,(H_2O)_{18}$	8/4	[50]
27	$\{[Na(H_2O)_4]_3[V_{12}B_{18}O_{54}(OH)_6(H_2O)]_2\}(H_4tren)_4\,(H_3O)\,41H_2O$	10/2	[53]
28	$Na_8(H_3O)\{[Ni(H_2O)_5][V_{12}B_{18}O_{60}H_6]\}\,12.5H_2O$	11/1	[52]
29	$Na_5(H_3O)_4\{[Ni(H_2O)_3(en)][V_{12}B_{18}O_{60}H_6]\}\,9H_2O$	11/1	[52]
30	$Na_9(H_3O)\{Zn_{0.5}[V_{12}B_{18}O_{60}H_6]\}\,11H_2O$	11/1	[52]
31	$[Hen][H_2en]\{[Zn(en)_2]_3[V_{12}B_{18}O_{60}H_6]\}\,3H_2O$	9/3	[52]
32	$\{[Na(H_2O)_3]_4Na_2V_{12}B_{18}O_{56}(OH)_4(H_2O)\}(H_3dien)_2$	10/2	[55]
33	$\{V_{12}B_{18}O_{54}(OH)_6(H_2O)[K_6(H_2O)_{12}]\}\,2(H_2dien)\,3H_2O$	10/2	[67]
34	$K_6(CH_3NH_3)_4[V_{12}B_{18}O_{54}(OH)_6(H_2O)]\,2en\,12H_2O$	10/2	[68]
35	$K(H_3O)(enH_2)_4[V_{12}B_{18}O_{60}H_6]\,9.60H_2O$	10/2	[58]
36	$K_5(H_3O)_2(1,3\text{-}diapH_2)_2[V_{12}B_{18}O_{60}H_6]\,10.8H_2O$	11/1	[59]
37	$K_2(H_3O)_7(enH_2)[V_{12}B_{18}O_{60}H_6]\,9.0H_2O$	11/1	[59]
38	$[Zn(H_3tepa)V_{12}B_{18}O_{54}(OH)_6][H_2en]_2\,H_3O\,3H_2O$	10/2	[63]
39	$[V_{12}B_{18}Zn_3O_{63}H_{12}]\,3(C_4N_3H_{16})\,10NH_4\,5H_2O$	-	[56]
40	$[V_{12}B_{18}Mn_3O_{63}H_{12}]\,3(C_4N_3H_{16})\,10NH_4\,5H_2O$	-	[56]
41	$[V_{12}B_{18}Ni_3O_{63}H_{12}]\,3(C_4N_3H_{16})\,10NH_4\,5H_2O$	-	[56]

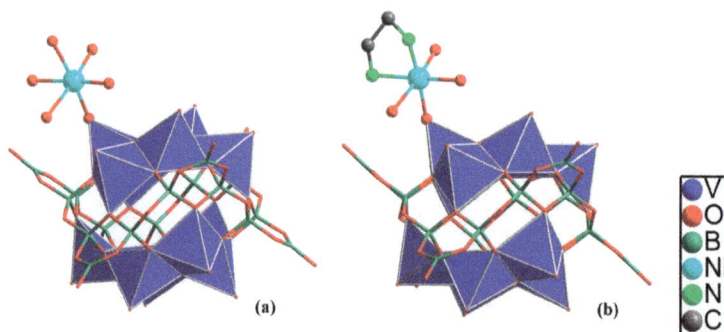

Figure 5. The $[V_{12}B_{18}O_{60}]$ core bonded to: (a) $Ni(H_2O)_5^{2+}$ in (28) and (b) $[Ni(H_2O)_3(en)]^{2+}$ in (29).

Unlike the crystalline systems previously mentioned within this section that contain protonated amine and alkaline cations, (**28**), (**29**), (**30**), (**31**), and (**38**) include coordination complexes of a secondary transition metal ion. Hermosilla-Ibáñez *et al.* in 2012 reported four new compounds based on the $[V_{12}B_{18}O_{60}]$ core, whose negative charge is stabilized by $Ni(H_2O)_5^{2+}$, Na^+ and H_3O^+ in (**28**), by $[Ni(H_2O)_3(en)]^{2+}$, Na^+ and H_3O^+ in (**29**), by Zn^{II}, Na^+ and H_3O^+ in (**30**), and by $Zn(en)_2^{2+}$, Hen^+ and H_2en^{2+} in (**31**) [52]. Each of the $Ni(H_2O)_5^{2+}$ and $[Ni(H_2O)_3(en)]^{2+}$ complexes in (**28**) and (**29**), respectively, are coordinated to the $[V_{12}B_{18}O_{60}]$ cluster through only one oxygen atom from a vanadyl group (Figure 5). This connection type differs from system (**8**) described above and reported by Lin *et al.* [45], in which six $Ni(en)_2^{2+}$ complexes are connected to two oxygen atoms from the borate fragment (Ni-(μ_3-O)-B$_2$) of the VBO polyanion (Figure 4b). Each zinc cation with partial occupancy (0.5) in (**30**) is four-coordinated to two vanadyl oxygen atoms and to two B-O-B oxygen bridges from two adjacent polyoxovanadoborates (Zn-O distances of 2.337(7) and 2.457(7) Å). Lattices (**31**) and (**38**) contain six-coordinated zinc complexes bonded to the $[V_{12}B_{18}O_{60}]$ clusters. Two en molecules and two oxygen-bridging atoms (B-O-B) from the same polyanion are bonded to the Zn^{II} in (**31**), while two tetraethylenepentamine (tepa) molecules, in addition to two vanadyl oxygen atoms from adjacent polyanions, are coordinated to the Zn^{II} in (**38**). The electroneutrality of both compounds is attained by protonated ethylenediamine ions, which allow the stabilization of the crystalline frameworks by hydrogen bonds. Additionally, (**31**) can be compared with compound (**6**) due to the fact that the latter contains six $Zn(en)_2^{2+}$ complexes connected to one $[V_{12}B_{18}O_{60}]$ core, while (**31**) contains only three $Zn(en)_2^{2+}$ complexes coordinated to the same polyoxometalate.

8. Coordination Geometry Analysis of the Counterbalancing Alkaline and Secondary Transition Metal Ions

On the basis of the crystallographic data included in the literature, we have calculated the best geometry for the alkaline and transition metal ions included in the corresponding crystalline lattices based on the $[V_{12}B_{18}O_{60}]$ core, using the SHAPE 2.1 program [69]. To carry out this study, the maximum M–O distance used to defined the coordination sphere for M = Na^+ and K^+ is 3.1 Å. In (**16**) Zhou *et al.* considered a longer K–O coordination distance of 3.4 Å, thus defining a 10-coordinate mode for some of the potassium ions, which is not included in our analysis [67].

Among all the counterbalancing alkaline ions of the $[V_{12}B_{18}O_{60}]$ polyanions, Li^+ is found to be five-coordinated, with a square pyramidal geometry (SPY-5) [58], while Cs^+ is eight-coordinated in a hexagonal bipyramidal geometry (HBPY-8) [57]. On the other hand, Na^+ and K^+ are found with more than one coordination number. As we reported earlier for (**14**), hexa-coordinated Na^+ ions are found with octahedral (OC-6) and trigonal prismatic (TPR-6) geometries [58]. Nevertheless, extra geometries for the $[NaO_x]$ are determined by the different coordination numbers (x = 5 and 6) in the other systems based on the $[V_{12}B_{18}O_{60}]$ cluster that contain sodium ions in their crystalline packing. As expected, potassium ions with a bigger ionic radius than sodium ions present coordination numbers from six to nine, as has been previously found in the literature [70–72]. Three different geometries are found when K^+ are six-coordinated, two when seven-coordinated, two when eight-coordinated, and one when nine-coordinated. The corresponding geometries are listed in Table 5.

In the case of Mn^{II}, Ni^{II}, Cu^{II}, and Zn^{II}, four- and six-coordination is found, as can be seen in Table 5. Only the six-coordinated manganese ions, which occupy only one crystallographic site in framework (40), adopt the trigonal prismatic geometry (TPR-6). The six-coordinated Ni^{II} ((8), (28), (29) and (41)), Cu^{II} ((7), (11), (24), (25), and (26)) and Zn^{II} ((9), (10), (13), (31), and (38)) share an octahedral geometry (OC-6). The square planar geometry mode (SP-4) is only found for Cu^{II} included in lattices (24) and (25). Zn^{II} presents the highest plasticity among the other transition metal ions, with coordination numbers four, five, and six. The four-coordinated Zn^{II} ions are in a tetrahedral geometry (T-4), whereas the five-coordinated Zn^{II} ions are found to be in a square pyramidal geometry (SPY-5) and vacant octahedral (vOC-5) geometries.

Table 5. List of the best geometries estimated for the alkaline and transition metal ions, using SHAPE 2.1.

Cation	Geometry	Geometry Symbol
Li	Square Pyramid	(SPY-5)
Na	Vacant Octahedron	(vOC-5)
	Trigonal Bipyramid	(TBPY-5)
	Square Pyramid	(SPY-5)
	Pentagonal Pyramid	(PPY-6)
	Octahedron	(OC-6)
	Trigonal Prism	(TPR-6)
K	Pentagonal Pyramid	(PPY-6)
	Octahedron	(OC-6)
	Trigonal Prism	(TPR-6)
	Capped Octahedron	(COC-7)
	Capped Trigonal Prism	(CTPR-7)
	Square Antiprism	(SAPR-8)
	Triangular Dodecahedron	(TDD-8)
	Tricaped Trigonal Prism	(TCTPR-9)
Cs	Hexagonal Bipyramid	(HBPY-8)
Mn	Trigonal Prism	(TPR-6)
Ni	Octahedron	(OC-6)
Cu	Square	(SP-4)
	Octahedron	(OC-6)
Zn	Tetrahedron	(T-4)
	Square Pyramid	(SPY-5)
	Vacant Octahedron	(vOC-5)
	Octahedron	(OC-6)

9. Spectroscopic Properties

The FT-IR fingerprint region characteristic of the $[V_{12}B_{18}O_{60}]$ core is observed between *ca.* 640 and 1420 cm^{-1}. The asymmetric and symmetric V–O–V stretching vibrations appear in the low energy region between 640 and 880 cm^{-1}, whereas the bands observed in the range of 900 and

960 cm^{-1} are assigned to the terminal V–O stretching vibrations of the vanadyl group. On the other hand, the borate fragments are characterized by B–O asymmetrical stretching vibrations for both the [BO$_3$] and [BO$_4$] units, appearing between 1020 and 1150 cm^{-1} for the trigonal and between 1300 and 1420 cm^{-1} for the tetrahedral units [34,43,44,48–50,52–55,57,58,63,67,68]. Müller *et al.* reported that the energy and shape of the vanadyl stretching bands depend on the oxidation state and on the existing interactions of the vanadyl groups of the polyoxovanadate anions in the crystalline packing [26]. In the IR spectra reported for all the studied systems, only (11) presents a sharp stretching vibration of the terminal V–O group. This fact can be rationalized considering that in this framework the polyanion has all the vanadyl groups equally connected to [Cu(dien)(H$_2$O)]$^{2+}$ complexes.

The optical properties of compounds (4), (5), (7), (9), (10), (12), (15), (16), (18), (19), (20), (21), (27), (29), (30), (35), and (38) have been studied by solid-state diffuse reflectance spectroscopy in the UV-visible region, since they are all insoluble in most common organic solvents and water. Three bands in the studied UV-visible region are reported for almost all the investigated systems: two bands in the high energy region, between 243 and 230 nm, and 344 and 310 nm, depending on the species, and one band that appears between 590 to 517 nm. In general, the two absorption bands in the high energy region are assigned as O→V and O→B charge transfer transitions, respectively. The less intense band in the low energy region has been assigned to Intervalence Charge Transfer Transitions (IVCT) and to d–d transitions [34,48,52–55,57,58,63,67]. With respect to the low energy absorption bands (*ca.* 500 nm), most authors consider that these arise from "presumably d-d electronic transitions" [34,53,54,63]. With respect to the bands in this same visible region of the polymetallic vanadium species, Robin and Day consider that they should be assigned to mixed valence absorptions [73]. However, the real meaning of these bands should become apparent once a more complete electronic description has been attained from quantum mechanical calculations. We are currently calculating the electronic spectra of these species by DFT methods.

Considering the similarity of the UV-visible spectra of these systems even when the polyanions are functionalized with secondary transition metal atoms [52], we can deduce that the crystalline lattices have a negligible effect on the electronic properties of the [V$_{12}$B$_{18}$O$_{60}$] core.

10. Magnetic Properties

Among all the studied compounds included in this review, the magnetic properties of only (5), (11), (12), (15), (18), (19), (27), and (38) have been reported [53–55,57,63]. All of these systems have a mixed valence ratio of 10V^{IV}/2V^V, and present a bulk antiferromagnetic behavior. The χT values at 300 and 2 K for the abovementioned compounds are listed in Table 6. (11) presents the highest χT value of 4.81 emu K mol^{-1}, which was explained by Zhou *et al.*, assuming that this value is very close to the theoretical χT value of 4.88 emu K mol^{-1}, considering 10 uncoupled V^{IV} plus three uncoupled Cu^{II} centers (g = 2.00 for both atoms) [55]. However, when the χT value of the three uncoupled Cu^{II} centers is subtracted (g = 2.00), the resultant χT value for the [V$_{12}$B$_{18}$O$_{60}$] polyanion of (11) is 3.68 emu K mol^{-1}, thus presenting the same trend followed by (5), (12), (15), (18), (19), and (27) (Figure 6). From the χT(T) graph reported by Zhou *et al.* [53] it is possible to infer that (27) is almost magnetically uncoupled at room temperature. On the other hand, the χT values of the rest of compounds, (5), (12), (15), (18), and (19) show that all of them are magnetically coupled at room temperature. Within the observed tendency of 3.34 to 3.83 emu K mol^{-1}, there is no clear correlation between the magnetic properties and the nature of the frameworks, which include different cations interacting with the polyanions, even though it is clear that the interactions between the different cations and the polyanion in the lattice affect the magnitude of the exchange phenomena in the cluster.

It is interesting to point out that Hermosilla-Ibáñez et al. reported in 2014 that it is possible to show by DFT calculations that the alkaline ions in compounds (18) and (19) quench the intracluster antiferromagnetic coupling, in comparison with compound (5) [57]. In this study, the results indicated that the presence of alkaline ions perturbs the extent of the spin density of the magnetic orbitals (d_{xy}); this perturbation is dependent on the distance between the alkaline cation and the oxygen of the vanadyl groups. Thus, the obtained modification of the orbital overlap due to the presence of the alkaline cations influences the magnitude of the antiferromagnetic interactions. Nevertheless, the existence of additional hydrogen bonds and/or covalent bonds should also influence the global magnetic properties.

As can be seen, the most coupled system is (38), which includes Zn(H$_3$tepa)$^{2+}$ and (enH$_2$)$^{2+}$ as counterbalancing ions. In this system the Zn(H$_3$tepa)$^{2+}$ complexes are coordinated to the [V$_{12}$B$_{18}$O$_{60}$] polyanions through two oxygen atoms of vanadyl groups from adjacent polyoxovanadoborates, i.e., acting as bridges between two polyanions. As discussed above, the coordination of the vanadyl groups with a cation influences the electronic properties, i.e. stretching vibrations and exchange interactions. In (38) the presence of the zinc(II) cations bridging the polyanion clearly increases the antiferromagnetic behavior of the material.

Table 6. List of the reported compounds with magnetic property studies.

Compound	Auxiliary cations	V^{IV}/V^V Ratio	χT (emu K mol^{-1}) (300 K)	χT (emu K mol^{-1}) (2 K)	Ref.
5	NH$_4^+$, (1,3-diapH$_2$)$^{2+}$	10/2	3.34	0.33	[57]
11	[Cu(dien)(H$_2$O)]$^{2+}$	10/2	4.81 (3.68)*	0.56	[55]
12	Cd(H$_2$O)$_2^{2+}$	10/2	3.60	0.10	[55]
15	Na$^+$	10/2	3.53	0.23	[54]
18	K$^+$, NH$_4^+$	10/2	3.57	0.40	[57]
19	K$^+$	10/2	3.58	0.38	[57]
27	[Na(H$_2$O)$_4$]$^+$, (H$_4$tren)$^{4+}$	10/2	3.83	0.15	[53]
38	Zn(H$_3$tepa)$^{2+}$, (enH$_2$)$^{2+}$	10/2	1.54	0.11	[63]

* The χT value in parentheses is the χT value of the polyanion, which was calculated by subtracting the χT value for the three uncoupled Cu^{II} centers, considering a g = 2 (χT = 1.13 emu K mol^{-1}).

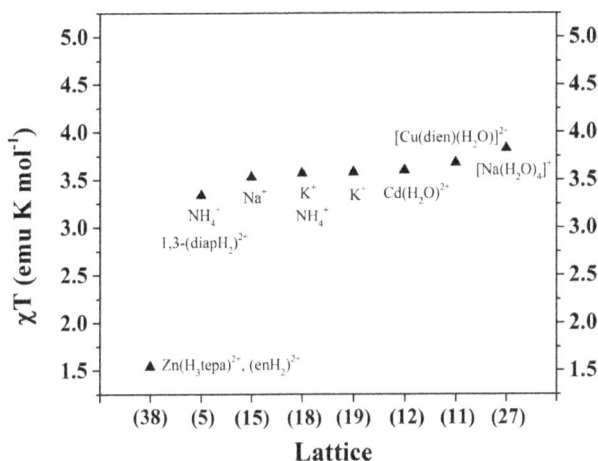

Figure 6. χT values for each lattice.

11. Final Remarks

The structural stability of the $[V_{12}B_{18}O_{60}]$ core allows the formation of polyoxometalate species with different crystalline lattices, depending on the cations present in the synthesis. This polyanion is potentially able to share bridging B-O-B oxygen atoms, both vanadyl and bridging B-O-B oxygen atoms, and in some cases, the 12 oxygen atoms from the vanadyl groups, thus permitting one-, two-, or three-dimensional frameworks to be obtained. The presence of auxiliary cations may be responsible for the alignment of the organic amines in only one direction, since they can fill hindered nucleophilic sites around the polyanions.

Hermosilla-Ibáñez *et al.* demonstrated that the organic diamines act as reducing agents in the reactions, as the presence of nitrate ions in the final mother liquors was detected by ionic liquid chromatography [57]. The FT-IR results show that the coordination of a cation to each of the existing vanadyl groups of the polyanion produces a single and sharp stretching band for the vanadyl group. Thus, the coordination of the vanadyl groups with cations influences the electronic properties, *i.e.*, stretching vibrations and exchange interactions. However, the similarity of the solid state reflectance spectra indicates that the crystalline lattices have a negligible effect on the electronic spectra of the $[V_{12}B_{18}O_{60}]$ core.

From the reported magnetic data it is clear to conclude that the $[V_{12}B_{18}O_{60}H_6]$ cluster with a $10V^{IV}/2V^V$ mixed-valence ratio presents a global antiferromagnetic exchange among the 10 spin carriers. It can also be concluded that the interactions of the cations in the crystal packing with the polyanion can modify the global antiferromagnetic interaction in the polyanion. Further studies must be done in order to reach a deeper understanding of the magnetic behavior in these compounds. At this time our group is working on the rationalization of the magnetic properties of the $[V_{12}B_{18}O_{60}]$ family.

Acknowledgments

The authors acknowledge financial support from projects FONDECYT 1120004 and Financiamiento Basal FB0807 (CEDENNA). This work was done under the international collaborative project LIA-MIF 836. Powered@ NLHPC: This research was partially supported by the super-computing infrastructure of the NLHPC (ECM-02), Centre for Mathematical Modelling CMM, Universidad de Chile. KMB thanks 21100772 and AT-24121391 doctoral scholarships and the USACH-French Embassy agreement for a doctoral mobility grant. PHI thanks Vicerrectoría de Investigación, Desarrollo e Innovación (USACH) for the POSTDOC_DICYT 001316 scholarship.

Author Contributions

The manuscript was written with equal contributions from all authors. DVY owns the general idea and plan of the publication. Most of the presented results are part of the Doctoral thesis of PHI and KMB. ES and VPG were part of the discussion and analysis of the presented data. ELF collaborated with the structural determinations.

Conflicts of Interest

The authors declare no conflict of interest.

References

1. Pope, M.T.; Müller, A. Polyoxometalate Chemistry From Topology via Self-Assembly to Applications. Kluwer Academic Publishers: Dordrecht, The Netherlands, 2001.
2. Yamase, T.; Pope, M.T. *Polyoxometalate Chemistry for Nano-Composite Design*; Kluwer Academic: Dordrecht, The Netherlands; Plenum Publishers: New York, NY, USA, 2002.
3. Borrás-Almenar, J.J.; Coronado, E.; Müller, A.; Pope, M.T. *Polyoxometalate Molecular Science*; Kluwer Academic Publishers: Dordrecht, The Netherlands, 2003.
4. Müller, A.; Peters, F.; Pope, M.T.; Gatteschi, D. Polyoxometalates: Very large clusters-nanoscale magnets. *Chem. Rev.* **1998**, *98*, 239–271.
5. Clemente-Juan, J.M.; Coronado, E.; Gaita-Ariño, A. Magnetic polyoxometalates: from molecular magnetism to molecular spintronics and quantum computing. *Chem. Soc. Rev.* **2012**, *41*, 7464–7478.
6. Suzuki, K.; Sato, R.; Mizuno, N. Reversible switching of single-molecule magnet behaviors by transformation of dinuclear dysprosium cores in polyoxometalates. *Chem. Sci.* **2013**, *4*, 596–600.
7. Mizuno, N.; Misono, M. Heterogeneous catalysis. *Chem. Rev.* **1998**, *98*, 199–217.
8. Lv, H.; Geletii, Y.V.; Zhao, C.; Vickers, J.W.; Zhu, G.; Luo, Z.; Song, J.; Lian, T.; Musaev, D.G.; Hill, C.L. Polyoxometalate water oxidation catalysts and the production of green fuel. *Chem. Soc. Rev.* **2012**, *41*, 7572–7589.
9. Rhule, J.T.; Hill, C.L.; Judd, D.A.; Schinazi, R.F. Polyoxometalates in medicine. *Chem. Rev.* **1998**, *98*, 327–357.

10. Wang, Y.; Weinstock, I.A. Polyoxometalate-decorated nanoparticles. *Chem. Soc. Rev.* **2012**, *41*, 7479–7496.

11. Miras, H.N.; Yan, J.; Long, D.-L.; Cronin, L. Engineering polyoxometalates with emergent properties. *Chem. Soc. Rev.* **2012**, *41*, 7403–7430.

12. Song, Y.-F.; Tsunashima, R. Recent advances on polyoxometalate-based molecular and composite materials. *Chem. Soc. Rev.* **2012**, *41*, 7384–7402.

13. Long, D.-L.; Tsunashima, R.; Cronin, L. Polyoxometalates: Building blocks for functional nanoscale systems. *Angew. Chem. Int. Ed.* **2010**, *49*, 1736–1758.

14. Streb, C.; Long, D.-L.; Cronin, L. Influence of organic amines on the self-assembly of hybrid polyoxo-molybdenum(V) phosphate frameworks. *CrystEngComm* **2006**, *8*, 629–634.

15. Yi, Z.; Yu, X.; Xia, W.; Zhao, L.; Yang, C.; Chen, Q.; Wang, X.; Xu, X.; Zhang, X. Influence of the steric hindrance of organic amines on the supramolecular network based on polyoxovanadates. *CrystEngComm.* **2010**, *12*, 242–249.

16. Long, D.-L.; Kögerler, P.; Farrugia, L.J.; Cronin, L. Restraining symmetry in the formation of small polyoxomolybdates: Building blocks of unprecedented topology resulting from "shrink-wrapping" $[H_2Mo_{16}O_{52}]^{10-}$-type clusters. *Angew. Chem. Int. Ed.* **2003**, *42*, 4180–4183.

17. Abbas, H.; Pickering, A.L.; Long, D.-L.; Kögerler, P.; Cronin, L. Controllable growth of chains and grids from polyoxomolybdate building blocks linked by Silver(I) dimers. *Chem. Eur. J.* **2005**, *11*, 1071–1078.

18. Streb, G.; McGlone, T.; Brücher, O.; Long, D.-L.; Cronin, L. Hybrid host-guest complexes: Directing the supramolecular structure through secondary host-guest interactions. *Chem. Eur. J.* **2008**, *14*, 8861–8868.

19. Mcglone, T.; Streb, C.; Long, D.-L.; Cronin, L. Guest-directed supramolecular architectures of $\{W_{36}\}$ polyoxometalate crowns. *Chem. Asian J.* **2009**, *4*, 1612–1618.

20. Day, V.W.; Klemperer, W.G.; Yaghi, O.M. A new structure type in polyoxoanion chemistry: synthesis and structure of the $V_5O_{14}{}^{3-}$ anion. *J. Am. Chem. Soc.* **1989**, *111*, 4518–4519.

21. Evans, H.T. The molecular structure of the isopoly complex ion, decavanadate ($V_{10}O_{28}{}^{16-}$). *Inorg. Chem.* **1966**, *5*, 967–977.

22. Day, V.W.; Klemperer, W.G.; Yaghi, O.M. Synthesis and characterization of a soluble oxide inclusion complex, $[CH_3CN \subset (V_{12}O_{12}{}^{4-})]$. *J. Am. Chem. Soc.* **1989**, *111*, 5959–5962.

23. Hou, D.; Hagen, K.S.; Hill, C.L. Tridecavanadate, $[V_{13}O_{34}]^{3-}$, a new high-potential isopolyvanadate. *J. Am. Chem. Soc.* **1992**, *114*, 5864–5866.

24. Hou, D.; Hagen, K.S.; Hill, C.L. Pentadecavanadate, $V_{15}O_{42}{}^{9-}$, a new highly condensed fully oxidized isopolyvanadate with kinetic stability in water. *J. Chem. Soc. Chem. Commun.* **1993**, *20*, 426–428.

25. Müller, A.; Krickemeyer, E.; Penk, M.; Walberg, H.-J.; Bögge, H. Spherical mixed-valence $[V_{15}O_{36}]^{5-}$, an example from an unusual cluster family. *Angew. Chem. Int. Ed. Engl.* **1987**, *26*, 1045–1046.

26. Müller, A.; Sessoli, R.; Krickemeyer, E.; Bögge, H.; Meyer, J.; Gatteschi, D.; Pardi, L.; Westphal, J.; Hovemeier, K.; Rohlfing, R.; Döring, J.; Hellweg, F.; Beugholt, C.; Schmidtmann, M. Polyoxovanadates: High-nuclearity spin clusters with interesting host–guest systems and different electron populations. synthesis, spin organization, magnetochemistry, and spectroscopic studies. *Inorg. Chem.* **1997**, *36*, 5239–5250.

27. Suber, L.; Bonamico, M.; Fares, V. Synthesis, magnetism, and X-ray molecular structure of the mixed-valence vanadium(IV/V)–oxygen cluster [$VO_4 \subset (V_{18}O_{45})$]$^{9-}$. *Inorg. Chem.* **1997**, *36*, 2030–2033.

28. Müller, A.; Krickemeyer, E.; Penk, M.; Rohlfing, R.; Armatage, A.; Bögge, H. Template-controlled formation of cluster shells or a type of molecular recognition: Synthesis of [$HV_{22}O_{54}(ClO_4)$]$^{6-}$ and [$H_2V_{18}O_{44}(N_3)$]$^{5-}$. *Angew. Chem. Int. Ed.* **1991**, *30*, 1674–1677.

29. Müller, A.; Rohlfing, R.; Döring, J.; Penk, M. Formation of a cluster sheath around a central cluster by a"self-organization process": the mixed valence polyoxovanadate[$V_{34}O_{82}$]$^{10-}$. *Angew. Chem. Int. Ed.* **1991**, *30*, 588–590.

30. Warren, C.J.; Rijssenbeek, J.T.; Rose, D.J.; Haushalter, R.C.; Zubieta, J. Hydrothermal synthesis and characterization of an unusual polyoxovanadium borate cluster: Structure of Rb_4[$(VO)_6\{\{B_{10}O_{16}(OH)_6\}\}_2$] 0.5$H_2O$. *Polyhedron* **1998**, *17*, 2599–2605.

31. Williams, I.D.; Wu, M.; Sung, H.H.-Y.; Zhang, X.X.; Yu, J. An organotemplated vanadium(IV) borate polymer from boric acid "flux" synthesis, [H_2en]$_4$[Hen]$_2$[$V_6B_{22}O_{53}H_8$] 5H_2O. *Chem. Commun.* **1998**, 2463–2464.

32. Cao, Y.-N.; Zhang, H.-H.; Huang, C.-C.; Sun, Y.-X.; Chen, Y.-P.; Guo, W.-J.; Zhang, F.-L. Synthesis and structural characterization of a new polyoxovanadium borate: ($H_3NCH_2CH_2NH_3$)$_4$[$(VO)_6(B_{10}O_{22})_2$](H_3O)$_7$. *Chin. J. Struct. Chem.* **2005**, *24*, 525–530.

33. Cai, Q.; Lu, B.; Zhang, J.; Shan, Y. Synthesis, structure and properties of ($H_2NCH_2CH_2NH_3$)$_3$ $\{(VO)_6[B_{10}O_{16}(OH)_6]_2\}$ 11H_2O. *J. Chem. Crystallogr.* **2008**, *38*, 321–325.

34. Liu, X.; Zhou, J.; An, L.; Chen, R.; Hu, F.; Tang, Q. Hydrothermal syntheses, crystal structures and characterization of new vanadoborates: The novel decorated cage cluster [$V_6B_{22}O_{44}(OH)_{10}$]. *J. Solid State Chem.* **2013**, *201*, 79–84.

35. Wu, M.; Law, T.S.-C.; Sung, H.H.-Y.; Cai, J.; Williams, I.D. Synthesis of elliptical vanadoborates housing bimetallic centers [$Zn_4(B_2O_4H_2)(V_{10}B_{28}O_{74})$]$^{8-}$ and [$Mn_4(C_2O_4)(V_{10}B_{28}O_{74}H_8)$]$^{10-}$. *Chem. Commun.* **2005**, *4*, 1827–1829.

36. Cao, Y.; Zhang, H.; Huang, C.; Yang, Q.; Chen, Y.; Sun, R.; Zhang, F.; Guo, W. Synthesis, crystal structure and two-dimensional infrared correlation spectroscopy of a layer-like transition metal (TM)-oxalate templated polyoxovanadium borate. *J. Solid State Chem.* **2005**, *178*, 3563–3570.

37. Chen, H.; Yu, Z.-B.; Bacsik, Z.; Zhao, H.; Yao, Q.; Sun, J. Construction of mesoporous frameworks with vanadoborate clusters. *Angew. Chem. Int. Ed.* **2014**, *53*, 3608–3611.

38. Warren, C.J.; Haushalter, R.C.; Rose, D.J.; Zubieta, J. A bimetallic main group oxide cluster of the oxovanadium borate system: ($H_3NCH_2CH_2NH_3$)$_4$[$(VO)_{12}O_4\{B_8O_{17}(OH)_4\}_2\{Mn(H_2O)_2\}_2$] H_2O. *Inorg. Chim. Acta* **1998**, *282*, 123–129.

39. Cao, Y.; Zhang, H.; Huang, C.; Chen, Y.; Sun, R.; Guo, W. Hydrothermal synthesis and crystal structure of a novel 1D polyoxovanadium borate: $(H_3NCH_2CH_2NH_3)_3[(VO)_{12}O_4\{B_8O_{17}(OH)_4\}_2\{Na(H_2O)\}_2(H_3O)_2(H_2O)_{6.5}$. *J. Mol. Struct.* **2005**, *733*, 211–216.

40. Sun, Y.-Q.; Li, G.-M.; Chen, Y.-P. A novel polyoxovanadium borate incorporating an organic amine ligand: synthesis and structure of $[V_{12}B_{16}O_{50}(OH)_7(en)]^{7-}$. *Dalton Trans.* **2012**, *41*, 5774–5777.

41. Liu, X.; Zhou, J.; Xiao, H.-P.; Kong, C.; Zou, H.; Tang, Q.; Li, J. Two new 3-D boratopolyoxovanadate architectures based on the $[V_{12}B_{16}O_{50}(OH)_8]^{12-}$ cluster with different metal linkers. *New J. Chem.* **2013**, *37*, 4077–4082.

42. Rijssenbeek, J.T.; Rose, D.J.; Haushalter, R.C.; Zubieta, J. Novel clusters of transition metals and main group oxides in the alkylamine/oxovanadium/borate system. *Angew. Chem. Int. Ed.* **1997**, *36*, 1008–1010.

43. Zhang, L.; Shi, Z.; Yang, G.; Chen, X.; Feng, S. Hydrothermal synthesis and X-ray single crystal structure of $[Zn(en)_2]_6[(VO)_{12}O_6B_{18}O_{39}(OH)_3]$ $13H_2O$. *J. Solid State Chem.* **1999**, *148*, 450–454.

44. Lin, Z.-H.; Zhang, H.-H.; Huang, C.-C.; Sun, R.-Q.; Chen, Y.-P.; Wu, X.-Y. Hydrothermal synthesis, crystal structure and properties of polyoxovanadium borate $H_3\{[Cu(en)_2]_5[(VO)_{12}O_6B_{18}O_{42}]\}[B(OH)_3]_2$ $16H_2O$. *Acta Chim. Sin.* **2004**, *62*, 391–398.

45. Lin, Z.-H.; Zhang, H.-H.; Huang, C.-C.; Sun, R.-Q.; Yang, Q.-Y.; Wu, X.-Y. Hydrothermal synthesis and crystal structure of $[Ni(en)_2]_6H_2[(VO)_{12}O_6B_{18}O_{42}]$ $15H_2O$. *Chin. J. Struct. Chem.* **2004**, *23*, 83–86.

46. Lin, Z.-H.; Yang, Q.-Y.; Zhang, H.-H.; Huang, C.-C.; Sun, R.-Q.; Wu, X.-Y. Hydrothermal synthesis and crystal structure of $(enH_2)_4Na_4H_3[(VO)_{12}O_6B_{18}O_{42}]$ $8H_2O$. *Chin. J. Struct. Chem.* **2004**, *23*, 590–595.

47. Lu, B.; Wang, H.; Zhang, L.; Dai, C.-Y.; Cai, Q.-H.; Shan, Y.-K. Hydrothermal synthesis and structure of $K_3Na_5(H_2NCH_2CH_2NH_3)_2\{(VO)_{12}O_6[B_3O_6(OH)]_6\}(H_2O)$ $12H_2O$. *Chin. J. Chem.* **2005**, *23*, 137–143.

48. Liu, X.; Zhou, J. The new vanadoborate-supported hexanuclear zinc complex $[Zn(teta)]_6[(VO)_{12}O_6B_{18}O_{36}(OH)_6](H_2O)$ $8H_2O$. *Z. Naturforsch.* **2011**, *66b*, 115–118.

49. Liu, X.; Zhou, J.; Zhou, Z.; Zhang, F. Hydrothermal syntheses and crystal structures of two new heteropolyoxovanadoborates containing $\{(VO)_{12}O_6[B_3O_6(OH)]_6(H_2O)\}$ cluster. *J. Clust. Sci.* **2011**, *22*, 65–72.

50. Li, G.-M.; Mei, H.-X.; Chen, X.-Y.; Chen, Y.-P.; Sun, Y.-Q.; Zhang, H.-H.; Chen, X.-P. A porous organic-inorganic hybrid compound constructed from polyoxovanadium borate anions, dinuclear Na sites and metal-organic units. *Chin. J. Struct. Chem.* **2011**, *30*, 785–792.

51. Brown, K.; Car, P.-E.; Vega, A.; Venegas-Yazigi, D.; Paredes-García, V.; Vaz, M.G.F.; Allao, R.A.; Pivan, J.-Y.; Le Fur, E.; Spodine, E. Polyoxometalate cluster $[V_{12}B_{18}O_{60}H_6]$ functionalized with the copper(II) bis-ethylenediamine complex. *Inorg. Chim. Acta* **2011**, *367*, 21–28.

52. Hermosilla-Ibáñez, P.; Car, P.E.; Vega, A.; Costamagna, J.; Caruso, F.; Pivan, J.-Y.; Le Fur, E.; Spodine, E.; Venegas-Yazigi, D. New structures based on the mixed valence polyoxometalate cluster $[V_{12}B_{18}O_{60}H_6]^{n-}$. *CrystEngComm* **2012**, *14*, 5604–5612.

53. Zhou, J.; Liu, X.; Hu, F.; Zou, H.; Li, X. A new 1-D extended vanadoborate containing triply bridged metal complex units. *Inorg. Chem. Commun.* **2012**, *25*, 51–54.

54. Zhou, J.; Liu, X.; Hu, F.; Zou, H.; Li, R.; Li, X. One novel 3-D vanadoborate with unusual 3-D Na–O–Na network. *RSC Adv.* **2012**, *2*, 10937–10940.

55. Zhou, J.; Liu, X.; Chen, R.; Xiao, H.-P.; Hu, F.; Zou, H.; Zhou, Y.; Liu, C.; Zhu, L. New 3-D polyoxovanadoborate architectures based on $[V_{12}B_{18}O_{60}]^{16-}$ clusters. *CrystEngComm* **2013**, *15*, 5057–5063.

56. Chen, H.; Deng, Y.; Yu, Z.; Zhao, H.; Yao, Q.; Zou, X.; Bäckvall, J.E.; Sun, J. 3D open-framework vanadoborate as a highly effective heterogeneous pre-catalyst for the oxidation of alkylbenzenes. *Chem. Mater.* **2013**, *25*, 5031–5036.

57. Hermosilla-Ibáñez, P.; Cañon-Mancisidor, W.; Costamagna, J.; Vega, A.; Paredes-García, V.; Garland, M.T.; Le Fur, E.; Cador, O.; Spodine, E.; Venegas-Yazigi, D. Crystal lattice effect on the quenching of the intracluster magnetic interaction in $[V_{12}B_{18}O_{60}H_6]^{10-}$ polyoxometalate. *Dalton Trans.* **2014**, *43*, 14132–14141.

58. Hermosilla-Ibáñez, P.; Costamagna, J.; Vega, A.; Paredes-García, V.; Le Fur, E.; Spodine, E.; Venegas-Yazigi, D. Coordination interactions in the crystalline lattice of alkaline ions with the polyoxometalate $[V_{12}B_{18}O_{60}H_6]^{10-}$ ligand. *J. Coord. Chem.* **2014**, *67*, 3940–3952.

59. Hermosilla-Ibáñez, P.; Costamagna, J.; Vega, A.; Paredes-García, V.; Garland, M.T.; Le Fur, E.; Spodine, E.; Venegas-Yazigi, D. Protonated diamines as linkers in the supramolecular assemblies based on the $[V_{12}B_{18}O_{60}H_6]$ polyoxovanadoborate anion. *J. Struct. Chem.* **2014**, *55*, 1453–1465.

60. Yamase, T.; Suzuki, M.; Ohtaka, K. Structures of photochemically prepared mixed-valence polyoxovanadate clusters: oblong $[V_{18}O_{44}(N_3)]^{14-}$, superkeggin $[V_{18}O_{42}(PO_4)]^{11-}$ and doughnut-shaped $[V_{12}B_{32}O_{84}Na_4]^{15-}$ anions. *Dalton Trans.* **1997**, *44*, 2463–2472.

61. Warren, C.J.; Rose, D.J.; Haushalter, R.C.; Zubieta, J. A new transition metal–main group oxide cluster in the oxovanadium–borate system: hydrothermal synthesis and structure of $(H_3O)_{12}[(VO)_{12}\{B_{16}O_{32}(OH)_4\}_2]\cdot 28H_2O$. *Inorg. Chem.* **1998**, *37*, 1140–1141.

62. Williams, I.D.; Wu, M.; Sung, H.H.-Y.; Law, T.S.-C.; Zhang, X.X. *Contemporary Boron Chemistry: Synthesis and Properties of Vanadoborate Cluster Materials*; Davidson, M.G., Hughes, A.K., Marder, T.B., Wade, K., Eds.; Royal Society of Chemistry: Cambridge, UK, **2000**.

63. Liu, X.; Zhao, R.; Zhou, J.; Liu, M. Three new vanadoborates functionalized with zinc complexes. *Inorg. Chem. Commun.* **2014**, *43*, 101–104.

64. Christ, C.L.; Clark, J.R. A crystal-chemical classification of borate structures with emphasis on hydrated borates. *Phys. Chem. Miner.* **1977**, *2*, 59–87.

65. Hagrman, P.J.; Finn, R.C.; Zubieta, J. Molecular manipulation of solid state structure: Influences of organic components on vanadium oxide architectures. *Solid State Sci.* **2001**, *3*, 745–774.

66. Zhang, Y.-N.; Zhou, B.-B.; Sha, J.-Q.; Su, Z.-H.; Cui, J.-W. Assembly of two layered cobalt–molybdenum phosphates: Changing interlayer distances by tuning the lengths of amine ligands. *J. Solid State Chem.* **2011**, *184*, 419–426.

67. An, L.; Zhou, J.; Xiao, H.-P.; Liu, X.; Zou, H.; Pan, C.-Y.; Liu, M.; Li, J. A series of new 3-D boratopolyoxovanadates containing five types of $[K_xO_y]_n$ building units. *CrystEngComm* **2014**, *16*, 4236–4244.

68. Meng, Q.; He, H.; Yang, B.-F.; Zhao, J.-W.; Yang, G.-Y. Synthesis and characterization of a 3-D framework constructed from $[V_{12}B_{18}O_{54}(OH)_6(H_2O)]^{10-}$ clusters and K^+ cations. *J. Clust. Sci.* **2014**, *25*, 1273–1282.

69. Llunell, M.; Casanova, D.; Cirera, J.; Alemany, P.; Alvarez, S. *SHAPE*, version 2.1; Universitat de Barcelona, Barcelona, Spain, 2013.

70. Delgado, F.S.; Ruiz-Pérez, C.; Sanchiz, J.; Lloret, F.; Julve, M. Versatile supramolecular self-assembly. Part I. Network formation and magnetic behaviour of the alkaline salts of the bis(malonate)cuprate(II) anion. *CrystEngComm* **2006**, *8*, 507–529.

71. Askarinejad, A.; Morsali, A. Potassium(I)thallium(I) heterometallic 3D polymeric mixed-anions compound, succinate-nitrate $[K_2Tl(MU-C_4H_4O_4)(MU-NO_3)]_n$. *Inorg. Chim. Acta* **2006**, *359*, 3379–3383.

72. Nagasubramanian, S.; Jayamani, A.; Thamilarasan, V.; Aravindan, G.; Ganesan, V.; Sengottuvelan, N. Hetero-metallic trigonal cage-shaped dimeric Ni_3K core complex of L-proline ligand: Synthesis, structural, electrochemical and DNA binding and cleavage activities. *J. Chem. Sci.* **2014**, *126*, 771–781.

73. Robin, M.B.; Day, P. Mixed Valence Chemistry-A Survey and Classification. *Adv. Inorg. Chem. Radiochem.* **1967**, *10*, 247–422.

Activity and Stability of the Tetramanganese Polyanion [Mn$_4$(H$_2$O)$_2$(PW$_9$O$_{34}$)$_2$]$^{10-}$ during Electrocatalytic Water Oxidation

Sara Goberna-Ferrón, Joaquín Soriano-López and José Ramón Galán-Mascarós

Abstract: In natural photosynthesis, the oxygen evolving center is a tetranuclear manganese cluster stabilized by amino acids, water molecules and counter ions. However, manganese complexes are rarely exhibiting catalytic activity in water oxidation conditions. This is also true for the family of water oxidation catalysts (WOCs) obtained from POM chemistry. We have studied the activity of the tetranuclear manganese POM [**Mn$_4$**(H$_2$O)$_2$(PW$_9$O$_{34}$)$_2$]$^{10-}$ (**Mn$_4$**), the manganese analog of the well-studied [Co$_4$(H$_2$O)$_2$(PW$_9$O$_{34}$)$_2$]$^{10-}$ (**Co$_4$**), one of the fastest and most interesting WOC candidates discovered up to date. Our electrocatalytic experiments indicate that **Mn$_4$** is indeed an active water oxidation catalysts, although unstable. It rapidly decomposes in water oxidation conditions. Bulk water electrocatalysis shows initial activities comparable to those of the cobalt counterpart, but in this case current density decreases very rapidly to become negligible just after 30 min, with the appearance of an inactive manganese oxide layer on the electrode.

Reprinted from *Inorganics*. Cite as: Goberna-Ferrón, S.; Soriano-López, J.; Galán-Mascarós, J. Activity and Stability of the Tetramanganese Polyanion [Mn$_4$(H$_2$O)$_2$(PW$_9$O$_{34}$)$_2$]$^{10-}$ during Electrocatalytic Water Oxidation. *Inorganics* **2015**, *3*, 332–340.

1. Introduction

Water oxidation catalysis is currently one of the hot topics in chemistry research [1–3], since it is regarded as the bottleneck in the development of an artificial photosynthesis scheme [4–6]. Fuels production by harvesting solar energy to reduce a substrate (as protons to hydrogen) needs to occur concomitant to the oxidation of water into oxygen. If this second process is not fast enough and energy efficient, cost-effective solar fuels production will never become a technological reality.

Polyoxometalates (POMs) are some of the most promising electrocatalysts to promote water oxidation [7,8] due to their dual nature. Being molecular species, POMs possess all advantages of homogeneous catalysts, being fast, monodispersed and easy to process. At the same time, as fragments of metal oxides, POMs also exhibit some of the advantages of heterogeneous catalysts, as they possess the right catalytic active sites, and appear to be highly robust and stable in water oxidation conditions [9].

The first POM reported as a water oxidation catalyst (WOC) was the tetraruthenium polyanion [(SiW$_{10}$O$_{36}$)$_2$Ru$_4$O$_5$(OH)(H$_2$O)$_4$)]$^{9-}$ [10,11]. Although very fast and rugged [12–15], its noble metal content precluded to envision technological impact since realization of artificial photosynthesis will require inexpensive and readily-available catalysts.

The breakthrough in the area appeared when the tetracobalt POM $[Co_4(H_2O)_2(PW_9O_{34})_2]^{10-}$ (**Co₄**), obtained from earth abundant metals, was reported to be one of the fastest WOCs up to date [16,17]. Some studies suggested that this POM may not be stable in water oxidation conditions [18–20]. Although this is true for certain conditions, and it can put into question the quantification of its catalytic activity, it is important to note that the genuine WOC activity of **Co₄** has been confirmed beyond any reasonable doubt [21–23], and corroborated even by the most (initially) critical research teams [24]. Following these results, several related POMs have shown WOC activity [7,25–29], most of them containing Co^{II} as water-ligated active center.

Artificial photosynthesis takes inspiration from natural photosynthesis in green plants and algae, where a tetramanganese cluster is responsible for the catalysis of the water oxidation reaction [30–33]. However, Mn-containing POMs have not been so successful. There is only one Mn-based example, reported very recently [34]. This trend of few Mn examples is not exclusive of POMs [1], and even if one considers heterogeneous catalysts, Mn oxide is much less active than the corresponding Co or Ni analogs [35]. Still, the use of Mn is attractive because it is earth abundant and relatively nontoxic when compared with the other more active metal-based catalysts.

In the search for Mn-containing POMs which could exhibit WOC activity we decided to test the Mn^{II} structural equivalent of the tetra cobalt WOC reported by Hill *et al.* [16]. The water electrochemistry of this polyanion $[Mn_4(H_2O)_2(PW_9O_{34})_2]^{10-}$ (**Mn₄**) confirms its activity as a water oxidation catalysts, albeit its intrinsic instability in water oxidation conditions.

2. Results and Discussion

Mn₄ was prepared as a mixed Na/K salt following literature procedure. The structure of the polyanion consists of rhomblike tetranuclear cluster of MnO6 octahedra sharing edges encapsulated by two lacunary Keggin-type $[PW_9O_{34}]^{9-}$ units (Figure 3b). All bridges between Mn atoms are oxo groups from WO6 or PO4 polyhedra. Two terminal water molecules complete the coordination positions of the Mn centers in the long diagonal of the rhomb.

We carried out cyclic voltammetry experiments in a pH 7 sodium phosphate buffer (NaP_i, 50 mM) solution with a amorphous carbon disc anode, a Pt wire cathode and a Ag/AgCl (NaCl 3 M) reference electrode (Figure 1). In the presence of **Mn₄** (1 mM) a significant catalytic water oxidation wave is observed above 1.0 V, which is accompanied by significant gas bubbling at the anode. This suggests that **Mn₄** is indeed a WOC at neutral pH. The catalytic activity appears to be slower In comparison with the activity of **Co₄** in the same conditions, since the **Mn₄**-catalyzed wave appears below the **Co₄** cyclic voltametry at all potentials. It is also worthy to mention that no pre-catalytic features are observed in the data, indicating that no additional redox processes occur before catalysis.

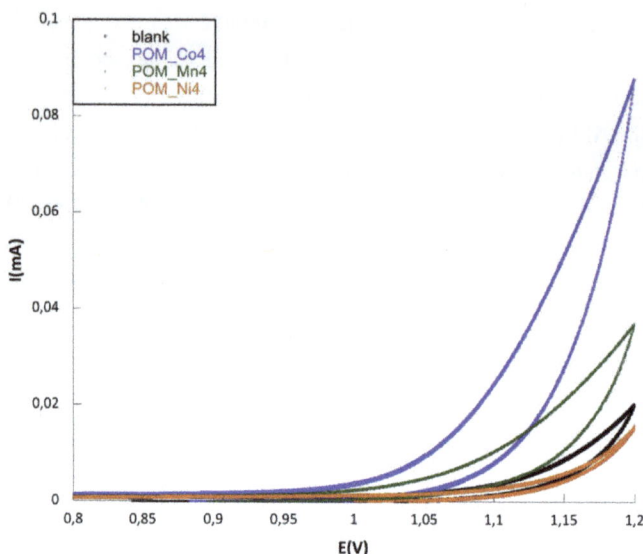

Figure 1. Cyclic voltammetry in a pH 7 sodium phosphate buffer (50 mM) water solution as electrolyte: 1.0 mM **Co₄** (**blue line**); 1.0 mM **Mn₄** (**green line**); blank (**black line**). E reported *vs.* Ag/AgCl (3 M) reference electrode.

We carried out bulk water electrolysis experiments under stirring in a two-chamber cell with both chambers connected through a glass frit (Figure 2). As anode and cathode we used FTO-coated glass (1 cm × 1 cm) and Pt mesh electrodes, respectively. The Ag/AgCl (NaCl 3 M) reference electrode was located close the anode, in the anodic compartment. Negligible current values were obtained when an anodic overpotential of ≈ 600 mV was applied (1.40 V *vs.* NHE) to this set-up with a pH 7 sodium phosphate buffer (NaP$_i$, 50 mM) solution with NaNO₃ (1 M) as electrolyte in the absence of a catalyst. The addition of **Mn₄** (1 mM) to the anodic compartment resulted in rapid gas evolution, with a fast current increase reaching typical currents (i) over 0.03 mA. After a short induction time, i keeps decreasing slowly to reach negligible values after just 30 min. During this time, the solution maintains essentially its original orange color, but a thin brown film grows on the anode. Chemical analysis of this film, which exhibited amorphous X-ray powder patterns, showed that manganese is the only metal significantly present. This suggests that, while the POM is stable and active in solution, there is a second process where it rapidly decomposes under an oxidation potential to yield a manganese oxide phase that growing on the electrode. This phase appears to be inactive as a WOC, and catalysis stops when the access of the remaining **Mn₄** in solution to the electrode is blocked by this film. Indeed, the as-used electrode does not show any significant remnant catalytic activity. A similar process was observed for other Co-containing POM-based WOCs [18], although the deposited Co-oxide is a highly active WOC. The inactivity of the Mn-oxide film observed in our case makes easier to identify the genuine catalytic activity of **Mn₄**. Although mixed valence MnO$_x$ are active WOCs [36–38], the MnO₂ phase is inactive [39]. Thus, this should be the major species in the deposited films. The formation of this oxidized MnO_x film precludes quantitative oxygen production during this water electrolysis.

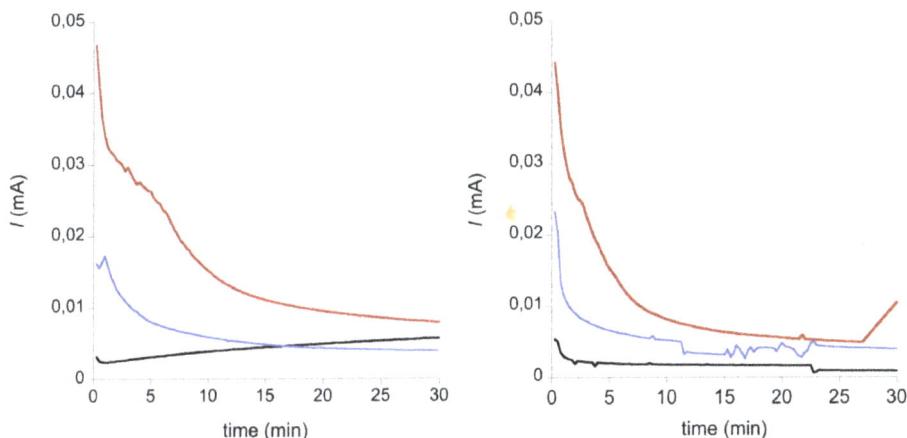

Figure 2. (a) Bulk water electrolysis under an applied anodic potential of 1.40 V/*vs.* NHE) with an fluorine-doped tin oxide (FTO) anode and Pt mesh cathode in a pH 7 sodium phosphate buffer (50 mM) water solution as electrolyte with 1.0 mM **Mn₄** (red line); a consecutive experiment with the as-used electrode in a **Mn₄**-free electrolyte solution (blue); blank (black line). (b) analogous experiment with addition of bpy(10 mM).

In Co-containing POMs, the formation of metal-oxide films has been assigned to the solution equilibria of the POMs, and not to a redox instability at the electrode [23]. Chelating agents able to trap traces of aqueous $[M(H_2O)_6]^{2+}$ $(M^{2+} = Co^{2+})$ generated in solution have been very effective to prevent the formation of such oxides. Unfortunately, the same strategy was not successful to prevent formation of the corresponding oxide in this case. Analogous bulk water electrolysis experiment in the presence of 2,2'-bipyridyl showed even faster catalyst decomposition (Figure 2b), suggesting this tetra manganese POM is not redox stable.

In other homogeneous WOCs their incorporation into a heterogeneous matrix has significantly improved their stability, while maintaining their catalytic activity [40]. Following the same strategy, we incorporated the insoluble $Cs_{10}[Mn_4(H_2O)_2(PW_9O_{34})_2]xH_2O$ salt into an amorphous carbon paste anode. However, the incorporation of the Cs-**Mn₄** component showed no effect in the catalytic current, indicating that **Mn₄** is also unstable in such conditions.

3. Experimental Section

3.1. Materials and Instrumentation

All reagents were purchased from Sigma-Aldrich (Madrid, Spain)(>99% purity) and used without further purification. Metal content in POM salts was analyzed with an Environmental Scanning

Electron Microscope JEOL-JMS6400 equipped with an Oxford Instruments X-ray elemental analyzer. Thermogrammetry was performed with power samples using a TGA/SDTA851 Mettler Toledo with MT1 microbalance.

3.2. Synthesis and Characterization

Co$_4$ and **Mn$_4$** were prepared from optimized literature methods as alkali salts [41,42]. The compounds were recrystallized from water, collected by filtration, dried in vacuum and characterized by IR spectroscopy. The counter cations and solvent content were determined by EDAX microanalysis and thermogravimetry, respectively. The molecular formulas are: $Na_5K_5[Co_4(H_2O)_2(PW_9O_{34})_2]$ 31H_2O (M_w = 5599.04) and $Na_5K_5[Mn_4(H_2O)_2(PW_9O_{34})_2]$ 31H_2O (M_w = 5583.08). The molecular structure of these polyanions is represented in Figure 3.

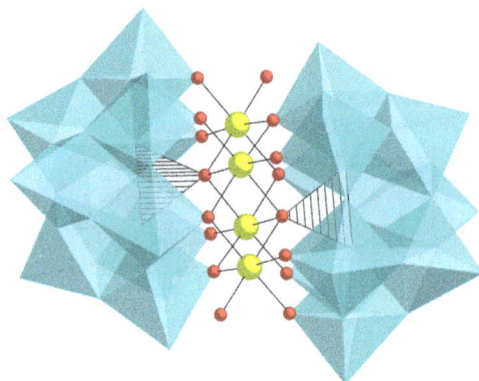

Figure 3. Molecular structure of the $[Mn_4(H_2O)_2(PW_9O_{34})_2]^{10-}$ (**Mn$_4$**) polyoxometalate.

3.3. Electrochemistry

Bulk water electrolysis were carried out with stirring in a two-chamber cell, with a porous frit connecting both chambers. In one chamber we placed a Pt mesh counter electrode, and in the other chamber a fluorine-doped tin oxide (FTO)-coated glass working electrode and a Ag/AgCl (NaCl 3 M) reference electrode. All potentials are reported *versus* NHE ($E = E_{obs} + 0.208$). Data were collected with a Biologic SP-150 potentiostat. Typical electrolysis experiments were carried out in a sodium phosphate (NaP_i) buffer pH = 7 solution with NaNO$_3$ (1 M) as electrolyte. Ohmic drop was compensated using the positive feedback compensation implemented in the instrument.

4. Conclusions

Electrochemistry data of a $[Mn_4(H_2O)_2(PW_9O_{34})_2]^{10-}$ at pH = 7 confirms its genuine activity of this POM to promote water oxidation. Nevertheless, its activity is lower when compared with the Co counterpart and, more significant, its stability is poor. During electrocatalytic water splitting, **Mn$_4$**

decomposition products (oxides) deposit on the anode precluding the catalytic activity. After a few minutes, catalysis becomes negligible, reaching values identical to those of the bare electrodes.

These results suggest that, although chosen by natural evolution, Mn may not be the best option for the development of water oxidation catalysts. In POMS, and also in other homogeneous catalysts, Mn-based WOCs appear to be intrinsically unstable during water oxidation [43]. In living entities, the instability of Mn WOCs may be an advantage, since it allows for easier self-repair mechanism in photosystem II. However, this is an important disadvantage when developing working devices.

Acknowledgments

We thank the financial support of the EU (CHEMCOMP ERC Stg grant 279313), the Spanish Ministerio de Economía y Competitividad (grant CTQ2012-34088) and through a Severo Ochoa Excellence Accreditation 2014–2018 (SEV-2013-0319) and the ICIQ Foundation.

Author Contributions

JRGM proposed the concept. JRGM and SGF designed the experiments. SGF and JSL performed the experiments. All authors analyzed the data and wrote the manuscript.

Conflicts of Interest

The authors declare no conflict of interest.

References

1. Spiccia, L.; Singh, A. Water oxidation catalysts based on abundant 1st row transition metals. *Coord. Chem. Rev.* **2013**, *257*, 2607–2622.
2. Dau, H.; Limberg, C.; Reier, T.; Risch, M.; Roggan, S.; Strasser, P. The mechanism of water oxidation: From electrolysis via homogeneous to biological catalysis. *ChemCatChem* **2010**, *2*, 724–761.
3. Galan-Mascaros, J.R. Water oxidation at electrodes modified with earth-abundant transition-metal catalysts. *ChemElectroChem* **2015**, *2*, 37–50.
4. McKone, J.R.; Lewis, N.S.; Gray, H.B. Will solar-driven water-splitting devices see the light of day? *Chem. Mater.* **2014**, *26*, 407–414.
5. Barber, J. Photosynthetic energy conversion: Natural and artificial. *Chem. Soc. Rev.* **2009**, *38*, 185–196.
6. Walter, M.G.; Warren, E.L.; McKone, J.R.; Boettcher, S.W.; Mi, Q.; Santori, E.A.; Lewis, N.S. Solar water splitting cells. *Chem. Rev.* **2010**, *110*, 6446–6473.
7. Lv, H.; Geletii, Y.V.; Zhao, C.; Vickers, J.W.; Zhu, G.; Luo, Z.; Song, J.; Lian, T.; Musaev, D.G.; Hill, C.L. Polyoxometalate water oxidation catalysts and the production of green fuel. *Chem. Soc. Rev.* **2012**, *41*, 7572–7589.
8. Streb, C. New trends in polyoxometalate photoredox chemistry: From photosensitisation to water oxidation catalysis. *Dalton Trans.* **2012**, *41*, 1651–1659.

9. Hill, C.L. Progress and challenges in polyoxometalate-based catalysis and catalytic materials chemistry. *J. Mol. Catal.* **2007**, *262*, 2–6.

10. Geletii, Y.V.; Botar, B.; Kögerler, P.; Hillesheim, D.A.; Musaev, D.G.; Hill, C.L. An all-inorganic, stable, and highly active tetraruthenium homogeneous catalyst for water oxidation. *Angew. Chem. Int. Ed.* **2008**, *47*, 3896–3899.

11. Sartorel, A.; Miro, P.; Salvadori, E.; Romain, S.; Carraro, M.; Scorrano, G.; di Valentin, M.; Llobet, A.; Bo, C.; Bonchio, M. Water oxidation at a tetraruthenate core stabilized by polyoxometalate ligands: experimental and computational evidence to trace the competent intermediates. *J. Am. Chem. Soc.* **2009**, *131*, 16051–16052.

12. Guo, S.X.; Liu, Y.; Lee, C.Y.; Bond, A.M.; Zhang, J.; Geletii, Y.V.; Hill, C.L. Graphene-supported $[Ru_4O_4(OH)_2(H_2O)_4(\gamma\text{-}SiW_{10}O_{36})_2]^{10-}$ for highly efficient electrocatalytic water oxidation. *Energy Environ. Sci.* **2013**, *6*, 2654–2663.

13. Orlandi, M.; Argazzi, R.; Sartorel, A.; Carraro, M.; Scorrano, G.; Bonchio, M.; Scandola, F. Ruthenium polyoxometalate water splitting catalyst: Very fast hole scavenging from photogenerated oxidants. *Chem. Commun.* **2010**, *46*, 3152–3154.

14. Toma, F.M.; Sartorel, A.; Iurlo, M.; Carraro, M.; Parisse, P.; Maccato, C.; Rapino, S.; Rodriguez Gonzalez, B.; Amenitsch, H.; da Ros, T.; *et al.* Efficient water oxidation at carbon nanotube-polyoxometalate electrocatalytic interfaces. *Nat. Chem.* **2010**, *2*, 826–831.

15. Besson, C.; Huang, Z.; Geletii, Y.V.; Lense, S.; Hardcastle, K.I.; Musaev, D.G.; Lian, T.; Proust, A.; Hill, C.L. $Cs_9)[(\gamma\text{-}PW_{10}O_{36})_2Ru_4O_5(OH)(H_2O)_4)]$, a new all-inorganic, soluble catalyst for the efficient visible-light-driven oxidation of water. *Chem. Commun.* **2010**, *46*, 2784–2786.

16. Yin, Q.; Tan, J.M.; Besson, C.; Geletii, Y.V.; Musaev, D.G.; Kuznetsov, A.E.; Luo, Z.; Hardcastle, K.I.; Hill, C.L. A fast soluble carbon-free molecular water oxidation catalyst based on abundant metals. *Science* **2010**, *328*, 342–345.

17. Huang, Z.; Luo, Z.; Geletii, Y.V.; Vickers, J.W.; Yin, Q.; Wu, D.; Hou, Y.; Ding, Y.; Song, J.; Musaev, D.G.; *et al.* Efficient light-driven carbon-free cobalt-based molecular catalyst for water oxidation. *J. Am. Chem. Soc.* **2011**, *133*, 2068–2071.

18. Stracke, J.J.; Finke, R.G. Electrocatalytic water oxidation beginning with the cobalt polyoxometalate $[Co_4)(H_2O)_2(PW_9O_{34})_2]^{10-}$: Identification of Heterogeneous CoO_x as the dominant catalyst. *J. Am. Chem. Soc.* **2011**, *133*, 14872–14875.

19. Stracke, J.J.; Finke, R.G. Water oxidation catalysis beginning with 2.5 μM $[Co_4(H_2O)_2(PW_9O_{34})_2]^{10-}$: Investigation of the true electrochemically driven catalyst at $\geq$ 600 mV overpotential at a glassy carbon electrode. *ACS Catal.* **2013**, *3*, 1209–1219.

20. Natali, M.; Berardi, S.; Sartorel, A.; Bonchio, M.; Campagna, S.; Scandola, F. Is $[Co_4(H_2O)_2(\alpha\text{-}PW_9O_{34})_2]^{10-}$ a genuine molecular catalyst in photochemical water oxidation? Answers from time-resolved hole scavenging experiments. *Chem. Commun.* **2012**, *48*, 8808–8810.

21. Vickers, J.W.; Lv, H.; Sumliner, J.M.; Zhu, G.; Luo, Z.; Musaev, D.G.; Geletii, Y.V.; Hill, C.L. Differentiating homogeneous and heterogeneous water oxidation catalysis: Confirmation that $[Co_4(H_2O)_2(\alpha\text{-}PW_9O_{34})_2]^{10-}$ is a molecular water oxidation catalyst. *J. Am. Chem. Soc.* **2013**, *135*, 14110–14118.

22. Schiwon, R.; Klingan, K.; Dau, H.; Limberg, C. Shining light on integrity of a tetracobalt-polyoxometalate water oxidation catalyst by X-ray spectroscopy before and after catalysis. *Chem. Commun.* **2013**, *50*, doi:10.1039/c3cc46629a.

23. Goberna-Ferron, S.; Vigara, L.; Soriano-López, J.; Galán-Mascarós, J.R. Identification of a nonanuclear Co_9^{II} polyoxometalate cluster as a homogeneous catalyst for water oxidation. *Inorg. Chem.* **2012**, *51*, 11707–11715.

24. Stracke, J.J.; Finke, R.G. Water oxidation catalysis beginning with $Co_4(H_2O)_2(PW_9O_{34})_2^{10-}$ when driven by the chemical oxidant ruthenium(III)tris(2,2⁻bipyridine): Stoichiometry, kinetic, and mechanistic studies en route to identifying the true catalyst. *ACS Catal.* **2014**, *4*, 79–89.

25. Song, F.; Ding, Y.; Ma, B.; Wang, C.; Wang, Q.; Du, X.; Fua, S.; Song, J. $K_7[(CoCo^{II})Co^{III}(H_2O)W_{11}O_{39}]$: a molecular mixed-valence Keggin polyoxometalate catalyst of high stability and efficiency for visible light-driven water oxidation. *Energy Environ. Sci.* **2013**, *6*, 1170–1184.

26. Zhu, G.; Glass, E.N.; Zhao, C.; Lv, H.; Vickers, J.W.; Geletii, Y.V.; Musaev, D.G.; Song, J.; Hill, C.L. A nickel containing polyoxometalate water oxidation catalyst. *Dalton Trans.* **2012**, *41*, 13043–13049.

27. Cao, R.; Ma, H.; Geletii, Y.V.; Hardcastle, K.I.; Hill, C.L. Structurally characterized iridium(III)-containing polytungstate and catalytic water oxidation activity. *Inorg. Chem.* **2009**, *48*, 5596–5598.

28. Zhu, G.; Geletii, Y.V.; P., K.; Schilder, H.; Song, J.; Lense, S.; Zhao, C.; I., H.K.; Musaev, D.G.; Hill, C.L. Water oxidation catalyzed by a new tetracobalt-substituted polyoxometalate complex: $[\{Co_4(\mu\text{-}OH)(H_2O)_3\}(Si_2W_{19}O_{70})]^{11-}$. *Dalton Trans.* **2012**, *41*, 2084–2090.

29. Lv, H.; Song, J.; Geletii, Y.V.; Vickers, J.W.; Sumliner, J.M.; Musaev, D.G.; Kögerler, P.; Zhuk, P.F.; Bacsa, J.; Zhu, G.; *et al.* An exceptionally fast homogeneous carbon-free cobalt-based water oxidation catalyst. *J. Am. Chem. Soc.* **2014**, *136*, 9268–9271.

30. Suga, M.; Akita, F.; Hirata, K.; Ueno, G.; Murakami, H.; Nakajima, Y.; Shimizu, T.; Yamashita, K.; Yamamoto, M.; Ago, H.; *et al.* Native structure of photosystem II at 1.95 Å resolution viewed by femtosecond X-ray pulses. *Nature* **2015**, *517*, 99–103.

31. Kern, J.; Alonso-Mori, R.; Hellmich, J.; Tran, R.; Hattne, J.; Laksmono, H.; Glöckner, C.; Echols, N.; Sierra, R.G.; Sellberg, J.; *et al.* Room temperature femtosecond X-ray diffraction of photosystem II microcrystals. *Proc. Natl. Acad. Sci. USA* **2012**, *109*, 9721–9726.

32. Umena, Y.; Kawakami, K.; Shen, J.R.; Kamiya, N. Crystal structure of oxygen-evolving photosystem II at a resolution of 1.9 Å. *Nature* **2011**, *473*, 55–60.

33. Ferreira, K.N.; Iverson, T.M.; Maghlaoui, K.; Barber, J.; Iwata, S. Architecture of the photosynthetic oxygen-evolving center. *Science* **2004**, *303*, 1831–1838.

34. Al-Oweini, R.; Sartorel, A.; Bassil, B.S.; Natali, M.; Berardi, S.; Scandola, F.; Kortz, U.; Bonchio, M. Photocatalytic water oxidation by a mixed-valent $Mn_3^{III}Mn^{IV}O_3$ manganese oxo core that mimics the natural oxygen-evolving center. *Angew. Chem. Int. Ed.* **2014**, *53*, 11182–11185.

35. Zaharieva, I.; Chernev, P.; Risch, M.; Klingan, K.; Kohlhoff, M.; Fischer, A.; Dau, H. Electrosynthesis, functional, and structural characterization of a water-oxidizing manganese oxide. *Energy Environ. Sci.* **2012**, *5*, 7081–7089.

36. Wiechen, M.; Najafpour, M.M.; Allakhverdiev, S.I.; Spiccia, L. Water oxidation catalysis by manganese oxides: Learning from evolution. *Energy Environ. Sci.* **2014**, *7*, 2203–2212.

37. Kuo, C.H.; Mosa, I.M.; Poyraz, A.S.; Biswas, S.; El-Sawy, A.M.; Song, W.; Luo, Z.; Chen, S.Y.; Rusling, J.F.; He, J.; *et al.* Robust mesoporous manganese oxide catalysts for water oxidation. *ACS Catal.* **2015**, *5*, 1693–1699.

38. Najafpour, M.M.; Moghaddam, A.N.; Dau, H.; Zaharieva, I. Fragments of layered manganese oxide are the real water oxidation catalyst after transformation of molecular precursor on clay. *J. Am. Chem. Soc.* **2014**, *136*, 7245–7248.

39. Zaharieva, I.; Chernev, P.; Risch, M.; Klingan, K.; Kohlhoff, M.; Fischer, A.; Dau, H. Electrosynthesis, functional, and structural characterization of a water-oxidizing manganese oxide. *Energy Environ. Sci.* **2012**, *5*, 7081–7089.

40. Soriano-López, J.; Goberna-Ferron, S.; Vigara, L.; Carbó, J.J.; Poblet, J.M.; Galan-Mascaros, J.R. Cobalt polyoxometalates as heterogeneous water oxidation catalysts. *Inorg. Chem.* **2013**, *52*, 4753–4755.

41. Weakley, T.J.R.; Evans, H.T.; Showell, J.S.; Tourne, G.F.; Tourne, C.M. 18-Tungstotetracobaltato(II)diphosphate and related anions. A novel structural class of heteropolyanions. *Chem. Commun.* **1973**, 139–140.

42. Gómez Garcia, C.; Coronado, E.; Gomez-Romero, P.; Casan-Pastor, N. A tetranuclear rhomblike cluster of manganese(II). Crystal structure and magnetic properties of the heteropoly complex $K_{10}[Mn_4(H_2O)_2(PW_9O_{34})_2]$ 20H$_2$O. *Inorg. Chem.* **1993**, *32*, 3378–3381.

43. Hocking, R.K.; Brimblecombe, R.; Chang, L.Y.; Singh, A.; Cheah, M.H.; Glover, C.; Casey, W.H.; Spiccia, L. Water-oxidation catalysis by manganese in a geochemical-like cycle. *Nat. Chem.* **2011**, *3*, 461–466.

Synthesis and Characterisation of the Europium (III) Dimolybdo-Enneatungsto-Silicate Dimer, [Eu(α-SiW$_9$Mo$_2$O$_{39}$)$_2$]$^{13-}$

Loïc Parent, Pedro de Oliveira, Anne-Lucie Teillout, Anne Dolbecq, Mohamed Haouas, Emmanuel Cadot and Israël M. Mbomekallé

Abstract: The chemistry of polyoxometalates (POMs) keeps drawing the attention of researchers, since they constitute a family of discrete molecular entities whose features may be easily modulated. Often considered soluble molecular oxide analogues, POMs possess enormous potential due to a myriad of choices concerning size, shape and chemical composition that may be tailored in order to fine-tune their physico-chemical properties. Thanks to the recent progress in single-crystal X ray diffraction, new POMs exhibiting diverse and unexpected structures have been regularly reported and described. We find it relevant to systematically analyse the different equilibria that govern the formation of POMs, in order to be able to establish reliable synthesis protocols leading to new molecules. In this context, we have been able to synthesise the Eu^{3+}-containing silico-molybdo-tungstic dimer, [Eu(α-SiW$_9$Mo$_2$O$_{39}$)$_2$]$^{13-}$. We describe the synthesis and characterisation of this new species by several physico-chemical methods, such as single-crystal X-ray diffraction, ^{183}W NMR and electrochemistry.

Reprinted from *Inorganics*. Cite as: Parent, L.; de Oliveira, P.; Teillout, A.-L.; Dolbecq, A.; Haouas, M.; Cadot, E.; Mbomekallé, I.M. Synthesis and Characterisation of the Europium (III) Dimolybdo-Enneatungsto-Silicate Dimer, [Eu(α-SiW$_9$Mo$_2$O$_{39}$)$_2$]$^{13-}$. *Inorganics* **2015**, *3*, 341-354.

1. Introduction

Polyoxometalates (POMs) are oxo-metal clusters in which the metal element M is often in its highest oxidation state (M = W^{VI}, Mo^{VI}, V^{V}…). Berzélius was the first to isolate a POM in 1826: the ammonium salt of the 12-molybdophosphate [PMo$_{12}$O$_{40}$]$^{3-}$ anion [1], but the first description of the structure of one of the compounds of this family, the [PW$_{12}$O$_{40}$]$^{3-}$ anion, was made more than a century later by Keggin [2]. Then, several hundreds of structures have been described, especially from the second half of the 20th century onwards. Nowadays, a myriad of new molecules, ranging from the simplest ones to unexpected and rather complex structures, are reported every year. In fact, upon modulating the experimental conditions and selecting the nature of the metal elements and their molar ratios, the different substitution and addition reactions involving POM entities with respect to metal cations offer a wide range of possibilities regarding the synthesis of new molecules which may find applications in a variety of fields such as medicine [3,4], catalysis [5,6], nanotechnologies [7,8], magnetism [9,10], *etc.* It is, then, possible and more and more common to synthesise POM species that include in their structures some elements specially selected in order to impart the features required for certain targeted applications.

Lanthanide (Ln) cations, for example, have a partially filled 4f orbital. As a consequence, they possess photoluminescence properties. In fact, they exhibit a high purity colour luminescence in the visible or in the near IR range [11], being employed in TV and computer screens, in optical fibres, *etc.* [12]. The incorporation of lanthanide cations in POMs rose a considerable interest due to the possibility of creating synergy between their electronic properties. The first POM containing a lanthanide cation as the hetero-element was the compound $[CeMo_{12}O_{42}]^{8-}$, described in 1914 [13]. Several decades later, in the beginning of the 70s, Peacock and Weakley reported the synthesis of Ln-containing POMs which formed monomers and dimers of the (POM)Ln and (POM)$_2$Ln types [14]. Afterwards, the family of Ln-containing POMs never ceased to grow, concomitantly becoming more and more diversified [15], on the basis of a synthesis protocol that varies little. It consists of two major steps: 1) creation of mono- or multi-lacunary POM structures which will behave as ligands; 2) co-ordination of the lanthanide cation by those ligands via oxo bridges. POM:Ln complexes are obtained, exhibiting one of the following different ratios: 1:1 [16,17], 2:1 [17–20] or 2:2 [21–23]. The most common POM structures known as Linqvist, Anderson, Keggin and Dawson may be used as lacunary species, and in some cases the co-ordination sphere of the lanthanide cation is completed by an organic ligand [24,25]. Finally, it is not unusual to come across less common structures combining fragments from different families or involving "giant" POMs [26–28]. As previously stated, the major interest of these POM:Ln compounds is to obtain a synergistic effect between the electronic properties of the POM, often considered an electron reservoir, and the lanthanide, whose 4f orbitals are partially empty. The studies of these systems are quite often focused on their luminescence properties and rarely on their electrochemical behaviour [29]. The present study concerns the synthesis, structural characterisations, electrochemical studies and electro-catalytic properties of a POM containing the Keggin fragment $[SiW_9Mo_2O_{39}]^{8-}$ and the Eu^{3+} cation, obtained as the potassium salt of the respective dimer: $K_{13}[Eu(\alpha\text{-}SiW_9Mo_2O_{39})_2]$ 21H_2O. The formation constants of this complex had already been calculated by Choppin *et al.* [30], but this is the first time that the compound has been isolated and characterised.

2. Results and Discussion

2.1. Synthesis of $K_{13}[Eu(\alpha\text{-}SiW_9Mo_2O_{39})_2]$ 21H_2O

The compound $K_{13}[Eu(\alpha\text{-}SiW_9Mo_2O_{39})_2]$ 21H_2O (**1**) was prepared according to a method different from that described in the literature for its homologous species $K_{13}[Eu(\alpha\text{-}SiW_{11}O_{39})_2]$ 18H_2O [16,19]. In fact, the lacunary compound $[\alpha\text{-}SiW_9Mo_2O_{39}]^{8-}$ being less robust than its molybdenum-free homologue, the synthesis has to be carried out at room temperature and in a buffered medium (acetate buffer, pH ~4.7) in order to prevent it from decaying. Likewise, the sequence of reagent addition is reversed. The compound $[\alpha\text{-}SiW_9Mo_2O_{39}]^{8-}$ is added in small aliquots to a solution containing the Eu^{3+} ions. It may be assumed that the reaction rate between the lacunary species $[\alpha\text{-}SiW_9Mo_2O_{39}]^{8-}$ and the Eu^{3+} cation is high enough. In addition, the formation equilibrium constants determined by Choppin *et al.* [30] for POM-Eu complexes indicate that the compound $[Eu(\alpha\text{-}SiW_9Mo_2O_{39})_2]^{13-}$ is sufficiently stable in solution. This implies that the lacunary fragments $[\alpha\text{-}SiW_9Mo_2O_{39}]^{8-}$ remain free in solution for a relatively short time, which should be enough to prevent them from decomposing.

2.2. Elemental Analysis, TGA and IR Spectroscopy

The elemental analysis and the TGA results are in good agreement with the formula derived from single crystal X-ray diffraction. It consists of a potassium salt that crystallises with 21 water molecules. The main IR absorption bands observed correspond to the stretching vibrations of the X–O bonds, with X = Si or W [15,16]. The presence of Mo results in slight shifts when compared to the values reported in the study by Pope et $al.$ or that by Balula and Freire cited before [15,16]. The formula of compound **1** corresponds well to that of the parent dimer $K_{13}[Eu(\alpha-SiW_{11}O_{39})_2]$ $18H_2O$.

2.3. Structure

The dimeric anion $[Eu(SiW_9Mo_2O_{39})_2]^{13-}$ is built of two monovacant Keggin moieties $[\alpha-SiW_9Mo_2O_{39}]^{8-}$ connected to a central Eu^{3+} cation (Figure 1A). The Mo and W centres are localised in the structure, the two MoO_6 octahedra belonging to two different trinuclear fragments and sharing a corner. The rare-earth cation is slightly off-centred of the vacant site of the $[\alpha-SiW_9Mo_2O_{39}]^{8-}$ sub-units (Figure 1B), as shown by the longer Si$\cdots$Eu distance (4.52 Å) when compared to the Si$\cdots$M (M = Mo, W) distances (3.48 – 3.58 Å). It may be noticed that the POM has no symmetry element and in particular is non-centro-symmetric. The local symmetry around the Eu^{3+} ion, co-ordinated by eight oxygen atoms of the Keggin fragments, is approximately square anti-prismatic (Figure 1C), as also observed, for example, in the well-known $[EuW_{10}O_{36}]^{9-}$ species [31] and also in $[Eu(\alpha-SiW_{11}O_{39})_2]^{13-}$ [29] and $[Eu(\beta_2-SiW_{11}O_{39})_2]^{13-}$ [19]. However, a pseudo C_2 axis can be identified passing through the Eu^{3+} ion (Figure 1D). Overall, the structure of $[Eu(\alpha-SiW_9Mo_2O_{39})_2]^{13-}$ is quite close to that of $[Eu(\alpha-SiW_{11}O_{39})_2]^{13-}$ [29], except for the presence of four localised Mo^{VI} centres in the former complex. However, $K_{13}[Eu(\alpha-SiW_{11}O_{39})_2]$ $18H_2O$ [29] and $K_{13}[Eu(\alpha-SiW_9Mo_2O_{39})_2]$ $21H_2O$ (**1**) are not isostructural, due to small differences in the crystal packing.

2.4. NMR Spectroscopy

Tungsten-183 (^{183}W) NMR spectroscopy allows the characterisation of the structure of $[Eu(SiW_9Mo_2O_{39})_2]^{13-}$ in solution. Figure 2 shows the ^{183}W NMR spectrum taken at 29 °C. It shows nine resonances (the monomer would exhibit five), each integrating for one W atom. This multiplicity is rather consistent with a point-group symmetry C_2 and thus a staggered conformation similar to that observed in $[Ln(SiW_{11}O_{39})_2]^{13-}$, Ln = Yb or Lu, and in opposition to the eclipsed conformation (C_{2v}) observed for Ln = La [32]. Indeed, the latter gave rise to six lines in the ^{183}W NMR spectrum, with five integrating for two W atoms and one integrating for one W atom, whereas the spectrum of the former, with either of the heavy lanthanides, exhibited eleven lines of equal intensity [32]. Thus, in the present case of the $[Eu(SiW_9Mo_2O_{39})_2]^{13-}$ species, all nine W atoms in a POM unit are not equivalent to each other and each lacunary fragment is related to the other by a two-fold rotation. Consistently, the ^{29}Si NMR spectrum shows one resonance at $\delta = -71.9$ ppm (Figure S3 in Supporting Information). Interestingly, among the nine ^{183}W lines, two of them undergo strong up-field shifts ($\sim$−1200 ppm) that allow them to be assigned to W atoms close to the paramagnetic Eu centre. The other two metal centres bridged to Eu through metal-oxo junctions are then the two Mo atoms, in agreement with X-ray diffraction analysis. Also, six of the remaining ^{183}W lines are relatively broad (line width ~5–8 Hz), presumably as a consequence of residual paramagnetic interaction. Finally, the last line at $\delta = -127.4$ ppm is very narrow (line width <1 Hz) and even exhibits $^2J_{W-O-W}$ coupling satellites ($^2J = 16$ Hz). It should, thus, be attributed to the W atom positioned at the longest distance from the Eu centre.

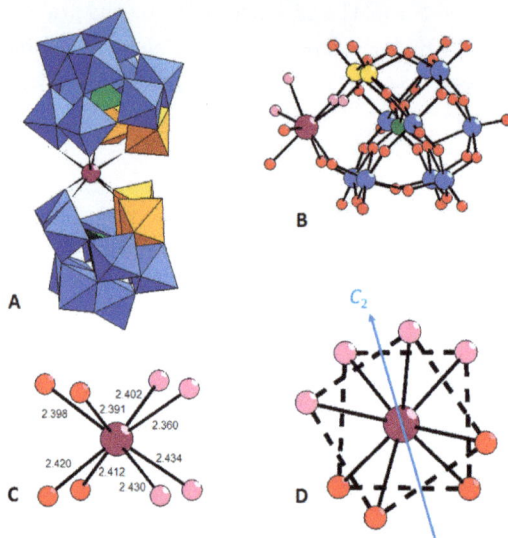

Figure 1. **(A)** Mixed ball-and-stick and polyhedral representation of $[Eu(SiW_9Mo_2O_{39})_2]^{13-}$; **(B)** ball and stick representation of the $\{(SiW_9Mo_2O_{39})EuO_4\}$ fragment; **(C)** environment around the Eu^{3+} ion, with the Eu-O distances (Å) indicated for each bond; **(D)** position of the pseudo C_2 axis passing through the Eu^{3+} ion. Blue octahedra: WO_6; orange octahedra: MoO_6; green tetrahedra: SiO_4; plum spheres: Eu; blue spheres: W; orange spheres: Mo; red spheres: O; pink spheres: O bound to Mo ions.

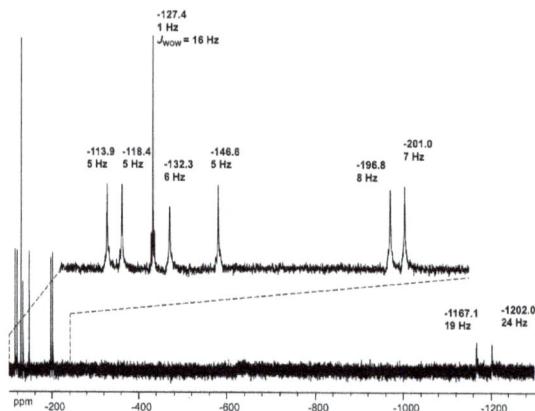

Figure 2. ^{183}W NMR spectrum of the lithium salt of $[Eu(SiW_9Mo_2O_{39})_2]^{13-}$ in D_2O/H_2O solution.

2.5. Electrochemistry

Compound **1** is stable at pH 3 in lithium sulphate solution. For lower pH values (see SI), it slowly decomposes to give rise to the saturated species $[SiW_9Mo_3O_{40}]^{4-}$. Figure 3 shows the CVs of **1** and of $[SiW_9Mo_2O_{39}]^{8-}$ recorded in the same experimental conditions and restricted to the redox processes attributed to the Mo^{VI} centres of the two compounds. The CV of **1** shows a quasi-reversible wave with $E_{pc} = -0.170$ V $vs.$ SCE and $\Delta E = 0.070$ V ($\Delta E = E_{pa} - E_{pc}$). The CV of $[SiW_9Mo_2O_{39}]^{8-}$ reveals a slower process taking place at far more negative potentials, -0.396 V $vs.$ SCE. It is a rather irreversible process when compared to that of compound **1**.

We were taken aback by the shape of the re-oxidation wave of **1**, which suggests that there is a desorption step subsequent to the formation of a film on the surface of the working electrode. In order to sort this out, we studied the dependence of the oxidation peak current intensity, I_{pa}, as a function of the scan rate, v (Figure S4-A). The plot of I_{pa} as a function of v does not give a straight line, as should be the case for a simple adsorbed species mechanism (Figure S4-B). Likewise, the plot of I_{pa} as a function of the square root of the scan rate, $v^{1/2}$, is not a straight line either (Figure S4-C), as expected from the shape of the re-oxidation wave. It is highly likely that the overall process is simultaneously controlled by diffusion and adsorption phenomena. This sort of mechanism has been previously described by Amatore et $al.$ for certain electro-catalytic processes [33].

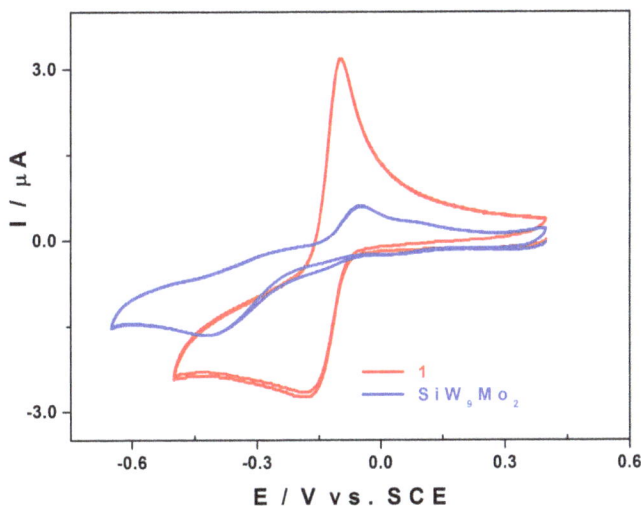

Figure 3. CVs of **1** (red) and of $[SiW_9Mo_2O_{39}]^{8-}$ (blue) in 0.5 M $Li_2SO_4 + H_2SO_4$/pH 3. Scan rate: 10 mV.s^{-1}. POM concentration: 2.5×10^{-4} M; working electrode: EPG; counter electrode: Pt; reference electrode: SCE.

A controlled potential coulometry experiment carried out at -0.40 V $vs.$ SCE yielded a consumption of 4 moles of electrons per mole of **1**, meaning that the Mo^{VI} centres in compound **1** are all reduced to Mo^V by the uptake of an electron each, the solution becoming violet (Figure S5-A). Upon resetting the potential to $+0.30$ V $vs.$ SCE, a re-oxidation step ensues in which over 95% of the previously consumed charge is recovered, and the solution reverts to its initial colourless aspect. The CVs recorded between these different steps show that compound **1** does not decompose when undergoing its reduction followed by its re-oxidation (Figure S5-B). Beyond this first redox wave, there is a large irreversible process having a reduction peak potential at −0.860 V *vs.* SCE (Figure S6).

We found it interesting to compare the response of the species **1** with that of its parent compound, the Keggin-type silico-tungstic derivative possessing three Mo^{VI} centres, $[SiW_9Mo_3O_{40}]^{4-}$, as far as their redox and electro-catalytic properties are concerned.

2.5.1. Compared Redox Behaviour of **1** and of $[SiW_9Mo_3O_{40}]^{4-}$ at pH 3.0

Figure 4 shows the CVs of **1** and of $[SiW_9Mo_3O_{40}]^{4-}$ in 0.5 M Li_2SO_4 + H_2SO_4/pH 3. Even if the three Mo^{VI} centres of the compound $[SiW_9Mo_3O_{40}]^{4-}$ are equivalent, their overall reduction does not take place in a single step, as is the case for compound **1**. In fact, the CV of $[SiW_9Mo_3O_{40}]^{4-}$ reveals three successive single-electron waves corresponding to the one-electron reduction of each of the Mo^{VI} centres. We recall that we observed a single four-electron redox wave for compound **1** in the same experimental conditions. The symmetry of compound **1** implies that the four Mo^{VI} centres are divided into two equivalent groups, each comprising two Mo^{VI} centres whose molecular orbitals have the same energy when the molecule is in its oxidised state. The uptake of four electrons, despite being fast, corresponds to a multi-step process, as is the case for all multi-electron reductions. We would expect to observe a decrease of the level of degeneracy after the uptake of the first electron, resulting in the splitting of the initial single wave into several waves: a sequence of either two two-electron steps or four single-electron steps. The latter phenomenon is observed upon the reduction of the three Mo^{VI} centres of the compound $[SiW_9Mo_3O_{40}]^{4-}$, which are equivalent when the molecule is in its highest oxidation state, but whose degree of degeneracy decreases when the first electron is added. There are three successive single-electron reduction waves in the CV of $[SiW_9Mo_3O_{40}]^{4-}$, whereas that of compound **1** reveals a single four-electron wave in the same experimental conditions. The degree of degeneracy is not affected in the latter, the four Mo centres remaining equivalent throughout the addition of the four electrons.

This behaviour should have a beneficial influence on the electro-catalytic properties of **1**. Indeed, the most promising compounds for electro-catalysis are those capable of exchanging a large number of electrons in one go.

2.5.2. Electro-Catalytic Activity of **1** towards O_2 and H_2O_2 Reduction

We were also interested in the electro-catalytic efficiency of compound **1** with respect to the reduction of both dioxygen (O_2) and hydrogen peroxide (H_2O_2). In fact, it is important to check these two processes and to compare their respective onset potentials, that is, the potential values at which the electro-catalytic reactions start. In the case of the species **1**, the onset potential for the reduction of H_2O_2 is less negative than that for O_2. This means that the electro-reduction of H_2O_2 on an EPG electrode is easier than that of O_2 in the presence of compound **1**. When we concentrate on the electro-reduction of O_2, the overall process will be pursued beyond the formation of H_2O_2 up to H_2O as the final product.

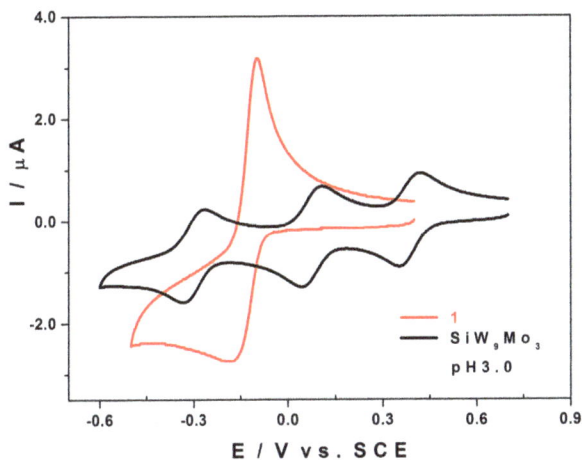

Figure 4. CVs of **1** (red) and of $[SiW_9Mo_3O_{40}]^{4-}$ (black) in 0.5 M Li_2SO_4 + H_2SO_4/pH 3. Scan rate: 10 mV.s^{-1}. POM concentration: 2.5×10^{-4} M. Working electrode: EPG; counter electrode: Pt; reference electrode: SCE.

Figure 5A confirms that the reduction of H_2O_2 is easier than that of O_2. Also, the efficiency of H_2O_2 reduction increases upon successive cycling (Figure 5B). The working electrode surface activation is probably due to the formation of a deposit that does not totally re-dissolve in the presence of H_2O_2. In the case of the electro-reduction of O_2, the activation phenomenon is not observed, the successive CVs being superposable (Figure S7). This suggests that the process of electro-catalytic reduction of O_2 by compound **1** either does not include a H_2O_2 formation step or H_2O_2 is a transient species whose lifespan is so short that no activation film forms on the electrode surface, meaning that the response shown in Figure 5B is not observed.

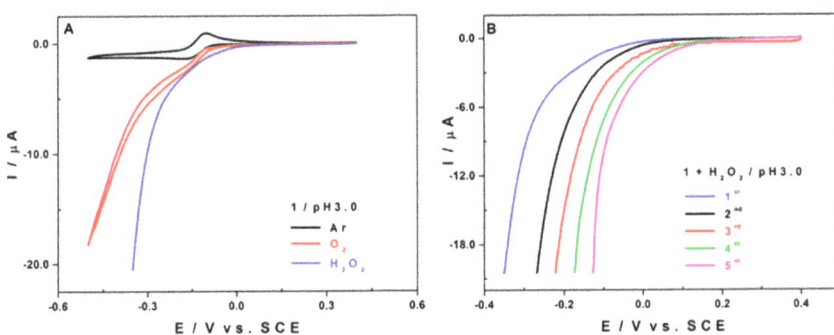

Figure 5. (A) CVs of **1** alone in an argon saturated solution (black), in a O_2 saturated solution (red), and in the presence of 0.25 M H_2O_2 (blue) in 0.5 M Li_2SO_4 + H_2SO_4/pH 3. **(B)** Successive CVs of **1** in the presence of 0.25 M H_2O_2 in 0.5 M Li_2SO_4 + H_2SO_4/pH 3. Scan rate: 2 mV.s^{-1}. POM concentration: 2.5×10^{-4} M. Working electrode: EPG; counter electrode: Pt; reference electrode: SCE.

This activation phenomenon is not observed with the species $[SiW_9Mo_3O_{40}]^{4-}$ and its onset potential for the reduction of H_2O_2 is more negative than that for the reduction of O_2 (Figure 8S-A). It is also important to point out that for the electro-catalytic reduction processes of both O_2 and H_2O_2, compound **1** is always more efficient than the parent species $[SiW_9Mo_3O_{40}]^{4-}$ (Figure 8S-B).

3. Experimental Section

3.1. Synthesis of $K_{13}[Eu(\alpha-SiW_9Mo_2O_{39})_2]$ $21H_2O$

0.18 g of $EuCl_3$ $6H_2O$ (0.5 mmol) are dissolved in 30 mL of 0.5M $NaCH_3COO$ + 0.5M CH_3COOH/pH 4.7, and then 3.0 g of $K_8[\alpha-SiW_9Mo_2O_{39}]$ nH_2O (1 mmol) are added in small aliquots. This procedure is adapted from the method described by Cadot *et al.* [34]. The mixture is stirred at room temperature until the dissolution is complete, and then filtered. After a few days, needle-like white crystals form in the colourless filtrate, which are recovered by filtration and dried in the open air. A mass of 1.24 g of the white crystals of the compound $K_{13}[Eu(\alpha-SiW_9Mo_2O_{39})_2]$ $21H_2O$ was obtained, corresponding to a yield of 41%. IR (cm^{-1}): 1003(w), 940 (m), 884 (s), 813 (m), 755 (m), 713 (m), 528 (w).

3.2. TGA, and IR Spectroscopy

Hydration water contents were determined by thermogravimetric analysis performed in a nitrogen atmosphere between 25 and 600 °C, with a heating speed of 5 °C.min^{-1}, using a Metler Toledo TGA/DSC1.. Infrared spectra were recorded on a Nicolet 6700 FT spectrometer driven by a PC with the OMNIC E.S.P. 8.2 software.

3.3. X-ray Crystallography

Data collection was carried out using a Siemens SMART three-circle diffractometer equipped with a CCD bi-dimensional detector using the monochromatised wavelength λ(Mo $K\alpha$) = 0.71073 Å. Absorption correction was based on multiple and symmetry-equivalent reflections in the data set using the SADABS program [35] based on the method of Blessing [36]. The structure was solved by direct methods and refined by full-matrix least-squares using the SHELX-TL package [37]. There is a discrepancy between the formulae determined by elemental analysis and that deduced from the crystallographic atom list because of the difficulty in locating all the disordered water molecules and counter-ions. These disordered water molecules and counter-ions, when located, were refined isotropically and with partial occupancy factors. Crystallographic data are given in Table 1. Further details of the crystal structure investigation may be obtained from the Fachinformationszentrum Karlsruhe, D-76344 Eggenstein-Leopoldshafen (Germany), on quoting the depository number CSD-429499.

Table 1. Crystallographic data.

Parameters	1
Formula	$EuK_{13}Mo_4O_{94}Si_2W_{18}$
Formula weight, g	5913.50
Crystal system	triclinic
Space group	P-1
a/Å	12.2447(10)
b/Å	12.7803(10)
c/Å	33.762(3)
α/°	83.799(2)
β/°	85.106(2)
γ/°	66.214(2)
V/Å^3	4801.4(6)
Z	2
D_{calc}/g cm^{-3}	4.090
μ/mm^{-1}	23.307
Data/parameters	37169/26808
R_{int}	0.0560
GOF	0.946
R_1 ($I > 2\sigma(I)$)	0.0720
wR_2	0.1786

3.4. NMR Spectroscopy

The NMR spectra were obtained on a Bruker AVANCE-400 spectrometer using a 10 mm broad-band probe. The [183]W spectrum at 16.7 MHz was recorded with a spectral width of 100 kHz, an acquisition time of 0.7 s, a pulse delay of 0.2 s, a pulse width of 50 µs (90° tip angle) and a number of scans of 131072. The ^{29}Si spectrum at 79.5 MHz was recorded with a spectral width of 4 kHz, an acquisition time of 0.5 s, a pulse delay of 0.1 s, a pulse width of 14 µs (90° tip angle) and a number of scans of 16384. For all spectra, the temperature was controlled and fixed at 29 °C. The [183]W spectrum was referenced to 2.0 mol.dm^{-3} Na$_2$WO$_4$ and the ^{29}Si spectrum to SiMe$_4$. For the ^{29}Si and the [183]W chemical shifts, the convention used is that the more negative chemical shifts denote up-field resonances.

3.5. Electrochemistry

Pure water was obtained with a Milli-Q Intregral 5 purification set. All reagents were of high-purity grade and were used as purchased without further purification: H$_2$SO$_4$ (Sigma Aldrich) and Li$_2$SO$_4$ H$_2$O (Acros Organics). The composition of the electrolyte was 0.5 M Li$_2$SO$_4$ + H$_2$SO$_4$/pH 3.0.

The stability of polyanions **1**, $[\alpha\text{-}SiW_9Mo_2O_{39}]^{8-}$ and $[SiW_9Mo_3O_{40}]^{4-}$ in solution in this medium was long enough to allow their characterisation by cyclic voltammetry and control potential coulometry. The UV-visible spectra were recorded on a Perkin-Elmer 750 spectrophotometer with 10^{-4} M solutions of the polyanion. Matched 2.00 mm optical path quartz cuvettes were used.

Electrochemical data were obtained using an EG & G 273 A potentiostat driven by a PC with the M270 software. A one-compartment cell with a standard three-electrode configuration was used for cyclic voltammetry experiments. The reference electrode was a saturated calomel electrode (SCE) and the counter electrode a platinum gauze of large surface area; both electrodes were separated from the bulk electrolyte solution via fritted compartments filled with the same electrolyte. The working electrodes were a 3 mm OD pyrolytic carbon disc or a *ca.* $10 \times 10 \times 2$ mm^3 glassy carbon stick (Le Carbone-Lorraine, France). The pre-treatment of the first electrode before each experiment is adapted from a method described elsewhere [38]. The stick is polished twice with SiC paper, grit 500 (Struers). After each polishing step, which lasts for about 5 min, the stick is rinsed and sonicated twice in Millipore water for a total of 10 minutes. Prior to each experiment, solutions were thoroughly de-aerated for at least 30 minutes with pure Ar. A positive pressure of this gas was maintained during subsequent work. All cyclic voltammograms were recorded at a scan rate of 10 mV.s^{-1} and potentials are quoted against the saturated calomel electrode (SCE) unless otherwise stated. The polyanion concentration was $\times 10^{-4}$ M. All experiments were performed at room temperature, which is controlled and fixed for the laboratory at 20°C. Results were very reproducible from one experiment to the other and slight variations observed over successive runs are rather attributed to the uncertainty associated with the detection limit of our equipment (potentiostat, hardware and software) and not to the working electrode pre-treatment nor to possible fluctuations in temperature.

4. Conclusions

In an acetate buffer medium, the mono-lacunary polyanion $[\alpha\text{-}SiW_9Mo_2O_{39}]^{8-}$ reacts with the Eu^{3+} cation, giving rise to the dimer species $[Eu(\alpha\text{-}SiW_9Mo_2O_{39})_2]^{13-}$. This new compound was characterised by several physico-chemical methods, such as TGA and IR spectroscopy, which allowed us to infer its chemical formula, $K_{13}[Eu(\alpha\text{-}SiW_9Mo_2O_{39})_2].21H_2O$. Single crystal X ray diffraction, ^{183}W and ^{29}Si NMR results are coherent with the fact that the compound is made of two $[\alpha\text{-}SiW_9Mo_2O_{39}]^{8-}$ fragments which co-ordinate the Eu^{3+} cation via 8 atoms.

The structure of compound **1** shows that there are two sorts of equivalent Mo centres, but the cyclic voltammetry revealed that they are all simultaneously reduced, contrary to the results observed with the parent compound $[\alpha\text{-}SiW_9Mo_3O_{40}]^{4-}$, for which there are three reduction steps corresponding to the three Mo centres. Compound **1** exhibits a good catalytic efficiency towards the reduction of O_2 and H_2O_2.

Supplementary Materials

Additional experimental data: UV-visible spectra of **1**, SiW_9Mo_2 and $EuCl_3$ 6H$_2$O (Figure S1), TGA curve of **1** (Figure S2), ^{29}Si NMR spectrum of **1** (Figure S3), additional electrochemical data (Figures S4-S8).

Acknowledgments

This work was supported by the Centre National de la Recherche Scientifique (UMR 8000, UMR 8180), the University of Paris-Sud and the University of Versailles.

Author Contribution

The preparation of the manuscript was made by all authors. L.P.: Syntheses, electrochemical and Electrocatalytic studies. P.d.O.: Careful follow-up and improvement of the manuscript. A.-L.T.: Electrochemical and electrocatalytic studies. A.D.: Single crystal X-ray diffraction and wrote that part. M.H.: NMR studies and wrote that part. E.C.: Careful follow-up and improvement of the manuscript. I.M.M.: General idea and plan for the publication.

Conflicts of Interest

The authors declare no conflict of interest.

References

1. Berzelius, J.J. Beitrag zur näheren Kenntniss des Molybdäns. *Ann. Phys.* **1826**, *82*, 369–392 (In German).
2. Keggin, J.F. Structure of the Molecule of 12-Phosphotungstic. *Nature* **1933**, *131*, 908–909.
3. Rhule, J.T.; Hill, C.L.; Judd, D.A.; Schinazi, R.F. Polyoxometalates in Medicine. *Chem. Rev.* **1998**, *98*, 327–358.
4. Prudent, R.; Moucadel, V.; Laudet, B.; Barette, C.; Lafanechère, L.; Hasenknopf, B.; Li, J.; Bareyt, S.; Lacôte, E.; Thorimbert, S.; Malacria, M.; Gouzerh, P.; Cochet, C. Identification of polyoxometalates as nanomolar noncompetitive inhibitors of protein kinase CK2. *Chem. Biol.* **2008**, *15*, 683–692.
5. Lv, H.; Geletii, Y.V; Zhao, C.; Vickers, J.W.; Zhu, G.; Luo, Z.; Song, J.; Lian, T.; Musaev, D.G.; Hill, C.L. Polyoxometalate water oxidation catalysts and the production of green fuel. *Chem. Soc. Rev.* **2012**, *41*, 7572–7589.
6. Mizuno, N.; Kamata, K. Catalytic oxidation of hydrocarbons with hydrogen peroxide by vanadium-based polyoxometalates. *Coord. Chem. Rev.* **2011**, *255*, 2358–2370.
7. Mitchell, S.G.; de la Fuente, J.M. The synergistic behavior of polyoxometalates and metal nanoparticles: from synthetic approaches to functional nanohybrid materials. *J. Mater. Chem.* **2012**, *22*, 18091–18100.
8. Mondloch, J.E.; Bayram, E.; Finke, R.G. A review of the kinetics and mechanisms of formation of supported-nanoparticle heterogeneous catalysts. *J. Mol. Catal. A* **2012**, *355*, 1–38.
9. Oms, O.; Dolbecq, A.; Mialane, P. Diversity in structures and properties of 3d-incorporating polyoxotungstates. *Chem. Soc. Rev.* **2012**, *41*, 7497–7536.
10. Zheng, S.-T.; Yang, G.-Y. Recent advances in paramagnetic-TM-substituted polyoxometalates (TM = Mn, Fe, Co, Ni, Cu). *Chem. Soc. Rev.* **2012**, *41*, 7623–7646.
11. Bünzli, J.-C.G.; Piguet, C. Taking advantage of luminescent lanthanide ions. *Chem. Soc. Rev.* **2005**, *34*, 1048–1077.
12. Binnemans, K. Lanthanide-based luminescent hybrid materials. *Chem. Rev.* **2009**, *109*, 4283–4374.

13. Barbieri, G.A. The position of cerium in the periodic system and the complex molybdates of tetravalent cerium, Atti della Accademia Nazionale dei Lincei, Classe di Scienze Fisiche. *Mat. Nat. Rend.* **1914**, *23*, 805–812 (In Italian).

14. Peacock, R.D.; Weakley, T.J.R. Heteropolytungstate complexes of the lanthanide elements. Part I. Preparation and reactions. *J. Chem. Soc. A* **1971**, 1836–1839.

15. Granadeiro, C.M.; de Castro, B.; Balula, S.S.; Cunha-Silva, L. Lanthanopolyoxometalates: From the structure of polyanions to the design of functional materials. *Polyhedron* **2013**, *52*, 10–24.

16. Sadakane, M.; Dickman, M.; Pope, M. Controlled Assembly of Polyoxometalate Chains from Lacunary Building Blocks and Lanthanide-Cation Linkers. *Angew. Chem. Int. Ed.* **2000**, *39*, 2914–2916.

17. Luo, Q.; Howell, R.C.; Bartis, J.; Dankova, M.; Horrocks, W.D.; Rheingold, A.L.; Francesconi, L.C. Lanthanide Complexes of $[\alpha\text{-}2\text{-}P_2W_{17}O_{61}]^{10-}$: Solid State and Solution Studies. *Inorg. Chem.* **2002**, *41*, 6112–6117.

18. Gaunt, A.J.; May, I.; Sarsfield, M.J.; Collison, D.; Helliwell, M.; Denniss, I.S. A rare structural characterisation of the phosphomolybdate lacunary anion, $[PMo_{11}O_{39}]_7$?. Crystal structures of the Ln(III) complexes, $(NH_4)_{11}[Ln(PMo_{11}O_{39})_2]$ 16H$_2$O (Ln = Ce^{III}, Sm^{III}, Dy^{III} or Lu^{III}) Electronic supplementary information (ESI) available: IR. *Dalton Trans.* **2003**, 2767–2771.

19. Bassil, B.S.; Dickman, M.H.; von der Kammer, B.; Kortz, U. The monolanthanide-containing silicotungstates $[Ln(\beta_2\text{-}SiW11O_{39})_2]^{13-}$ (Ln = La, Ce, Sm, Eu, Gd, Tb, Yb, Lu): a synthetic and structural investigation. *Inorg. Chem.* **2007**, *46*, 2452–2458.

20. Ostuni, A.; Bachman, R.E.; Pope, M.T. Multiple Diastereomers of $[M^{n+}(\alpha_m\text{-}P_2W_{17}O_{61})_2]^{(20-n)-}$ (M = U^{IV}, Th^{IV}, Ce^{III}; m = 1, 2). *Syn-* and *Anti*-Conformations of the Polytungstate Ligands in $\alpha_1\alpha_1$, $\alpha_1\alpha_2$ and $\alpha_2\alpha_2$ Complexes. *J. Clust. Sci.* **2003**, *14*, 431–446.

21. Niu, J.; Wang, K.; Chen, H.; Zhao, J.; Ma, P.; Wang, J.; Li, M.; Bai, Y.; Dang, D. Assembly Chemistry between Lanthanide Cations and Monovacant Keggin Polyoxotungstates: Two Types of Lanthanide Substituted Phosphotungstates $[\{(\alpha\text{-}PW_{11}O_{39}H)Ln(H_2O)_3\}_2]^{6-}$ and $[\{(\alpha\text{-}PW_{11}O_{39})Ln(H_2O)(\eta2,\mu\text{-}1,1)\text{-}CH_3COO\}_2]^{10-}$. *Cryst. Growth Des.* **2009**, *9*, 4362–4372.

22. Sadakane, M.; Ostuni, A.; Pope, M.T. Formation of 1:1 and 2:2 complexes of Ce(III) with the heteropolytungstate anion $\alpha_2\text{-}[P_2W_{17}O_{61}]^{10-}$, and their interaction with proline. The structure of $[Ce_2(P_2W_{17}O_{61})_2(H_2O)_8]^{14-}$. *Dalt. Trans.* **2002**, 63–67.

23. Zhang, C.; Bensaid, L.; McGregor, D.; Fang, X.; Howell, R.C.; Burton-Pye, B.; Luo, Q.; Todaro, L.; Francesconi, L.C. Influence of the Lanthanide Ion and Solution Conditions on Formation of Lanthanide Wells–Dawson Polyoxotungstates. *J. Clust. Sci.* **2006**, *17*, 389–425.

24. Kortz, U. Rare-Earth Substituted Polyoxoanions: $[\{La(CH_3COO)(H_2O)_2 (2\text{-}P_2W_{17}O_{61})\}_2]^{16-}$ and $[\{Nd(H_2O)_3(\alpha_2\text{-}P_2W_{17}O_{61})\}_2]^{14-}$. *J. Clust. Sci.* **2003**, *14*, 205–214.

25. Mialane, P.; Dolbecq, A.; Marrot, J.; Sécheresse, F. Oligomerization of Yb(III)-substituted Dawson polyoxotungstates by oxalato ligands. *Inorg. Chem. Commun.* **2005**, *8*, 740–742.

26. Dickman, M.H.; Gama, G.J.; Kim, K.; Pope, M.T. The structures of europium(III)- and uranium(IV) derivatives of $[P_5W_{30}O_{110}]^{15-}$: Evidence for "cryptohydration". *J. Clust. Sci.* **1996**, *7*, 567–583.

27. Naruke, H.; Yamase, T. Crystal structure of $K_{18.5}H_{1.5}[Ce_3(CO_3)(SbW_9O_{33})(W_5O_{18})_3]$ $14H_2O$. *J. Alloys Compd.* **1998**, *268*, 100–106.

28. Müller, A.; Beugholt, C.; Bögge, H.; Schmidtmann, M. Influencing the Size of Giant Rings by Manipulating Their Curvatures: $Na_6[Mo_{120}O_{366}(H_2O)_{48}H_{12}\{Pr(H_2O)_5\}_6]$ ($\sim200H_2O$) with Open Shell Metal Centers at the Cluster Surface. *Inorg. Chem.* **2000**, *39*, 3112–3113.

29. Julião, D.; Fernandes, D.M.; Cunha-Silva, L.; Ananias, D.; Balula, S.S.; Freire, C. Sandwich lanthano-silicotungstates: Structure, electrochemistry and photoluminescence properties. *Polyhedron* **2013**, *52*, 308–314.

30. VanPelt, C.E.; Crooks, W.J.; Choppin, G.R. Stability constant determination and characterization of the complexation of trivalent lanthanides with polyoxometalates. *Inorg. Chim. Acta* **2003**, *346*, 215–222.

31. Yamase, T.; Ozeki, T.; Ueda, K. Structure of $NaSr_4[EuW_{10}O_{36}]$ $34.5H_2O$. *Acta Cryst.* **1993**, *C49*, 1572–1574.

32. Bartis, J.; Sukal, S.; Dankova, M.; Kraft, E.; Kronzon, R.; Blumenstein, M.; Francesconi, L.C. Lanthanide complexes of polyoxometalates: characterization by tungsten-183 and phosphorus-31 nuclear magnetic resonance spectroscopy. *Dalton Trans.* **1997**, 1937–1944.

33. Klymenko, O.V.; Svir, I.; Amatore, C. Molecular electrochemistry and electrocatalysis: a dynamic view. *Mol. Phys.* **2014**, *112*, 1273–1283.

34. Cadot, E.; Thouvenot, R.; Teze, A.; Herve, G. Syntheses and multinuclear NMR characterizations of alpha-$[SiMo_2W_9O_{39}]^{8-}$ and alpha-$[SiMo_{3-x}V_xW_9O_{40}]^{(4+x)-}$ (x = 1, 2) heteropolyoxometalates. *Inorg. Chem.* **1992**, *31*, 4128–4133.

35. Sheldrick, G.M. *SADABS*, program for scaling and correction of area detector data; University of Göttingen: Germany 1997.

36. Blessing, R.H. An empirical correction for absorption anisotropy. *Acta Cryst.* **1995**, *C51*, 33–38.

37. Sheldrick, G.M. *SHELX-TL*, version 5.03; Software Package for the Crystal Structure Determination; Siemens Analytical X-ray Instrument Division: Madison, WI, USA, 1994.

38. Vilà, N.; Aparicio, P.A.; Sécheresse, F.; Poblet, J.M.; López, X.; Mbomekallé, I.M. Electrochemical behavior of α_1/α_2-$[Fe(H_2O)P_2W_{17}O_{61}]^{7-}$ isomers in solution: experimental and DFT studies. *Inorg. Chem.* **2012**, *51*, 6129–6138.

Vanadium(V)-Substitution Reactions of Wells–Dawson-Type Polyoxometalates: From $[X_2M_{18}O_{62}]^{6-}$ (X = P, As; M = Mo, W) to $[X_2VM_{17}O_{62}]^{7-}$

Tadaharu Ueda, Yuriko Nishimoto, Rie Saito, Miho Ohnishi and Jun-ichi Nambu

Abstract: The formation processes of V(V)-substituted polyoxometalates with the Wells–Dawson-type structure were studied by cyclic voltammetry and by ^{31}P NMR and Raman spectroscopy. Generally, the vanadium-substituted heteropolytungstates, $[P_2VW_{17}O_{62}]^{7-}$ and $[As_2VW_{17}O_{62}]^{7-}$, were prepared by mixing equimolar amounts of the corresponding lacunary species—$[P_2W_{17}O_{61}]^{10-}$ and $[As_2W_{17}O_{61}]^{10-}$—and vanadate. According to the results of various measurements in the present study, the tungsten site in the framework of $[P_2W_{18}O_{62}]^{6-}$ and $[As_2W_{18}O_{62}]^{6-}$ without defect sites could be substituted with V(V) to form the $[P_2VW_{17}O_{62}]^{7-}$ and $[As_2VW_{17}O_{62}]^{7-}$, respectively. The order in which the reagents were mixed was observed to be the key factor for the formation of Dawson-type V(V)-substituted polyoxometalates. Even when the concentration of each reagent was identical, the final products differed depending on the order of their addition to the reaction mixture. Unlike Wells–Dawson-type heteropolytungstates, the molybdenum sites in the framework of $[P_2Mo_{18}O_{62}]^{6-}$ and $[As_2Mo_{18}O_{62}]^{6-}$ were substituted with V(V), but formed Keggin-type $[PVMo_{11}O_{40}]^{4-}$ and $[AsVMo_{11}O_{40}]^{4-}$ instead of $[P_2VMo_{17}O_{62}]^{7-}$ and $[As_2VMo_{17}O_{62}]^{7-}$, respectively, even though a variety of reaction conditions were used. The formation constant of the $[PVMo_{11}O_{40}]^{4-}$ and $[AsVMo_{11}O_{40}]^{4-}$ was hypothesized to be substantially greater than that of the $[P_2VMo_{17}O_{62}]^{7-}$ and $[As_2VMo_{17}O_{62}]^{7-}$.

Reprinted from *Inorganics*. Cite as: Ueda, T.; Nishimoto, Y.; Saito, R.; Ohnishi, M.; Nambu, J. Vanadium(V)-Substitution Reactions of Wells–Dawson-Type Polyoxometalates: From $[X_2M_{18}O_{62}]^{6-}$ (X = P, As; M = Mo, W) to $[X_2VM_{17}O_{62}]^{7-}$. *Inorganics* **2015**, *3*, 355-369.

1. Introduction

Metal-substituted or metal-incorporated polyoxometalates (POMs) based on the Keggin and Wells–Dawson structures are of great interest in catalysis and other applications, as well as in fundamental studies, because they exhibit specific and excellent properties depending on the incorporated metals [1–9]. Cobalt- and ruthenium-incorporated POMs, such as $[Co_4(H_2O)_2(XW_9O_{34})_2]^{n-}$ (X = P, Si) and $Rb_8K_2[\{Ru_4O_4(OH)_2(H_2O)_4\}-(\gamma-SiW_{10}O_{36})_2]$, in particular, exhibit excellent catalytic activity toward water oxidation and water splitting [10–15]. Among the various metal-substituted POMs, vanadium-substituted POMs have been extensively investigated in terms of synthesis, characterization and applications. In particular, they have been used as oxidation catalysts in various organic syntheses, because they can be reduced at a more positive potential than the corresponding parent, non-substituted POMs [16–26]. Interestingly, vanadate can also be incorporated as the central ion of POMs [27–32]. Generally, metal-substituted or metal-incorporated POMs have been prepared with transition-metal atoms incorporated into the vacant sites of the lacunary species, such as $[PW_{11}O_{39}]^{7-}$ and $[P_2W_{17}O_{61}]^{10-}$ [1–9]. A variety of Wells–Dawson-type molybdenum- and/or vanadium-substituted tungstophosphates have been prepared from the parent $[P_2W_{18}O_{62}]^{6-}$ via a decomposition and re-building method by controlling the pH (Figure 1). Large-sized POMs have also been prepared by mixing metal cations and the lacunary POMs as building blocks. Currently, the synthesis of metal-substituted or metal-incorporated POMs can be controlled to some extent through the use of the lacunary POMs. In addition, details of the formation of some of POMs have been investigated using NMR and Raman spectroscopy and cyclic voltammetry [33–43]. However, many aspects of the formation of POMs in solution remain ambiguous.

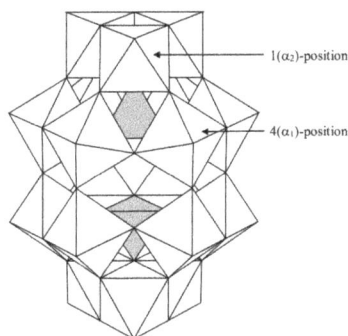

Figure 1. Structure of the Well–Dawson-type polyoxometalate.

Recently, we investigated the formation of Keggin-type V(V)-substituted POMs, $[XV_aM_{12-a}O_{40}]^{(3+a)-}$ (X = P, As, Si, Ge; M = Mo, W; a = 1, 2) (XV_aM_{12-a}), in both aqueous and aqueous-organic solutions using cyclic voltammetry, Raman spectroscopy and ^{31}P NMR [44,45]. Unlike other metal-substituted POMs, vanadate ions could be substituted, along with a few molybdenum or tungsten units, into the framework of XM_{12}, which did not contain vacant sites, to form V(V)-substituted POMs. On the basis of this idea, mono-vanadium-substituted tungstosulfates 1- and 4-$[S_2VW_{17}O_{62}]^{5-}$ (S_2VW_{17}) were prepared by substituting vanadate for tungsten units in the framework of $[S_2W_{18}O_{62}]^{4-}$ (S_2W_{18}) [46].

In the present study, we investigated the formation of Wells–Dawson-type vanadium-substituted POMs, $[X_2VM_{17}O_{62}]^{7-}$ (X_2VM_{17}) (X = As, P; M = Mo, W) from the parent POMs, $[X_2M_{18}O_{62}]^{6-}$ (X_2M_{18}) using ^{31}P NMR and Raman spectroscopy and cyclic voltammetry.

2. Results and Discussion

2.1. Characterization of POMs Related to the Present Study

Solid-state Raman bands of all of the investigated POMs and the ^{31}P NMR signals of P_2W_{18}, P_2VW_{17}, P_2Mo_{18} and related phosphorus-containing POMs are listed in Tables 1 and 2, respectively. Characteristic bands and signals were used to assign peaks observed in the spectra of the products of the vanadium-substitution reaction in the solution phase.

Table 1. Raman band (cm^{-1}) of As_2VW_{17}, As_2W_{18}, P_2VW_{17}, P_2W_{18}, P_2Mo_{18} and As_2Mo_{18}.

Compounds	$\nu_{Raman}(M\text{-}O_d)$
As_2VW_{17}	982
As_2W_{18}	982 (995) [a]
P_2VW_{17}	985
P_2W_{18}	983 (993) [a]
P_2Mo_{18} [b]	971
As_2Mo_{18} [b]	971

O_a: oxygen bonded with arsenic atom; O_b: octahedral corner-sharing oxygen; O_c: octahedral edge-sharing oxygen; O_d: terminal oxygen; M: W or Mo; X: P or As. The counter cation is n-Bu_4N^+. [a,b] Raman band of $K_6As_2W_{18}O_{62}$ and $K_6P_2W_{18}O_{62}$ from [47,48], respectively.

Table 2. ^{31}P NMR signal of phosphorus-centered polyoxometalates (POMs).

Compounds	^{31}P NMR signals (vs 85% H_3PO_4)	Ref.
α-P_2W_{18}	−12.44	[49]
β-P_2W_{18}	−11.0, −11.7	[50]
1-P_2W_{17}	−6.79, −13.63	[49]
4-P_2W_{17}	−8.53, −12.86	[49]
P_2W_{15}	+0.1, −13.3	[50]
1-P_2VW_{17}	−10.84, −12.92	[49]
4-P_2VW_{17}	−11.83, −12.90	[49]
1,2,3-$P_2V_3W_{15}$	−6.25, −13.89	[49]
α-PW_{12}	−14.42	[45]
β-PW_{12}	−13.50	[45]
PW_{11}	−11.8 [a], −10.2 [b]	[41]
A-α-PW_9	−5.1	[41]
PVW_{11}	−14.11	[45]
P_2Mo_{18}	−2.45	[47]
α-PMo_{12}	−3.16	[47]
PMo_{11}	−0.79 [c], −1.05 [d]	[51]
A-PMo_9	−0.93	[47]
$PVMo_{11}$	−3.53	[44]

[a] Measured at pH 2.0; [b] measured at pH 4.0-7.5; [c] measured at pH 3.4; [d] measured at pH 2.5.

The major Raman lines of the As_2W_{18} and As_2VW_{17} complexes were observed at 995 and 982 cm^{-1}, respectively. The Raman band at approximately 990 cm^{-1} is due to the stretching mode of W-O_d and is affected by the cation in the solid state [52]. The Raman bands of the n-Bu_4N^+ salts of As_2W_{18} and P_2W_{18}, in which a small number of H^+ are included as counter cations, occur at lower wavenumbers than those of small cation salts, such as potassium. These Raman bands were used as reference signals to investigate the formation and conversion processes of the As_2W_{18} and As_2VW_{17} complexes in the reaction mixtures, while taking into consideration the effect of the proton.

Cyclic voltammograms of 5.0×10^{-4} M $X_2V_aW_{18-a}$ (X = P, As; a = 0, 1) were measured in 95% (v/v) CH_3CN containing 0.1 M $HClO_4$ (Figure A1). Details are provided in the Supporting Information. The voltammetric behaviors of X_2VW_{17} were used to determine whether the vanadium-substitution reaction occurred on the basis of the appearance of a new redox wave at a more positive potential than that of the corresponding parent non-substituted species.

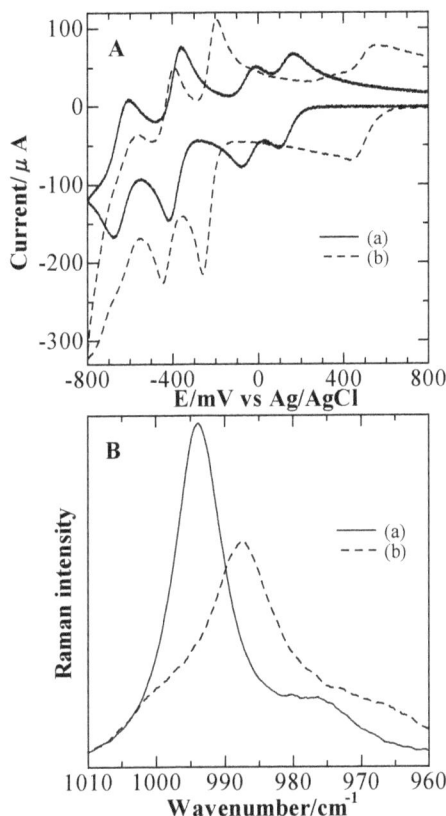

Figure 2. Cyclic voltammograms (**A**) and Raman spectra (**B**) of a 100 mM W(VI)–200 mM As(V)–10 mM V(V)–pH 2.0 system collected (a) after a solution without V(V) was heated at 80 °C for one week and (b) after 10 mM V(V) was added and the solution was heated again at 80 °C for one day.

2.2. Formation of Dawson-Type Vanadium-Substituted Tungstoarsenate and Tungstophosphate Complexes

Figures 2A(a) and 2B(a) show a cyclic voltammogram and a Raman spectrum, respectively, of a 100 mM W(VI)–200 mM As(V)–pH 2.5 system heated at 80 °C for one week. Four-step reduction waves were observed at 95, −83, −416 and −676 mV, which corresponded to one-, one-, two- and two-electron transfers, respectively. In addition, a Raman peak appeared at 994 cm^{-1}. These voltammetric waves and the Raman peak are due to the formation of the As_2W_{18}. After 10 mM V(V) was added, the pH was re-adjusted to 2.5 with conc. HCl and the solution was heated at 80 °C for one day; the original reduction waves and the Raman peak at 994 cm^{-1} disappeared, and new reduction waves at 437, −263 and −448 (Figure 2A(b)) and a new Raman band at 988 cm^{-1} appeared (Figure 2B(b)). The voltammetric and Raman spectral behaviors indicate that one of the tungsten units in the As$_2$W$_{18}$ was substituted with vanadate to form the As$_2$VW$_{17}$. The vanadium substitution of P$_2$W$_{18}$ leads to 1-P$_2$VW$_{17}$, whose vanadium is located at a polar site, as described below. In addition, the Wells–Dawson-type 1-S$_2$VW$_{17}$ was prepared by addition of V(V) to the reaction mixture, where the parent S$_2$W$_{18}$ is dominant [46]. The vanadium unit of the observed As$_2$VW$_{17}$ was located at the polar position, although Raman spectroscopy, cyclic voltammetry and other measurements provided no information on this perspective.

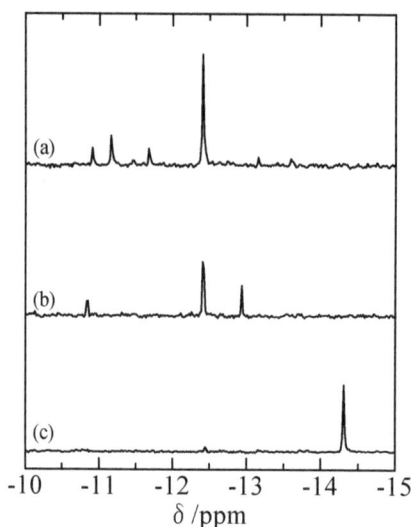

Figure 3. ^{31}P NMR spectrum of a 100 mM W(VI)–500 mM P(V)–V(V)–pH 2.0 system collected (**a**) after a solution without V(V) was heated at 80 °C for one week, (**b**) after 50 mM V(V) was added and the solution was heated again at 80 °C for one day and (**c**) after a solution of 100 mM W(VI)–500 mM P(V)–50 mM V(V)–pH 2 was heated at 80 °C for one week.

The formation of P$_2$VW$_{17}$ by the substitution reaction of P$_2$W$_{18}$ with V(V) was investigated using ^{31}P NMR. After a solution of 100 mM W(VI) and 500 mM P(V) was heated at pH 2 at 80 °C for one week, four main ^{31}P NMR peaks at −10.9, −11.2, −11.7 and −12.4 ppm were observed; these peaks were attributed to the formation of α-P$_2$W$_{18}$ (−12.4 ppm), β-P$_2$W$_{18}$ (−10.9, −11.7 ppm) and PW$_{11}$ (−11.2 ppm) (Figure 3a). Upon further heating at 80 °C for one day following the addition of 50 mM V(V) and adjustment of the pH to 2.0, the three peaks at −10.9, −11.2 and −11.7 ppm disappeared, whereas two new peaks appeared at −10.8 and −12.9 ppm; these peaks were ascribed to the formation of the 1-P$_2$VW$_{17}$. However, the peak at −12.4 ppm did not completely disappear despite the addition of 100 mM V(V). A similar result was obtained in the case of the substitution reaction of the Keggin-type XM$_{12}$ (X = P, As; M = Mo, W) with V(V). The PW$_{12}$ and PMo$_{12}$ were not fully converted into PV$_a$W$_{12-a}$ and PV$_a$Mo$_{12-a}$, respectively, in aqueous CH$_3$CN solution, although both

AsW$_{12}$ and AsMo$_{12}$ were fully converted into the corresponding vanadium-substituted species [44,45]. The relatively high stability of P$_2$W$_{18}$, compared to that of As$_2$W$_{18}$ would lead to incomplete conversion of the P$_2$W$_{18}$ into the P$_2$VW$_{17}$, similar to the conversion of the PM$_{12}$ (M = Mo, W) into the PVM$_{11}$.

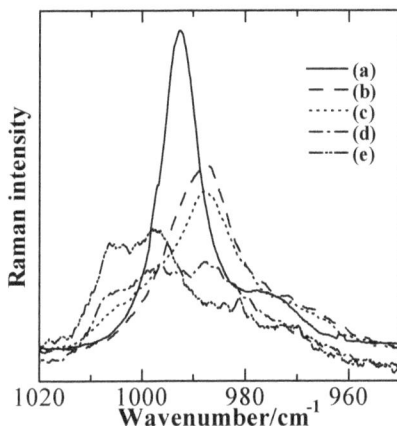

Figure 4. Raman spectra of a 100 mM W(VI)–200 mM As(V)–pH 2.0 system containing various concentrations of V(V) collected after each solution was heated at 80 °C for one week. [V(V)]/mM = (a) 0; (b) 5; (c) 10; (d) 15; and (e) 20.

To investigate the direct formation of X$_2$VW$_{17}$ (X = As, P), but not through X$_2$W$_{18}$ as an intermediate species, Raman spectra and ^{31}P NMR spectra were collected after a solution of W(VI)–V(V)-As(V) or –P(V) was heated at an appropriate pH. Figure 4 shows the Raman spectra collected after a solution of 100 mM W(VI), 200 mM As(V) and 0–20 mM V(V) were heated at pH 2 at 80 °C for one week. In the presence of 5–15 mM V(V), the Raman peak shifted from 993 cm^{-1} to 987 cm^{-1} when the concentration of V(V) was increased, thereby indicating the substitution of a tungsten unit in As$_2$W$_{18}$ with V(V) to form As$_2$VW$_{17}$. However, a broad, new Raman peak appeared at 1005 cm^{-1}, and its intensity increased depending on the concentration of V(V) (7–20 mM) (Figure 4, (d) and (e)). The new peak at approximately 1005 cm^{-1} was due to the formation of Keggin-type AsVW$_{11}$ [45]. Indeed, AsVW$_{11}$ was isolated along with As$_2$VW$_{17}$ when prepared in accordance with the literature [48], where the solution conditions used in the synthesis were similar to those used in our experiments, as previously described.

Figure 3c shows the ^{31}P NMR spectrum collected for a solution of 100 mM W(VI), 500 mM P(V) and 50 mM V(V) at pH 2 heated at 80 °C for one week. Only one peak (at –14.3 ppm) was observed that was attributable to the formation of the Keggin-type PVW$_{11}$. Even when the concentration of V(V) was increased or decreased, the two peaks at –10.8 and –12.9 ppm associated with the formation of P$_2$VW$_{17}$ were not observed. Notably, the occurrence of As$_2$W$_{18}$ or P$_2$W$_{18}$ in the reaction mixture appears to be essential for the formation of the corresponding V(V)-substituted species, As$_2$VW$_{17}$ and P$_2$VW$_{17}$.

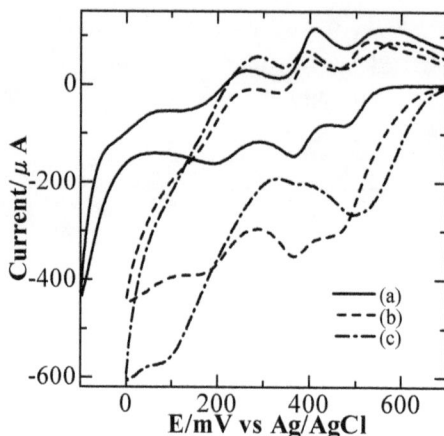

Figure 5. Cyclic voltammograms of a 100 mM Mo(VI)–10 mM As(V)–10 mM V(V)–0.5 M HCl system collected (a) after a solution without V(V) was heated at 80 °C for 7 h, (b) immediately after 10 mM V(V) was added to solution (a) and subsequently heated at 80 °C for 2 h and (c) after solution (b) was heated at 80 °C for one day.

2.3. Formation of Wells–Dawson-Type Vanadium-Substituted Molybdoarsenate and Molybdophosphate Complexes

The vanadium substitution reactions of P_2Mo_{18} and As_2Mo_{18} were also investigated using cyclic voltammetry and ^{31}P NMR spectroscopy. Three reduction peaks appeared at 477, 363 and 194 mV in the cyclic voltammogram collected after a solution of 100 mM Mo(VI), 10 mM As(V) and 0.5 M HCl was heated at 80 °C for 8 h; these peaks were due to the formation of As_2Mo_{18} (Figure 5(a)) [47]. After 10 mM V(V) was added and the solution was heated at 80 °C for 2 h, the original three peaks increased in magnitude of current, but the reduction potentials did not change (Figure 5(b)). Moreover, after the solution was heated at 80 °C for 15 h, a new reduction peak appeared at 500 mV in addition to a small peak at 375 mV, whereas the intensities of the original three reduction waves were diminished (Figure 5(c)). This new wave is ascribed to the formation of Keggin-type AsV_aMo_{12-a} [44]. In fact, only AsV_zMo_{12-z} ($z = 1,2$) could be isolated as a tetra-alkyl ammonium salt from this solution (see the Supporting Information). The occurrence of $AsVMo_{17}$ has not been confirmed, although we have extensively examined the solution under various conditions. In addition, we have investigated the direct formation of As_2VMo_{17} from a Mo(VI)–As(V)–V(V)–HCl system under various conditions. However, only Keggin-type vanadium-substituted molybdoarsenate was formed in the solution.

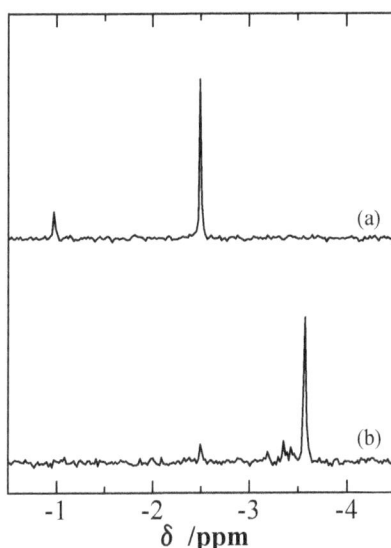

Figure 6. ^{31}P NMR spectra of a 100 mM Mo(VI)–100 mM P(V)–25 mM V(V)–1.0 M HCl system collected (**a**) after a solution without V(V) was heated at 80 °C for 30 h and (**b**) after V(V) was added and the solution was heated again at 80 °C for one day.

The substitution reaction of a molybdenum unit in P_2Mo_{18} with V(V) into P_2VMo_{17} was also extensively investigated. The ^{31}P NMR spectrum for a 100 mM Mo(VI)–100 mM P(V)–1.0 M HCl system showed two peaks at −0.97 and −2.49 ppm due to the formation of PMo_9 and P_2Mo_{18}, respectively (Figure 6a) [33,34,47]. After 10 mM V(V) was added and the solution was allowed to stand at room temperature for 1 h, new peaks appeared at −3.57 in addition several peaks at around −3.4 ppm, which are ascribed to the formation of mono-vanadium-substituted and multi-vanadium-substituted Keggin-type molybdophosphate, respectively. Meanwhile, the intensities of the peaks at −0.97 ppm and −2.49 ppm decreased in intensity, indicating that a small amount of P_2Mo_{18} remained in the solution. Although we have investigated the solutions under various conditions by changing the order of addition of reagents and the heating time, only Keggin-type vanadium-substituted molybdophosphates were formed. As with the Mo(VI)–As(V)–V(V) system, no P_2VMo_{17} appeared at any considerable concentration in the Mo(VI)–P(V)–V(V) system. For both the Mo(VI)–V(V)–As(V) and –P(V) systems, no X_2VMo_{17}(X = As, P) formed from X_2Mo_{18}. These results suggest that Keggin-type $XVMo_{11}$ should be substantially more stable than Wells–Dawson-type X_2VMo_{17} in the solution.

W(VI)-X(V)-V(V) systems

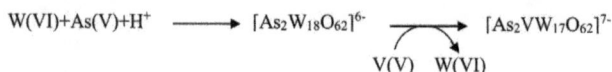

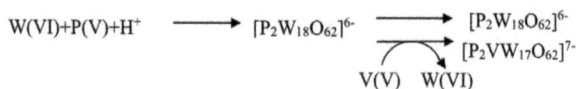

$W(VI)+As(V)+H^+ \longrightarrow [As_2W_{18}O_{62}]^{6-} \rightleftharpoons [As_2VW_{17}O_{62}]^{7-}$
$V(V)\quad W(VI)$

$W(VI)+As(V)+V(V)+H^+ \longrightarrow [As_2W_{18}O_{62}]^{6-}, [As_2VW_{17}O_{62}]^{7-}, [AsV_aW_{12-a}O_{40}]^{(3+a)-}$

$W(VI)+P(V)+H^+ \longrightarrow [P_2W_{18}O_{62}]^{6-} \rightleftharpoons \begin{matrix}[P_2W_{18}O_{62}]^{6-}\\ [P_2VW_{17}O_{62}]^{7-}\end{matrix}$
$V(V)\quad W(VI)$

$W(VI)+P(V)+V(V)+H^+ \longrightarrow [PV_aW_{12-a}O_{40}]^{(3+a)-}$

Mo(VI)-X(V)-V(V) systems

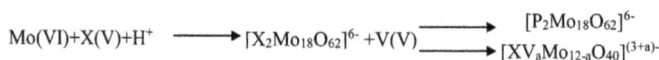

$Mo(VI)+X(V)+H^+ \longrightarrow [X_2Mo_{18}O_{62}]^{6-} +V(V) \begin{matrix}\longrightarrow [P_2Mo_{18}O_{62}]^{6-}\\ \longrightarrow [XV_aMo_{12-a}O_{40}]^{(3+a)-}\end{matrix}$

$Mo(VI)+X(V)+V(V)+H^+ \longrightarrow [XV_aMo_{12-a}O_{40}]^{(3+a)-}$

Scheme 1. Vanadium substitution reaction of Wells–Dawson-type POMs for the M(VI)–X(V)–V(V) (M = W, Mo; X = P, As) systems in aqueous solution.

3. Experimental Section

Voltammetric measurements were performed using a microcomputer-controlled system. A glassy carbon (GC) electrode (BAS GC-30S) with a surface area of 0.071 cm^2 was used as the working electrode, and a platinum wire served as the counter electrode. The reference electrodes were Ag/AgCl (saturated KCl) for aqueous solutions and Ag/Ag$^+$ (0.01 M (M = mol/dm^3) AgNO$_3$ in acetonitrile) for acetonitrile solutions. Prior to each measurement, the GC electrode was polished with 0.1-μm diamond slurry and washed with distilled water. The voltammograms were recorded at 25 ± 0.1 °C. Unless otherwise noted, the voltage scan rate was 100 mV/s. Raman spectra were recorded on a Horiba Jobin Yvon model HR-800 spectrophotometer. The argon line at 514.5 nm was used for excitation. The Raman measurements were conducted at 20 °C. For quantitative measurements, the Raman intensities were normalized by the 1048-cm^{-1} band associated with NO$_3^-$; 0.1 M NaNO$_3$ was added and used as an internal standard. The ^{31}P NMR measurements were performed on a JEOL model JNM-LA400 spectrometer at 161.70 MHz. An inner tube containing D$_2$O was used as an instrumental lock. Chemical shifts were referenced to 85% H$_3$PO$_4$. The preparation procedures of the n-Bu$_4$N$^+$ salts of the POMs used in the present study are described in the Appendix.

4. Conclusions

In the present study, the V(V)-substitution reactions of Wells–Dawson-type POMs were investigated in aqueous solution using ^{31}P NMR and Raman spectroscopy and cyclic voltammetry. The elucidated vanadium substitution reaction processes are summarized in Scheme 1. The addition of V(V) ions to a solution containing As$_2$W$_{18}$ or P$_2$W$_{18}$ led to the corresponding As$_2$VW$_{17}$ or 1-P$_2$VW$_{17}$, respectively, although not all P$_2$W$_{18}$ in the solution was completely transformed into

1-P_2VW_{17}, even after a large excess of V(V) was added. These results suggest that P_2W_{18} is more stable than As_2W_{18}. As_2VW_{17} and 1-P_2VW_{17} can be prepared directly from As_2W_{18} and P_2W_{18}, respectively, via the addition of V(V), a synthetic procedure that is simpler than those previously reported [48–51]. This simplified procedure enables the preparation of a large amount of X_2VW_{17} at industrial levels as oxidant catalysts at low cost. Interestingly, tungsten in the one-position of X_2W_{18} could be substituted with V(V) to form X_2W_{17}, while X_2W_{18} decomposed with a weak base to form 4-X_2W_{17}. The order of addition of the reagents to the reaction mixture was very important. Even when the concentrations of W(VI), V(V), H^+ and P(V) or As(V) were the same, mixing and then heating the reagents led to the Keggin-type XVW_{11}, although As_2VW_{17} was observed as a mixture in the case of the As(V) system. This result implies that the occurrence of X_2W_{18} in the reaction mixtures was essential for the formation of the corresponding V(V)-substituted species, X_2VW_{17}. In contrast, only $XVMo_{11}$ (X = As, P) formed, rather than X_2VMo_{17}, in the Mo(VI)–V(V)–P(V) and – As(V) systems, even when various concentrations of V(V), P(V), As(V) and acid were used and the order of the addition of the reagents was varied. The results suggested that the formation constants of $XVMo_{11}$ are likely much greater than those of X_2VMo_{17}. To the best of our knowledge, the literature contains few reports of the synthesis of Wells–Dawson-type vanadium-substituted phosphorus- and arsenic-centered heteropolymolybdates, X_2VMo_{17} (X = P, As). According to the results of the present study, X_2VMo_{17} cannot be synthesized using simple procedures.

Acknowledgements

This work was supported by a Grant-in-aid for Scientific Research (No. 25410095) from the Ministry of Education, Culture, Sports, Science and Technology of Japan, a Special Research Grant for Rare Metals and Green Technology and a Kochi University President's Discretionary Grant.

Author Contributions

T.U.: Writing the manuscript and discussion of the results; Y.N. and R.S.: voltammetry and Raman spectroscopy measurements; M.O. and J.N.: Synthesis of polyoxometalates and ^{31}P NMR spectroscopy measurement.

Conflicts of Interest

The authors declare no conflict of interest.

Appendix

A1. Cyclic Voltammograms of $X_2V_aW_{18-a}$ (X = P, As; a = 0, 1)

Figure A1 shows the cyclic voltammograms of 5.0×10^{-4} M $X_2V_aW_{18-a}$ (X = P, As; a = 0, 1) in 95% (v/v) CH_3CN containing 0.1 M $HClO_4$. The X_2W_{18} complexes exhibit two-step redox waves with midpoint-potentials (E_{mid}) of -440 and -610 mV vs. Fc/Fc^+ for P_2W_{18} and -400 and -550 mV for As_2W_{18}, with a current ratio of 1:1, where $E_{mid} = (E_{pc} + E_{pa})/2$, where E_{pc} is the cathodic peak potential and E_{pa} is the anodic peak potential. Each wave was diffusion-controlled. Coulometric analysis and normal-pulse voltammetric measurements confirmed that the two-step reduction waves corresponded to two two-electron transfers. In the case of cyclic voltammetry performed in aqueous acid solution (pH = 1.0: [HCl] = 0.1 M and [NaCl] = 0.9 M), the first two-electron transfer redox wave observed in Fig. 2 was split into two one-electron transfer waves at +0.04 and −0.14 V $vs.$ SCE for P_2W_{18}, and +0.08 and -0.10 V for As_2W_{18} [53,54]. Because the redox potential of the vanadium components of vanadium-substituted POMs XVM_{11} and S_2VW_{17} was more positive than those of the molybdenum or tungsten components in these complexes, the reduction waves at approximately 200 mV should be ascribed to the reduction of V(V) to V(IV) in the X_2VW_{17} complexes [46,55–58]. Each of the E_{mid} for PM_{12} (M = Mo, W) occurred at a more negative potential than that of the corresponding AsM_{12}. Similarly, each E_{mid} for P_2W_{18} and P_2VW_{17} also occurred at a more negative potential than that of the corresponding As_2W_{18} and As_2VW_{17}, although we could not elucidate the origin of the oxidation wave at 0 mV in the voltammograms of As_2VW_{17} and P_2VW_{17}. No oxidation wave was observed when the switch potential was set at a more positive potential than the tungsten reduction potential. On the basis of these voltammetry results, the occurrence of the vanadium substitution reaction can be determined from the appearance of a new peak at more positive potentials than the redox potentials of the Mo or W component in the POMs after the addition of V(V) to the reaction mixtures.

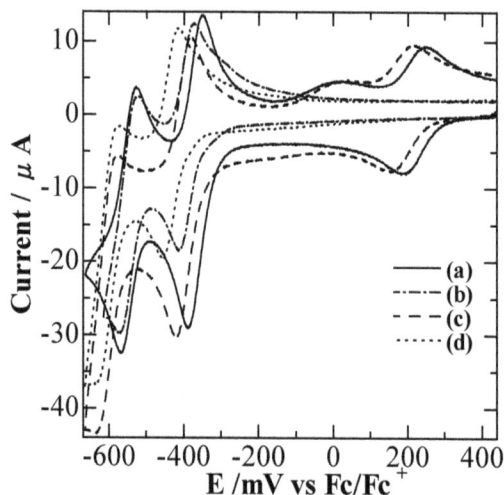

Figure A1. Cyclic voltammograms of (a) $[As_2VW_{17}O_{62}]^{7-}$, (b) $[As_2W_{18}O_{62}]^{6-}$, (c) $[P_2VW_{17}O_{62}]^{7-}$, and (d) $[P_2W_{18}O_{62}]^{6-}$ in 95% (v/v) CH_3CN containing 0.1 M HClO_4.

A2. Identification of POMs Isolated as a *tetra*-Butylammonium Salt from a 100 mM Mo(VI)–10 mM As(V)–0.5 M HCl System after the Addition of 5–20 mM V(V)

POMs were isolated as a tetra-butylammonium salt from a 100 mM Mo(VI)–10 mM As(V)–0.5 M HCl–5–20 mM V(V) system after measuring cyclic voltammograms as shown in Figure 5. The IR spectra of the isolated salts were measured using a Jasco 460-plus IR spectrometer; IR bands were observed at 950, 890, 847, and 786 cm^{-1}. These bands were compared with the reported data to identify the mixture of n-Bu$_4$N$^+$ salts of [AsVMo$_{11}$O$_{40}$]$^{4-}$ and [AsV$_2$Mo$_{10}$O$_{40}$]$^{5-}$ [44].

References

1. Pope, M.T. *Heteropoly and Isopoly Oxometalates*; Springer: Berlin, Gemany, 1983.
2. Hill, C.L. Introduction: Polyoxometalates s Multicomponent Molecular Vehicles To Probe Fundamental Issues and Practical Problems. *Chem. Rev.* **1998**, *98*, 1-390.
3. Pope, M.T.; Müller, A. *Polyoxometalate Chemistry From Topology via Self-Assembly to Applications*; Kluwer Academic Publishers: New York, NY, USA, 2001.
4. Yamase, T.; Pope, M.T. Eds., *Polyoxometalate Chemistry for Nano-Composite Design*, Kluwer Academic/Plenum Publishers: New York, NY, USA, 2002.
5. Borrás-Almenar, J.J.; Coronado, E.; Müller, A.; Pope, M.T. *Polyoxometalate Molecular Science*; Kluwer Academic Publishers: New York, NY, USA, 2003.
6. Mialane, P.; Dolbecq, A.; Sécheresse, F. Functionalization of polyoxometalates by carboxylato and azido ligands: macromolecular complexes and extended compounds. *Chem. Commun.* **2006**, *33*, 3477–3485.
7. Long, D.-L.; Burkholder, E.; Cronin, L. Polyoxometalate clusters, nanostructures and materials: From self assembly to designer materials and devices. *Chem. Soc. Rev.* **2007**, *36*, 105–121.
8. Proust, A.; Thouvenot, R.; Gouzerh, P. Functionalization of polyoxometalates: towards advanced applications in catalysis and materials science, *Chem. Commun.* **2008**, 1837–1852
9. Ueda, T.; Kotsuki, H. Heteropoly acids: green chemical catalysts in organic synthesis, *Heterocycle* **2008**, *76*, 73–97.
10. Huang, Z.; Luo, Z.; Geletii, Y.V.; Vickers, J.W.; Yin, Q.; Wu, D.; Hou, Y.; Ding, Y.; Song, J.; Musaev, D.G.; Hill, C.L.; Lian, T. Efficient light-driven carbon-free cobalt-based molecular catalyst for water oxidation, *J. Am. Chem. Soc.* **2011**, *133*, 2068–2071.
11. Car, P.-E.; Guttentag, M.; Baldridge, K.K.; Alberto, R.; Patzke, G.R. Synthesis and characterization of open and sandwich-type polyoxometalates reveals visible-light-driven water oxidation via POM-photosensitizer complexes, *Green Chem.* **2012**, *14*, 1680–1688.
12. Natali, M.; Orlandi, M.; Berardi, S.; Campagna, S.; Bonchio, M.; Sartorel, A.; Scandola, F. Photoinduced Water Oxidation by a Tetraruthenium Polyoxometalate Catalyst: Ion-pairing and Primary Processes with Ru(bpy)$_3$$^{2+}$ Photosensitizer. *Inorg. Chem.* **2012**, *51*, 7324–7331;
13. Guo, S.-X.; Liu, Y.; Lee, C.-Y.; Bond, A.M.; Zhang, J.; Geletii, Y.V.; Hill, C.L. Graphene-supported [{Ru$_4$O$_4$(OH)$_2$(H$_2$O)$_4$}(γ-SiW$_{10}$O$_{36}$)$_2$]$^{10-}$ for highly efficient electrocatalytic water oxidation, *Energy Environ. Sci.* **2013**, *6*, 2654–2663.
14. Liu, Y.; Guo, S.-X.; Bond, A.M.; Zhang, J.; Geletii, Y.V.; Hill, C.L. Voltammetric Determination of the Reversible Potentials for [{Ru$_4$O$_4$(OH)$_2$(H$_2$O)$_4$}(γ-SiW$_{10}$O$_{36}$)$_2$]$^{10-}$ over the pH Range of 2-12: Electrolyte Dependence and Implications for Water Oxidation Catalysis, *Inorg Chem* **2013**, *52*, 11986–11996.

15. Anwar, N.; Sartorel, A.; Yaqub, M.; Wearen, K.; Laffir, F.; Armstrong, G.; Dickinson, C.; Bonchio, M.; McCormac, T. Surface immobilization of a tetra-ruthenium substituted polyoxometalate water oxidation catalyst through the employment of conducting polypyrrole and the layer-by-layer (LBL) technique, *ACS Appl. Mater. Interfaces* **2014**, *6*, 8022–8031.

16. Hill, C.L. POMs in Catalysis. *J. Mol. Catal. A* **2007**, 262 1-242.

17. Kholdeeva, O.A.; Maksimchuk N.V.; Maksimov, G.M. Polyoxometalate-based heterogeneous catalysts for liquid phase selective oxidations: Comparison of different strategies. *Catal. Today* **2010**, *157*, 107–113.

18. Mizuno, N.; Kamata K.; Yamaguchi, K. Green Oxidation Reactions by Polyoxometalate-Based Catalysts: From Molecular to Solid Catalysts. *Top. Catal.* **2010**, *53*, 876–893.

19. Neumann, R. Activation of Molecular Oxygen, Polyoxometalates, and Liquid-Phase Catalytic Oxidation, *Inorg. Chem.* **2010**, *49*, 3594–3601.

20. Han,; Z.G. Bond A.M.; Zhao, C. Recent trends in the use of polyoxometalate-based material for efficient water oxidation. *Sci. China Chem.* **2011**, *54*, 1877–1887.

21. Maksimchuk, N.V.; Kholdeeva, O.A.; Kovalenko, K.A.; Fedin, V.P. MIL-101 Supported Polyoxometalates: Synthesis, Characterization, and Catalytic Applications in Selective Liquid-Phase Oxidation. *Isr. J. Chem.* **2011**, *51*, 281–289.

22. Mizuno N.; Kamata, K. Catalytic oxidation of hydrocarbons with hydrogen peroxide by vanadium-based polyoxometalates. *Coord. Chem. Rev.* **2011**, *255*, 2358–2370.

23. Deng, W.; Zhang Q.; Wang, Y. Polyoxometalates as efficient catalysts for transformations of cellulose into platform chemicals. *Dalton Trans.* **2012**, *41*, 9817–9831.

24. Lv, H.; Geletii, Y.V.; Zhao, C.; Vickers, J.W.; Zhu, G.; Luo, Z.; Song, J.; Lian, T.; Musaev D.G.; Hill, C.L. Polyoxometalate water oxidation catalysts and the production of green fuel. *Chem. Soc. Rev.* **2011**, *41*, 7572–7589.

25. Stracke, J.J.; Finke, R.G. Distinguishing Homogeneous from Heterogeneous Water Oxidation Catalysis when Beginning with Polyoxometalates, *ACS Catal.* **2014**, *4*, 909–933.

26. Sumliner, J.M.; Lv, H.; Fielden, J.; Geletii, Y.V.; Hill, C.L. Polyoxometalate Multi-Electron-Transfer Catalytic Systems for Water Splitting, *Eur. J. Inorg. Chem.* **2014**, 635–644.

27. Himeno, S.; Saito, A. Preparation of dodecamolybdovanadate(V). *Inorg. Chim. Acta* **1990**, *171*, 135–137.

28. Himeno, S.; Takamoto, M.; Higuchi, A.; Maekawa, M. Preparation and voltammetric characterization of Keggin-type tungstovanadate $[VW_{12}O_{40}]^{3-}$ and $[V(VW_{11})O_{40}]^{4-}$ complexes. *Inorg. Chim. Acta* **2003**, *348*, 57–62.

29. Himeno, S.; Kawasaki, K.; Hashimoto, M. Preparation and characterization of an α-Wells-Dawson-type $[V_2Mo_{18}O_{62}]^{6-}$ complex. *Bull. Chem. Soc. Jpn.* **2008**, *81*, 1465–1471.

30. Miras, H.N.; Ochoa, M.N. C.; Long, D.-L.; Cronin, L. Controlling transformations in the assembly of polyoxometalate clusters: $\{Mo_{11}V_7\}$, $\{Mo_{17}V_8\}$ and $\{Mo_{72}V_{30}\}$, *Chem. Commun.* **2010**, *46*, 8148–8150.

31. Miras, H.N.; Stone, D.; Long, D.-L.; McInnes, E.J. L.; Kogerler, P.; Cronin, L. Exploring the Structure and Properties of Transition Metal Templated $\{VM_{17}(VO_4)_2\}$ Dawson-Like Capsules, *Inorg. Chem.* **2011**, *50*, 8384–8391.

32. Miras, H.N.; Sorus, M.; Hawkett, J.; Sells, D.O.; McInnes, E.J. L.; Cronin, L. Oscillatory Template Exchange in Polyoxometalate Capsules: A Ligand-Triggered, Redox-Powered, Chemically Damped Oscillation. *J. Am. Chem. Soc.* **2012**, *134*, 6980–6983.

33. Pettersson, L.; Andersson, I.; Oehman, L.O. Multicomponent polyanions. 35. A phosphorus-31 NMR study of aqueous molybdophosphates. *Acta Chem. Scand., Ser. A* **1985**, *A39*, 53–58.

34. Pettersson, L.; Andersson, I.; Oehman, L.O. Multicomponent polyanions. 39. Speciation in the aqueous hydrogen ion-molybdate(MoO_4^{2-})-hydrogenphosphate(HPO_4^{2-}) system as deduced from a combined Emf-phosphorus-31 NMR study. *Inorg. Chem.* **1986**, *25*, 4726–4733.

35. Pettersson, L.; Andersson, I.; Grate, J.H.; Selling, A. Multicomponent Polyanions. 46. Characterization of the Isomeric Keggin Decamolybdodivanadophosphate Ions in Aqueous Solution by ^{31}P and ^{51}V NMR. *Inorg. Chem.* **1994**, *33*, 982–993.

36. Katano, H.; Osakai, T.; Himeno, S.; Saito, A. A kinetic study of the formation of 12-molybdosilicate and 12-molybdogermanate in aqueous solutions by ion transfer voltammetry with the nitrobenzene-water interface. *Electrochim. Acta* **1995**, *40*, 2935–2942.

37. Ueda, T.; Sano, K.; Himeno, S.; Hori, T. Formation and conversion of yellow molybdophosphonate complexes in aqueous-organic media. *Bull. Chem. Soc. Jpn.* **1997**, *70*, 1093–1099.

38. Himeno, S.; Ueda, T.; Shiomi, M.; Hori, T. Raman studies on the formation of 12-molybdopyrophosphate. *Inorg. Chim. Acta* **1997**, *262*, 219–223.

39. Himeno, S.; Sano, K.; Niiya, H.; Yamazaki, Y.; Ueda, T.; Hori, T. Formation and conversion of yellow heteropoly complexes in a Mo(VI)-Se(IV), Te(IV) system, *Inorg. Chim. Acta* **1998**, *281*, 214–220.

40. Selling, A.; Andersson, I.; Grate, J.H.; Pettersson, L. Multicomponent polyanions. 49. A potentiometric and (^{31}P, ^{51}V) NMR study of the aqueous molybdovanadophosphate system. *Eur. J. Inorg. Chem.* **2000**, 1509–1521.

41. Himeno, S.; Takamoto, M.; Ueda, T. Formation of α- and β-Keggin-Type $[PW_{12}O_{40}]^{3-}$ Complexes in Aqueous Media, *Bull. Chem. Soc. Jpn.* **2005**, *78*, 1463–1468.

42. Hashimoto, M.; Andersson, I.; Pettersson, L. An equilibrium analysis of the aqueous H^+-MoO_4^{2-}-$(HP)O_3^{2-}$ and H^+-MoO_4^{2-}-$(HP)O_3^{2-}$-HPO_4^{2-} systems. *Dalton Trans.* **2007**, 124–132.

43. Hashimoto, M.; Andersson, I.; Pettersson, L. A ^{31}P-NMR study of the H^+-$MoO4^{2-}$-$(HP)O_3^{2-}$-HPO_4^{2-}-$(C_6H_5P)O_3^{2-}$-$(CH_3P)O_3^{2-}$ system at low Mo_{tot}/P_{tot} ratio - Formation of mixed-hetero X_2M_5-type polyanions. *Dalton Trans.* **2009**, 3321–3327.

44. Ueda, T.; Wada, K.; Hojo, M. Voltammetric and Raman spectroscopic study on the formation of Keggin-type V(V)-substituted molybdoarsenate complexes in aqueous and aqueous-organic solution, *Polyhedron* **2001**, *20*, 83–89.

45. Ueda, T.; Nambu, J.; Yokota, H.; Hojo, M. The effect of water-miscible organic solvents on the substitution reaction of Keggin-type heteropolysilicates and -germanates with vanadium(V) ion. *Polyhedron* **2009**, *28*, 43–48.

46. Ueda, T.; Ohnishi, M.; Shiro, M.; Nambu, J.-i.; Yonemura, T.; Boas, J.F.; Bond, A.M. Synthesis and Characterization of Novel Wells-Dawson-Type Mono Vanadium(V)-Substituted Tungsto-polyoxometalate Isomers: 1- and 4-$[S_2VW_{17}O_{62}]^{5-}$. *Inorg. Chem.* **2014**, *53*, 4891–4898.

47. Himeno, S.; Hashimoto, M.; Ueda, T. Formation and conversion of molybdophosphate and -arsenate complexes in aqueous solution. *Inorg. Chim. Acta* **1999**, *284*, 237–245

48. Roy, S.K.; Day, K.C. Synthesis and characterization of sodium, potassium, guanidinium and tetramethylammonium salts of 17-tungstovanadodiarsenate(V). *Ind. J. Chem.* **1992**, 31A, 64–66.

49. Contant, R.; Abbessi, M.; Thouvenot, R.; Herve, G. Dawson Type Heteropolyanions. 3. Syntheses and ^{31}P, ^{51}V, and ^{183}W NMR Structural Investigation of Octadeca(molybdotungstovanado)diphosphates Related to the $[H_2P_2W_{12}O_{48}]^{12-}$ Anion. *Inorg. Chem.* **2004**, *43*, 3597–3604.

50. Contant, R.; Klemperer, W. G.; Yaghi, O. Potassium Octadecatungstodiphosphates(V) and Related Lacunary Compounds. *Inorg. Synth.* **1990**, *27*, 104–111.

51. Rob van Veen, J.A.; Sudmeijer, O.; Emeis, C.A.; Wit, H. On the identification of molybdophosphate complexes in aqueous solution. *Dalton Trans.* **1986**, 1825–1831.

52. Rocchiccioli-Deltcheff, C.; Fournier, M.; Franck, R.; Thouvenot, R. Vibrational investigations of polyoxometalates. 2. Evidence for anion-anion interactions in molybdenum(VI) and tungsten(VI) compounds related to the Keggin structure. *Inorg. Chem.* **1983**, *22*, 207–216.

53. Contant, R.; Thouvenot, R. Hétéropolyanions de type Dawson. 2. Synthèses de polyoxotungstoarsénates lacunaires dérivant de l'octadécatungstodiarsénate. Étude structurale par RMN du tungstène-183 des octadéca(molybdotungstovanado)diarsénates apparentés. *Can. J. Chem.* **1991**, *69*, 1498–1506 (In French).

54. Abbessi, M.; Contant, R.; Thouvenot, R.; Herve, G. Dawson type heteropolyanions. 1. Multinuclear (phosphorus-31, vanadium-51, tungsten-183) NMR structural investigations of octadeca(molybdotungstovanado)diphosphates .alpha.-1,2,3-$[P_2MM'_2W_{15}O_{62}]^{n-}$ (M, M' = Mo, V, W): syntheses of new related compounds. *Inorg. Chem.* **1991**, *30*, 1695.

55. Pope, M.T. Heteropoly and isopoly anions as oxo complexes and their reducibility to mixed-valence blues. *Inorg. Chem.* **1972**, *11*, 1973–1974.

56. Smith, D.P.; Pope, M.T. Heteropoly 12-metallophosphates containing tungsten and vanadium. Preparation, voltammetry, and properties of mono-, di-, tetra-, and hexavanado complexes. *Inorg. Chem.* **1973**, *12*, 331–336.

57. Altenau, J.J.; Pope, M.T.; Prados, R.A.; So, H. Models for heteropoly blues. Degrees of valence trapping in vanadium(IV)- and molybdenum(V)-substituted Keggin anions. *Inorg. Chem.* **1975**, *14*, 417–421.

58. Himeno, S.; Osakai, T.; Saito, A. Preparation and Properties of Heteropoly Molybdovanadate(V) Complexes. *Bull. Chem. Soc. Jpn.* **1991**, *64*, 21–28.

Water Oxidation by Ru-Polyoxometalate Catalysts: Overpotential Dependency on the Number and Charge of the Metal Centers

Simone Piccinin and Stefano Fabris

Abstract: Water oxidation is efficiently catalyzed by several Ru-based polyoxometalate (POM) molecular catalysts differing in the number, local atomistic environment and oxidation state of the Ru sites. We employ density functional theory calculations to rationalize the dependency of the reaction overpotential on the main structural and electronic molecular properties. In particular, we compare the thermodynamics of the water oxidation cycle for single-site Ru-POM and multiple-site Ru_4-POM complexes. For the Ru-POM case, we also investigate the reaction free energy as a function of the Ru oxidation state. We find that the overpotential of these molecular catalysts is primarily determined by the oxidation state of the metal center and is minimum for Ru(IV). In solution, the number of active sites is shown to play a minor role on the reaction energetics. The results are rationalized and discussed in terms of the local structure around the active sites and of the electrostatic screening due to the molecular structure or the solvent.

Reprinted from *Inorganics*. Cite as: Piccinin, S.; Fabris, S. Water Oxidation by Ru-Polyoxometalate Catalysts: Overpotential Dependency on the Number and Charge of the Metal Centers. *Inorganics* **2015**, *3*, 374–387.

1. Introduction

Replacing fossil fuels with renewable energy sources, like solar and wind, requires developing a strategy to cope with the intrinsic variability of these sources. To this end, Nature employs photosynthesis, which enables plants to convert sunlight into chemical energy through electrochemical reactions where H_2O and CO_2 are the reactants, while sugars and O_2 are the products. Mimicking this process with artificial devices would allow storing solar energy in the form of chemical fuels, which can be used at a later time to generate heat or electricity through combustion or fuel cells. Developing these technologies with high-enough efficiency is one of the current grand challenges in physical sciences [1,2].

One of the main bottlenecks for artificial solar fuel production is the lack of efficient, stable and inexpensive catalytic materials for the anodic reaction, the oxidation of water:

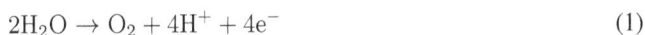

$$2H_2O \rightarrow O_2 + 4H^+ + 4e^- \tag{1}$$

The oxidation potential of this four-electron semi-reaction is 1.23 V, measured at pH = 0 against the normal hydrogen electron (NHE). This high oxidation potential reflects the high stability of water:

$$\Delta G^0(2H_2O \rightarrow O_2 + 4H_2) = 4.92 \text{ eV} \tag{2}$$

Using sunlight to promote the electrochemical splitting of water would therefore afford storing solar energy in the rearrangement of the chemical bonds among hydrogen and oxygen atoms, producing molecular H_2. Using the electrons released to the anode through the oxidation of water to reduce CO_2, rather than protons, would lead to the production of more manageable fuels, like, for example, methanol. Regardless of the fuel produced at the cathode, the source of electrons and protons at the cathode will be water, and suitable catalysts are needed to facilitate its oxidation.

RuO_2 and IrO_2 catalysts, although efficient, suffer from poor stability in non-alkaline conditions and are based on rare (and hence, expensive) elements, preventing their application on the massive scale required to replace fossil fuels. Third-row transition metals, such as Ni, Co, Fe and Mn, on the other hand, are much more abundant, and therefore, oxides of these elements are very attractive for the fabrication of anodic materials. The oxygen evolving complex in photosystem II, the catalyst used by green plants and some bacteria for oxidizing water, is indeed based on Earth-abundant elements, like Mn and Ca. It contains a Mn_4CaO_5 core with a distorted cubane structure [3], and recent experiments aimed at utilizing third-row elements within a similar cubane motif have achieved impressive results [4].

Molecular catalysts for water oxidation have also been intensively investigated [5], especially since the discovery of the "blue dimer" molecule (cis,cis-[(bpy)$^{2-}$ $(H_2O)Ru^{II}ORu^{II}(H_2O)(bpy)_2]^{4+}$) [6]. While in heterogeneous catalysts, it is usually difficult to identify the catalytically-active species, homogeneous catalysts are better suited to mechanistic studies [7]. Measuring the catalyst performance on a per-atom basis, molecular catalysts are in general more efficient than metal oxides, but they suffer from poor stability [8], since the organic ligands are quickly oxidized during the catalytic cycle.

Polyoxometalates (POMs) represent a special class of molecular catalysts, since they are fully-inorganic compounds that can serve as scaffolds to host different kinds of transition metal atoms or as ligands that stabilize multi-center metal-oxo cores. Recent experiments have shown that POMs with a metal-oxo core containing four transition metals atoms, like Ru [9,10], Co [11] or Mn [12], can be very efficient water oxidation catalysts. There is an obvious structural analogy between the metal-oxo core of these molecules, containing four transition metal atoms, and the Mn_4CaO_5 core of the natural catalyst for water oxidation [3], also containing four transition metal centers.

The discovery that single-site catalysts can also be efficient catalysts for water oxidation [13] has led to significant progress in identifying the reaction mechanism, mostly due to the structural simplicity of their active site [14]. Single-site POMs have also been shown to be active in water oxidation, in particular a Ru-based POMs with the Keggin structure [Ru^{III}(H$_2$O)XW$_{11}$O$_{39}$]$^{5-}$ (X = Si, Ge) [15]. The mechanism of water oxidation promoted by this single-site catalysis has been proposed to involve a nucleophilic attack of a water molecule on a Ru-oxo intermediate [15,16], in line with with what we have recently proposed for the four-atom analog Ru$_4$-POM [17] and for a variety of Ru-based molecular catalysts [18].

It is suggested that both the single-site and multiple-site POMs can promote water oxidation with a common reaction mechanism and that they have a very similar local structural environment around the Ru site. In this work, we focus on the following fundamental questions: How does the water oxidation thermodynamics and overpotential depend on the number of active sites in these molecular catalysts? How does the reaction thermodynamics depend on the charge state of the metal centers? We provide insight into these fundamental issues by means of density functional theory (DFT) calculations. Our DFT simulations investigate the catalytic mechanism of water oxidation promoted by the single-site $[Ru^{III}(H_2O)SiW_{11}O_{39}]^{5-}$ (Ru-POM). The effects of the Ru oxidation state are investigated by studying analogous Ru-POM systems in which the Ru center has higher oxidation states, namely Ru(IV) and Ru(V). To assess the dependence of the reaction thermodynamics on the number of active sites in the molecular catalyst, we compare the results for the Ru-POM against those obtained for the Ru_4-POM $[Ru_4(\mu\text{-}O)_4(\mu\text{-}OH)_2(H_2O)_4(\gamma\text{-}SiW_{10}O_{36})]^{10-}$ that we investigated in our previous works [17,19].

2. Computational Methods

2.1. Electronic Structure Calculations

The calculations are based on the spin-polarized DFT, and the exchange and correlation potential is approximated using either the Perdew–Burke–Ernzerhof (PBE) generalized gradient-corrected approximation [20] or the B3LYP hybrid functional [21,22]. This choice of exchange and correlation functionals is motivated by recent work [23] comparing the accuracy of DFT calculations using a variety of exchange and correlation functionals against highly accurate CCSD(T)calculations for water oxidation on a single-site Ru catalyst. This work has shown that hybrid functionals, such as B3LYP, PBE0 [24] and M06 [24], are the best-performing DFT methods, resulting in errors for the reaction energies and barriers for the rate-determining step of 1–2 kcal/mol. A similar conclusion was also recently drawn by computing the thermodynamics of water oxidation on Co-based catalysts [25].

The geometry optimizations of the single-center Ru-POM are performed with the LANL2DZ basis set, using the Los Alamos relativistic core potentials (EPCs) [26] for Ru and W atoms. The total energies are then computed with single-point (*i.e.*, fixed geometry) calculations using the Stuttgart–Dresden (SDD) basis set and EPC [27,28]. Solvation effects are included with the SMDsolvation model [29] through single-point calculations on the optimized geometries. All of the above calculations were performed with the Gaussian 09 software [30].

The comparison of the thermodynamics for the single-center Ru-POM presented in this work with the results for the multiple-center Ru_4-POM catalysts described in our previous work [17,19] requires using the same level of theory for the two systems. The available energetics for water oxidation promoted by Ru_4-POM was obtained with the CP2Kcode [31]. To benchmark the reproducibility of the results when changing the code, we have therefore calculated the energetics of the catalytic cycle promoted by Ru-POM also with CP2K, employing the same computational parameters of [19]. This benchmark is reported in Appendix A. In the CP2K calculations, the Ru and W atoms are modeled using the DZVP-SR-MOLOPTbasis set, while for H and O atoms, we used the the TZV2P-MOLOPT basis set. The cutoff for the plane wave representation of the charge density was set to 350 Ry. The molecules were simulated in vacuum, in cubic boxes with a side of 24 Å, using the Martyna–Tuckerman method to decouple the periodic replicas of the system. More details regarding this computational approach can be found in our previous work [19].

To compute the effects of solvation in the Ru_4-POM molecule, we used the COSMO (conductor-like screening model) method [32], as implemented in the NWChemcode [33]. Due to the high computational cost of modeling this large molecule, solvation effects were in this case evaluated with a single-point calculation using the PBE functional and the TZVPbasis set. The solvation energy was then used to correct the B3LYP results obtained in a vacuum. Since the reactions we model in this work involve the addition and removal of hydrogen atoms from adsorbates in contact with the liquid environment, the notorious difficulty of implicit solvent models to describe hydrogen bonds accurately can lead to inaccuracies. However, the magnitude of these inaccuracies, as will be shown below, are significantly smaller than the magnitude of the reaction energies and the difference in reaction energies, which are the quantities of interest in this work.

2.2. The Water Oxidation Cycle

In this work, we investigate the energetics of the water oxidation reaction focusing on the free energy differences between the intermediates visited by the system along the reaction cycle. Following previous works [17,19], here, we assume that (i) all oxidation steps are coupled to a proton transfer, resulting in four consecutive proton-coupled electron-transfer (PCET) steps, and (ii) the O-O bond formation takes place through a nucleophilic attack of a solvent water molecule on a Ru-oxo species (also known as the acid-base mechanism). The water oxidation cycle investigated in this work is schematically shown in Figure 1. Several studies indicate this mechanism to be active on a variety of single- and multi-center molecular catalysts, including the well-known "blue dimer" [34] and the Ru_4-POM complex [17], and it has also been suggested to be at play in the case of metal oxide surfaces, such as $RuO_2(110)$ [35].

Assuming we start with a metal center M in oxidation state n (M^n), the first step involves the removal of a proton from an incoming water and the oxidation of the metal center from oxidation state n to $n + 1$ (M^{n+1}), leading to the formation of a hydroxo ligand. The second step involves the formation of an oxo intermediate and the formal oxidation of the metal to M^{n+2}. The following step consists of the nucleophilic attachment, resulting in a hydroperoxo ligand (OOH^-) and in the reduction of the metal to M^{n+1}. The final step consists of the release of molecular oxygen and the reduction of metal back to the initial oxidation state (M^n). We stress that the oxidation states reported in Figure 1 are just formal oxidation states, while calculations show that, for example, in Ru(IV), the formation of the oxo ligand is accompanied by the reduction of the metal center and the formation of an oxyl ligand, leading to a Ru(V)-O˙, rather than a Ru(VI)-O intermediate, as shown in our previous work [17]. Similarly, the OOH ligand has a radical character, reducing the Ru metal in a similar fashion to the oxyl radical.

We note that the assumption that all oxidations are coupled to proton removal might not be valid under strongly acidic conditions. This is an important factor to consider in the comparison with the experiments. Indeed, Murakami et al. [15] showed that at low pH, the first oxidation does not involve a proton exchange, while the second one involves the loss of two protons.

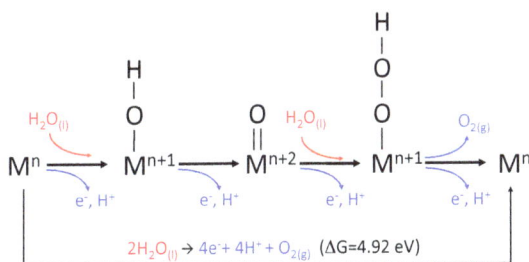

Figure 1. Scheme illustrating the water oxidation mechanism investigated in this work.

2.3. Reaction Thermodynamics

The thermodynamics of the electrochemical catalytic cycle was studied with the protocol proposed by Nørskov *et al.* [36]. This approach allows for calculating the free energy difference among the various intermediates. Energy barriers between different reaction steps are not included in the approach. The method has been extensively applied to study water oxidation [35,37,38] and oxygen reduction [36] reactions on metal and metal oxide surfaces.

In this approach, all oxidations are assumed to be PCET steps. The advantage of this assumption is that the energetics of the electrochemical reaction can be referenced to the normal hydrogen electrode (NHE), and the problematic calculation of the distinct chemical potentials of H^+ and e^- can be avoided. By referring the calculated electrode potentials to the NHE, the free energy of the pair $H^+ + e^-$ is equal to half the free energy of a hydrogen molecule at standard conditions, which can be easily and accurately calculated:

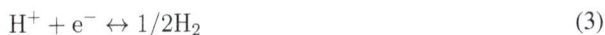

$$H^+ + e^- \leftrightarrow 1/2H_2 \qquad (3)$$

At zero bias and at pH $= 0$, the free energy change ΔG of a PCET step reaction is computed as:

$$\Delta G = \Delta E + \Delta ZPE + \Delta H - T\Delta S \qquad (4)$$

where ΔE is the difference of the DFT total energies, ΔZPE is the change in zero point energy, computed using the DFT vibrational frequencies, and ΔH and ΔS are the changes in enthalpy and entropy, computed using standard thermodynamic tables [39].

The energy of the various reaction intermediates has been computed in standard conditions of temperature and pressure and pH $= 0$. The energy difference between the initial and final states corresponds to the free energy change of the reaction Equation (1), *i.e.*, 4.92 eV. Computing this number from first principles would involve the calculation of the O_2 molecule, which is known to be poorly described in DFT (especially using GGAfunctionals), and would result in a significant underestimation of the free energy cost to split water. We therefore use the experimental value and fix the last point of the oxidation cycle, corresponding to the release of $O_{2(g)}$ to be 4.92 eV higher than the initial point.

The effect of a finite bias U on a state involving one electron in the electrode is modeled by lowering the energy of this state by $-eU$. In this approach, the overpotential for the water oxidation reaction is simply defined as the potential at which all steps become downhill in energy. The effect of pH can be accounted for correcting the free energy of H^+ ions by the concentration dependence of the entropy, resulting in a term $k_B T \ln 10 \times$ pH to be added to the value of ΔG computed at pH = 0 [36,40].

3. Results and Discussion

3.1. Electronic Structure of Ru-POM: Dependency on the Ru Oxidation State

We start by characterizing the electronic structure of the $[\text{Ru(III)-POM}]^{5-}$ complex ($[\text{Ru}^{III}(H_2O)SiW_{11}O_{39}]^{5-}$) that consists of a Keggin-type polyoxometalate with a surface W atom substituted by a Ru atom. We show in Figure 2 the relaxed geometry obtained using the PBE functional.

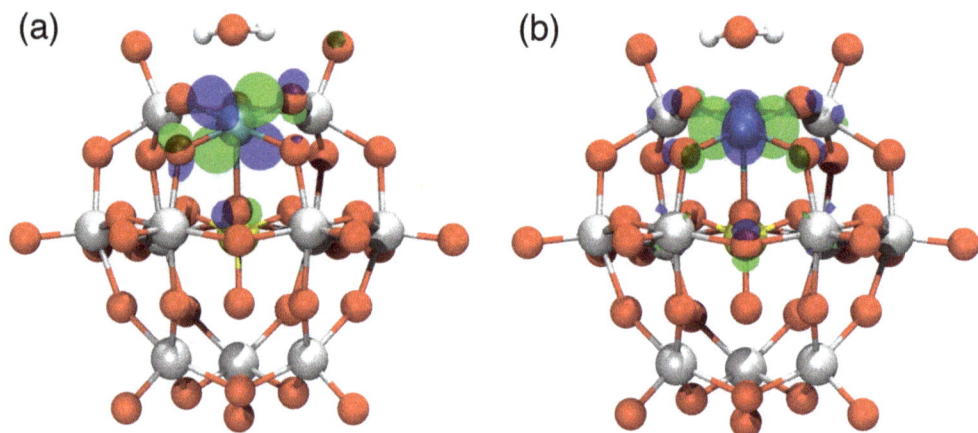

Figure 2. Relaxed geometry of the $[\text{Ru(III)-POM}]^{5-}$ molecule. Red, cyan, gray, yellow and white spheres represent O, Ru, W, Si and H atoms, respectively. The isosurface shows the spatial distribution of the HOMO (**a**) and LUMO (**b**) orbitals. POM, polyoxometalate.

The six-fold coordinated Ru ion is at the center of a distorted octahedron formed by five O ions of the POM molecule and by the O atom of the ligand water molecule. The octahedral ligand field induces a significant splitting of the t_{2g} and e_g levels. The charge and spin analysis of the ground state show that the Ru ion is in a low-spin d^5 electronic configuration, in which the five d electrons occupy the t_{2g} levels. This electronic configuration leaves one unpaired electron and, therefore, leads to a doublet state.

The highest occupied molecular orbital (HOMO) wavefunction is shown in Figure 2a: it is primarily localized on the Ru site and has a d_{xz} character. Similarly, we also find that the lowest unoccupied molecular orbital (LUMO) is mostly localized on the Ru ion (Figure 2b) and has a d_{xy} character. We find a HOMO-LUMO gap of 0.42 eV using the PBE functional and 2.85 eV using the B3LYP functional.

We now focus on the changes in the electronic structure resulting from oxidizing the [Ru(III)-POM]$^{5-}$ complex. In our computational setup, this can be achieved by reducing the number of electrons in the system, *i.e.*, by controlling the overall charge of the molecule.

The charge and spin analysis of the electronic ground states for the [Ru(IV)-POM]$^{4-}$ and [Ru(V)-POM]$^{3-}$ systems are reported in Table 1. The values of the Mulliken spin polarization at the Ru site were obtained with both the PBE and B3LYP functional. In each case, we report only the values corresponding to the lowest-energy spin configuration. The calculated Mulliken spin polarization follows the trend expected when going from a doublet in the case of Ru(III) to a triplet for Ru(IV) and to a doublet for Ru(V). In the case of a doublet, a single unpaired spin should result in μ equal to 1.0, while a triplet should result in μ equal to 2.0. We find lower values, indicating a certain degree of delocalization of the Mulliken spin density outside the Ru ion. We also find, in agreement with previous calculations [19], that hybrid functionals lead to a slightly larger localization of the spin density at the metal site, compared to the semi-local GGA-type of functionals.

Table 1. Mulliken spin polarization at the Ru ion as a function of the Ru oxidation state, evaluated in a vacuum. OS indicates the oxidation state of the Ru ion, "conf." the electronic configuration, Q the total charge of the anion, 2S + 1 the multiplicity and μ the Mulliken spin polarization in Bohr magnetons at the Ru ion obtained with the two functionals. PBE, Perdew–Burke–Ernzerhof.

OS	Conf.	Q	2S + 1	μ(PBE)	μ(B3LYP)
Ru(III)	d^5	−5	2	0.74	0.81
Ru(IV)	d^4	−4	3	1.31	1.48
Ru(V)	d^3	−3	2	0.76	0.67

3.2. Thermodynamics of the Water Oxidation Cycle

3.2.1. Ru-POM and the Effect of Ru Oxidation State

In Figure 3, we show the free energy cost to oxidize water along the reaction steps of the mechanism displayed in Figure 1. The reaction energetics is reported for the Ru-POM catalyst in both vacuum conditions (open symbols) and in solution (filled symbols and solid lines). Different colors of lines and symbols denote different charge states of the catalyst. The actual values of the free energy steps are also reported in Table 2.

The catalytic cycle promoted by Ru(III)-POM is denoted by the black symbols. The calculated energetics show that, when considering solvation effects, the step requiring the largest free-energy difference (2.06 eV) is determined by the formation of the hydroperoxo intermediate *OOH. The figure shows the free energy change along the oxidation cycle also for the Ru(IV)-POM and Ru(V)-POM catalysts. By increasing the oxidation state of the Ru atom, the free energy cost of the formation of hydroperoxo intermediate reduces while the cost for the second oxidation, the formation of the oxo intermediate, increases. In fact, we can see that the cost of those two steps is very similar for the Ru(IV)-POM catalyst (1.73 eV and 1.78 eV), while in the case of Ru(V)-POM, the formation of the *O intermediate becomes the most demanding step (1.83 eV).

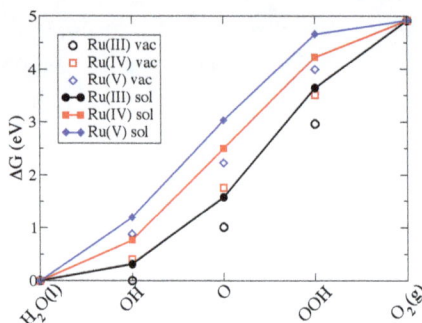

Figure 3. B3LYP free energy changes along the reaction cycle for single-center Ru-POMs with the Ru atom in different oxidation states (III, IV and V). Filled symbols represent the calculations performed accounting for the solvation effects, while empty symbols the calculations perfumed in a vacuum.

Table 2. Free energy changes along the catalytic cycle for the single-center Ru-POM and for Ru_4(IV)-POM computed with the B3LYP functional. Both results in a vacuum and in solution are reported.

Intermediate	In Vacuum				In Solution			
	Ru(III)	Ru(IV)	Ru(V)	Ru_4(IV)	Ru(III)	Ru(IV)	Ru(V)	Ru_4(IV)
H_2O(l)	0.00	0.00	0.00	0.00	0.00	0.00	0.00	0.00
*OH	0.00	0.40	0.89	0.78	0.31	0.77	1.20	0.72
*O	1.01	1.76	2.23	2.50	1.58	2.50	3.03	2.65
*OOH	2.97	3.01	4.00	4.20	3.64	4.28	4.66	4.35
O_2(g)	4.92	4.92	4.92	4.92	4.92	4.92	4.92	4.92

We can understand these results using the fact that the sum of the free energy cost to perform the second and third oxidation, *i.e.*, the transformation of a hydroxo ligand into a hydroperoxo ligand, is a universal constant independent of the catalyst. According to these results of Man *et al.* [41], the sum of these two steps has a free energy cost of 3.2 ($\pm$0.2 eV, 67% confidence interval, and $\pm$0.4 eV, 95% confidence interval), and in fact, we find values of 3.33, 3.51 and 3.46 eV for Ru(III)-, Ru(IV)- and Ru(V)-POM, respectively. The increase of the oxidation state of Ru leads to weaker bonding of the *O intermediate, and as a result, the free energy cost to form *OOH from *O decreases while the free energy cost to form *O from *OH increases. This is consistent with the fact that for Ru(III), where the *O binding energy is stronger (1.44 eV, where larger numbers indicate weaker binding), the most demanding step is the formation of *OOH, while for Ru(V), where the *O binding energy is weaker

(2.90 eV), the most demanding step is the formation of *O. Ru(IV) offers the best compromise, with a *O binding energy of 2.36 eV, leading to the free energy cost to form *O and *OOH being very similar.

According to the protocol proposed by Nørskov et al. [36], the overpotential for Ru(III)-POM is equal to $2.06 - 1.23 = 0.83$ V, while we obtain 0.55 V and 0.60 V for Ru(IV)- and Ru(V)-POM. An important result of our investigation is therefore that the best single-center Ru-POM catalyst with a Keggin structure is the one containing a Ru(IV) ion. Since the free energy cost of the two most demanding steps is very close (within 0.05 eV), this catalyst is almost optimal, within the constraint highlighted by Man et al. [41]. A similar conclusion was also reached in our previous work on the four-center Ru_4(IV)-POM [17].

3.2.2. Ru-POM vs. Ru_4-POM: The Effect of the Number of Ru Centers

In Figure 4a, we compare the energetics of the water oxidation cycle promoted by Ru-POM, considering Ru(III), Ru(IV) and Ru(V) single centers, with the one promoted by the four-center $Ru(IV)_4$-POM. In all cases, the calculations are performed using the B3LYP functional and accounting for solvation effects, as described in the Methodssection. The striking result of this comparison is that the values obtained for $Ru(IV)_4$-POM nicely match those for the single-center Ru(IV)-POM. This suggests that the most important factor in determining the energy cost of the four steps of the water oxidation cycle is the oxidation state of the metal center promoting the reaction, regardless of the number of metal centers. We stress here that the local atomistic environment around the Ru center, namely the octahedron formed by the six oxygen atoms neighboring the metal center, is similar in the single-site and multiple-site molecular catalysts.

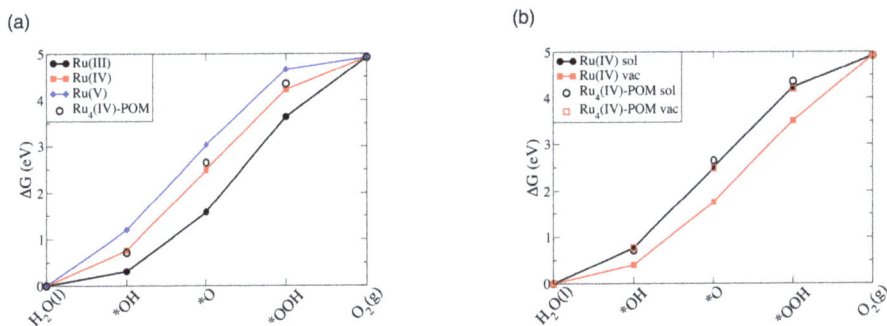

Figure 4. (a) B3LYP free energy changes along the reaction cycle computed in solution; (b) comparison of the effect of solvation on Ru(IV)-Keggin and $Ru(IV)_4$-POM.

Our results show that the molecular structure beyond the first O shell of the Ru active site has minor effects on the catalyst overpotential, which is instead primarily governed by the oxidation state of the metal site.

In addition, Figure 4b shows that solvation effects are much larger in the single-center Ru(IV)-POM compared to the four-center Ru_4(IV)-POM. This implies that the agreement discussed above is found only in solution, not in a vacuum. The reason for this behavior is likely due to electrostatic effects originating from the large charge of the the anions (-4 for Ru-POM and -10 for $Ru_4(IV)$-POM), which obviously cannot be screened when the calculations are performed in a vacuum. These effects manifest more clearly in the smaller catalyst, while is negligibly small for the larger $Ru_4(IV)$-POM, as already argued in our previous work [19].

4. Conclusions

In summary, we have used DFT calculations to investigate the water oxidation reaction catalyzed by single-center Ru-based polyoxometalates with the Keggin structure. By controlling the overall charge of the molecule, we were able to study the energetics of the reaction cycle as a function of the oxidation state of the Ru center. We found that the Ru(IV) ion leads to a lower overpotential compared to Ru(III) and Ru(V), which we rationalized on the basis of the strength of the Ru-Ointeraction. When compared against the $Ru_4(IV)$-POM catalyst, we found that the single-center and the multi-center catalysts based on the Ru(IV) ion display very similar energetics. For reaction mechanisms based on a single site, our results show that the electronic structure of the active metal ion controls the energy cost of the oxidation steps, while the local structure around the Ru sites beyond the first O shell does not contribute appreciably. Solvent effects are remarkably important for the smaller Keggin-type single-center molecules, while the $Ru_4(IV)$-POM molecule, likely due to its much larger size, is able to effectively screen its charge and, hence, display smaller effects of the solvent on its catalytic properties.

Acknowledgments

We acknowledge computational time on the Hermit supercomputer at HRLS, Germany, through the PRACEproject "ENCORASOL—Engineering multi-core transition metal catalysts for solar fuel production" and at CINECA HPC center, Italy. This work was partially supported by the EU FP7 COST action CM104 and by the FP7-NMP -2012 project chipCAT under Contract No. 310191. We are grateful to Andrea Sartorel and Marcella Bonchio for useful discussions.

Author Contributions

S.P. performed the calculations and the analysis. S.P. and S.F. planned the project and wrote the manuscript.

Conflicts of Interest

The authors declare no conflict of interest.

Appendix

A. Estimating the Error Bar in the Computed Thermodynamics

In this work, we employ different computational approaches to estimate the dependency of the reaction thermodynamics on the number and oxidation state of the active centers. This is necessary because the present data for the Ru-POM system are obtained with the Gaussian 09 code, while the available results for the Ru_4-POM system were obtained with the the CP2K code [17,19]. It is therefore useful to assess the error bar of this analysis by comparing the results obtained with the two approaches for the same system, namely the water oxidation cycle promoted by the single-center Ru(III)-POM in a vacuum.

In both cases, we use the B3LYP functional, while the choice of basis sets and pseudopotentials differ: in the case of CP2K calculations, we use DZVP-MOLOPT for Ru and W, TZVP-MOLOPT for O and H and GHTpseudopotentials [42], while in the case of Gaussian 09 calculations, we use SDD basis sets and effective core potentials. In both cases, the structures are optimized using the default thresholds on forces. In Table A1, we compare the free energy changes along the catalytic cycle obtained using the two approaches. The differences are in all cases small, the largest one being equal to 0.16 eV in the second step. While this difference is not negligible, indicating that our calculations are likely not fully converged with respect to the basis set, it is small enough to allow meaningful comparisons between the results obtained in our previous work on the four-center Ru_4-POM and the single-center Ru-POM.

Table A1. Comparison of the free energy changes along the catalytic cycle computed using CP2K(B3LYP/DZVP-TZV2P) and Gaussian (B3LYP/SDD-TZVP). Both sets of calculations are performed in a vacuum.

Intermediate	CP2K	Gaussian
$H_2O(l)$	0.00	0.00
*OH	−0.05	0.00
*O	1.17	1.01
*OOH	2.98	2.97
$O_2(g)$	4.92	4.92

One more source of error in our approach, when comparing solvation effects in Ru-POM and Ru_4-POM, is the use of two different solvation models, SMD and COSMO. We therefore evaluated the discrepancy between the two models on the smaller Ru(III)-POM molecule, for the case of a H_2O ligand. The SMD differences with respect to the vacuum case (*i.e.*, the solvation effects) in the free energy costs of the first three steps (the fourth step is set to 4.92 eV by construction) are 0.31 eV, 0.57 eV and 0.67 eV, respectively. The values obtained with the COSMO solvation model are 0.17 eV, 0.29 eV and 0.44 eV. The differences between the two solvation models are of the order of 0.2 eV, with the COSMO model resulting in smaller solvation effects. Furthermore, in this case, therefore, the errors are sufficiently small to allow for a meaningful comparison of the results obtained with these two different solvation models.

References

1. Lewis, N.S.; Nocera, D.G. Powering the planet: Chemical challenges in solar energy utilization. *Proc. Natl. Acad. Sci. USA* **2006**, *103*, 15729–15735.
2. Balzani, V.; Credi, A.; Venturi, M. Photochemical conversion of solar energy. *ChemSusChem* **2008**, *1*, 26–58.
3. Suga, M.; Akita, F.; Hirata, K.; Ueno, G.; Murakami, H.; Nakajima, Y.; Shimizu, T.; Yamashita, K.; Yamamoto, M.; Ago, H.; *et al.* Native structure of photosystem II at 1.95 A resolution viewed by femtosecond X-ray pulses. *Nature* **2015**, *517*, 99–103.
4. Kanan, M.W.; Nocera, D.G. *In situ* formation of an oxygen-evolving catalyst in neutral water containing phosphate and Co^{2+}. *Science* **2008**, *321*, 1072–1075.
5. Karkas, M.D.; Verho, O.; Johnston, E.V.; Pkermark, B. Artificial Photosynthesis: Molecular Systems for Catalytic Water Oxidation. *Chem. Rev.* **2014**, *114*, 11863–12001.
6. Gersten, S.W.; Samuels, G.J.; Meyer, T.J. Catalytic oxidation of water by an oxo-bridged ruthenium dimer. *J. Am. Chem. Soc.* **1982**, *104*, 4029–4030.
7. Dau, H.; Limberg, C.; Reier, T.; Risch, M.; Roggan, S.; Strasser, P. The Mechanism of Water Oxidation: From Electrolysis via Homogeneous to Biological Catalysis. *ChemCatChem* **2010**, *2*, 724–761.
8. Limburg, B.; Bouwman, E.; Bonnet, S. Molecular water oxidation catalysts based on transition metals and their decomposition pathways. *Coord. Chem. Rev.* **2012**, *256*, 1451–1467.
9. Sartorel, A.; Carraro, M.; Scorrano, G.; Zorzi, R.D.; Geremia, S.; McDaniel, N.; Bernhard, S.; Bonchio, M. Polyoxometalate embedding of a tetraruthenium(IV)-oxo-core by template-directed metalation of $[\gamma\text{-SiW}_{10}O_{36}]^{8-}$: a totally inorganic oxygen-evolving catalyst. *J. Am. Chem. Soc.* **2008**, *130*, 5006–5007.
10. Geletii, Y.V.; Botar, B.; Kögerler, P.; Hillesheim, D.A.; Musaev, D.G.; Hill, C.G. An all-inorganic, stable, and highly active tetraruthenium homogeneous catalyst for water oxidation. *Agew. Chem. Int. Ed.* **2008**, *47*, 3896–3899.
11. Yin, Q.; Tan, J.M.; Besson, C.; Geletii, Y.V.; Musaev, D.G.; Kuznetsov, A.E.; Luo, Z.; Hardcastle, K.I.; Hill, C.L. A Fast Soluble Carbon-Free Molecular Water Oxidation Catalyst Based on Abundant Metals. *Science* **2010**, *328*, 342–345.
12. Al-Oweini, R.; Sartorel, A.; Bassil, B.S.; Natali, M.; Berardi, S.; Scandola, F.; Kortz, U.; Bonchio, M. Photocatalytic Water Oxidation by a Mixed-Valent MnIII3MnIVO3 Manganese Oxo Core that Mimics the Natural Oxygen-Evolving Center. *Agew. Chem. Int. Ed.* **2014**, *53*, 11182–11185.
13. McDaniel, N.D.; Coughlin, F.J.; Tinker, L.L.; Bernhard, S. Cyclometalated iridium(III) Aquo complexes: efficient and tunable catalysts for the homogeneous oxidation of water. *J. Am. Chem. Soc.* **2008**, *130*, 210–217.

14. Wasylenko, D.J.; Palmer, R.D.; Berlinguette, C.P. Homogeneous water oxidation catalysts containing a single metal site. *Chem. Commun.* **2013**, *49*, 218–227.

15. Murakami, M.; Hong, D.; Suenobu, T.; Yamaguchi, S.; Ogura, T.; Fukuzumi, S. Catalytic mechanism of water oxidation with single-site ruthenium-heteropolytungstate complexes. *J. Am. Chem. Soc.* **2011**, *133*, 11605–11613.

16. Lang, Z.L.; Yang, G.C.; Ma, N.N.; Wen, S.Z.; Yan, L.K.; Guan, W.; Su, Z.M. DFT characterization on the mechanism of water splitting catalyzed by single-Ru-substituted polyoxometalates. *Dalton Trans.* **2013**, *42*, 10617–10625.

17. Piccinin, S.; Sartorel, A.; Bonchio, M.; Fabris, S. Water oxidation surface mechanisms replicated by a totally inorganic tetraruthenium-oxo molecular complex. *Proc. Natl. Acad. Sci. USA* **2013**, *110*, 4917–4922.

18. Romain, S.; Vigara, L.; Llobet, A. Oxygen-Oxygen Bond Formation Pathways Promoted by Ruthenium Complexes. *Acc. Chem. Res.* **2009**, *42*, 1944–1953.

19. Piccinin, S.; Fabris, S. First principles study of water oxidation catalyzed by a tetraruthenium-oxo core embedded in polyoxometalate ligands. *Phys. Chem. Chem. Phys.* **2011**, *13*, 7666–7674.

20. Perdew, J.P.; Burke, K.; Ernzerhof, M. Generalized Gradient Approximation Made Simple. *Phys. Rev. Lett.* **1996**, *77*, 3865.

21. Becke, A.D. Densityfunctional thermochemistry. III. The role of exact exchange. *J. Chem. Phys.* **1993**, *98*, 5648.

22. Stephens, P.; Devlin, F.J.; Chabalowski, C.F.; Frisch, M.J. Ab Initio Calculation of Vibrational Absorption and Circular Dichroism Spectra Using Density Functional Force Fields. *J. Phys. Chem.* **1994**, *98*, 11623.

23. Kang, R.; Yao, J.; Chen, H. Are DFT Methods Accurate in Mononuclear Ruthenium-Catalyzed Water Oxidation? An ab Initio Assessment. *J. Chem. Theory Comput.* **2013**, *9*, 1872–1879.

24. Perdew, J.P.; Burke, K.; Ernzerhof, M. Rationale for mixing exact exchange with density functional approximations. *J. Chem. Phys.* **1996**, *105*, 9982.

25. Kwapien, K.; Piccinin, S.; Fabris, S. Energetics of Water Oxidation Catalyzed by Cobalt Oxide Nanoparticles: Assessing the Accuracy of DFT and DFT+U Approaches against Coupled Cluster Methods. *J. Phys. Chem. Lett.* **2013**, *4*, 4223–4230.

26. Hay, P.J.; Wadt, W.R. Ab initio effective core potentials for molecular calculations. Potentials for the transition metal atoms Sc to Hg. *J. Chem. Phys.* **1985**, *82*, 270.

27. Andrae, D.; Haussermann, U.; Dolg, M.; Stoll, H.; Preuss, H. Energy-adjusted ab initio pseudopotentials for the second and third row transition elements. *Theor. Chem. Acc.* **1990**, *77*, 123–141.

28. Dolg, M.; Wedig, U.; Stoll, H.; Preuss, H. Energy-adjusted ab initio pseudopotentials for the first row transition elements. *J. Chem. Phys* **1987**, *86*, 866.

29. Marenich, A.V.; Cramer, C.J.; Truhlar, D.G. Universal Solvation Model Based on Solute Electron Density and on a Continuum Model of the Solvent Defined by the Bulk Dielectric Constant and Atomic Surface Tensions. *J. Phys. Chem. B* **2009**, *113*, 6378–6396.

30. Frisch, M.J.; Trucks, G.W.; Schlegel, H.B.; Scuseria, G.E.; Robb, M.A.; Cheeseman, J.R.; Scalmani, G.; Barone, V.; Mennucci, B.; Petersson, G.A.; *et al.* *Gaussian 09 Revision D.01*; Gaussian Inc.: Wallingford, CT, USA, 2009.

31. VandeVondele, J.; Krack, M.; Mohamed, F.; Parrinello, M.; Chassaing, T.; Hutter, J. Fast and accurate density functional calculations using a mixed Gaussian and plane waves approach. *Comp. Phys. Comm.* **2005**, *167*, 103–128.

32. Klamt, A.; Schuurmann, G. COSMO: a new approach to dielectric screening in solvents with explicit expressions for the screening energy and its gradient. *J. Chem. Soc. Perkin Trans. 2* **1993**, 799–805.

33. Valiev, M.; Bylaska, E.; Govind, N.; Kowalski, K.; Straatsma, T.; Dam, H.V.; Wang, D.; Nieplocha, J.; Apra, E.; Windus, T.; *et al.* NWChem: A comprehensive and scalable open-source solution for large scale molecular simulations. *Comp. Phys. Comm.* **2010**, *181*, 1477 – 1489.

34. Liu, F.; Concepcion, J.J.; Jurss, J.W.; Cardolaccia, T.; Templeton, J.L.; Meyer, T.J. Mechanisms of water oxidation from the blue dimer to photosystem II. *Inorg. Chem.* **2008**, *47*, 1727–1752.

35. Rossmeisl, J.; Qu, Z.W.; Zhu, H.; Kroes, G.J.; Nørskov, J.K. Electrolysis of water on oxide surfaces. *J. Electroanal. Chem.* **2007**, *607*, 83–89.

36. Nørskov, J.K.; Rossmeisl, J.; Logadottir, A.; Lindqvist, L.; Kitchin, J.R.; Bligaard, T.; Jonsson, H. Origin of the Overpotential for Oxygen Reduction at a Fuel-Cell Cathode. *J. Phys. Chem. B.* **2004**, *108*, 17886–17892.

37. Rossmeisl, J.; Logadottir, A.; Nørskov, J.K. Electrolysis of water on (oxidized) metal surfaces. *Chem. Phys.* **2005**, *319*, 178–184.

38. Rossmeisl, J.; Nørskov, J.K.; Taylor, C.D.; Janik, M.J.; Neurock, M. Calculated phase diagrams for the electrochemical oxidation and reduction of water over Pt(111). *J. Phys. Chem. B* **2006**, *110*, 21833–21839.

39. Stull, D.; Prophet, H. *JANAF Thermochemical Tables*, 2nd ed.; U.S. National Bureau of Standards: Washington, DC, USA, 1971.

40. Cheng, J.; Liu, X.; Kattirtzi, J.A.; VandeVondele, J.; Sprik, M. Aligning Electronic and Protonic Energy Levels of Proton-Coupled Electron Transfer in Water Oxidation on Aqueous TiO_2. *Angew. Chem.* **2014**, *126*, 12242–12246.

41. Man, I.C.; Su, H.Y.; Calle-Vallejo, F.; Hansen, H.A.; Martinez, J.I.; Inoglu, N.G.; Kitchin, J.; Jaramillo, T.F.; Norskov, J.K.; Rossmeisl, J. Universality in Oxygen Evolution Electrocatalysis on Oxide Surfaces. *ChemCatChem* **2011**, *3*, 1159–1165.

42. Goedecker, S.; Teter, M.; Hutter, J. Separable dual-space Gaussian pseudopotentials. *Phys. Rev. B* **1996**, *54*, 1703–1710.